New Thermodynamics: Say No to Entropy

(Second Edition of New Thermodynamics)

Kent W. Mayhew

ISBN: 1722479191
ISBN-13: 978-1722479190

DEDICATION

This book is dedicated to one of Canada's greatest physicists. He was a true thinker who would actually say what he believed. Importantly, he did not back down when faced with the illogical whims of those claiming accepted doctrine.

Sadly his peers wrongly chastised and then betrayed him. His removal from the University of Ottawa is a shameful rendition of the politics of office found within most institutions. Moreover, his expulsion remains a blight that all academics must share some responsibility for even allowing.

And in so doing, the very rights that so many before us fought for and too many died for were axed. That right being freedom of both thought and speech.

His name was Paul Marmet (1932-2005).

(see: https://www.Newtonphysics.on.ca/info/author.html)

TABLE OF CONTENTS

APPENDIX: TABLE OF CONTENTS

ACKNOWLEDGMENTS

I remain indebted to a few. People need to understand that humans do wonder and habitually blunder. Our past must be continually questioned. Valid opinions should be heard, especially when we actually have something to say. In that light I thank those who possess open minds i.e. Emilio Panarella (of *Physics Essays*), the group at *Progress in Physics* which includes Chifu E Ndikilar, Dmitri Rabounski and Andreas Ries. Open minds amonst those fully institutionalized is rare, hence often too hard to find.

There are also others whom I will tip my hat to, because they showed me more respect than I mayb have deserved: Herman Merte, Cynthia Whitney and Frank Lambert, all of whom I had several great email conversations. Also included in this is Arieh Ben Naim to whom I give credit for letting go of entropy but his refusal to digress from his probability-based edification has in my opinion has left him adhering to a different over complication. Simply put, probabilities give plausible results, rather than determine underlying theoretical logic.

Thanks, also goes out to Alexander Driega, who was my sometimes helpful, consummate, theoretical antagonist for various durations. And Bonnie Bonaventura, who helped me reformat my first paper. And, Chris Sheehan who helped with my third paper; it's the thought that counts. And Julie Vladimirovich for helping me write a tricky French version of an abstract. Certainly her linguistic talents are rarely matched.

Also thanks go out to my brother Lloyd and his wife Gail who at various times helped me with proof reading, and at times sort of believed. And William Humphries who also helped although his goals were very foreign to mine. Special thanks is also given to Nicholas Percival (of *John Chappells NPA*), who strived to help although making no claims of truly understanding my position nor its structure.

Least I forget my parents. I thank my mother Elizabeth, and my father Kenneth both of whom are no longer with us. I wish I could say that they fully believed in what I was doing but I cannot. Even so, they provided somewhere to land on whenever I crashed, and subsequently burned.

Of course, I was never completely alone in this endless adventure. I actually have discussed most of what I wrote with my various German Shepherds. First there was Ben who listened intently about DCI. Then Elsa, the brilliant one, who helped me figure out nucleation. Least I forget the impertinent one: Ruby who to the end believed that the Earth is flat. And now I live with "Q" the big friendly dufus who is actually way smarter than he looks.

My Apology

I apologize for not being a great writer and for the odd mistake. Accordingly, this book is to be taken as an insightful new beginning, clearly showing that a new simpler approach to thermodynamics is doable. And of course any final version will need the input of others, prior to presentation.

Moreover, given the choice of making the odd mistake while writing something that is fundamentally right, or being a perfectionist and writing with eloquent linguistic & mathematical purity something that is fundamentally erroneous. I will always take being *imperfectly right* over being *perfectly wrong*!

Preface: Entropy: All Beginnings Have an End!

Concerning my 2015 paper "Second law and lost work"[1], a reviewer put together a most remarkable tombstone worthy comment: "Although I think that the author is right, I cannot allow 150 yrs of indoctrination to be taken down by such a simple argument". Protected behind the sacred wall of peer review, the reviewer's commentary says everything. I of course retorted that simple arguments are the best arguments and eventually cooler heads prevailed and the paper was published with zero fanfare.

The above paper was followed by my paper: "Entropy: An ill-conceived mathematical contrivance"[2]. Together they clearly showed the possibility that entropy and her accomplice, the second law of thermodynamics are nothing short of one of sciences all-time greatest blunders.

Uffink[3] points out that Clausius's (1850) paper[4] represents for the first time, a clear rejection of the conservation of heat, while the validity of Carnot's theorem (*reductio ad absurdum*) is maintained. Clausius argues that if Carnot's theorem were false then a cyclic machine could be built, whose only effect would be that heat is transported from a cold to a hot reservoir. But this would be absurd, says Clausius[3,5], because: "[. . .]*this contradicts the further behaviour of heat, since it everywhere shows a tendency to smoothen any occurring temperature differences and therefore to pass from hotter to colder bodies*". Clausius's assertion is regarded as the initiation for second law, which can be construed as modern thermodynamic's birth.

Entropy's importance was inflated the next year by Kelvin (1851)[6] who accepted the validity of the second law, and similarly sought to put Carnot's theorem on this new footing. Kelvin eventual made a similar statement to Clauius's that being "*It is impossible for a self-acting machine, unaided by any external agency, to convey heat from one body to another at a higher temperature.*"[3,7]

Clausius seemingly initiated the universe's entropy always increases (time's arrow), only to seemingly regret this[3] by 1876. Whether or not Clausius had true misgivings, entropy became enshrined in thermodynamic dogma, and arguably has never been properly challenged until now.

Today entropy signifies many things to many people. It is readily applied to numerous situations, thus becoming fundamental to numerous scientific principles ranging from time's arrow to information theory. Yet its definition remains more muddled today than it was when it was first envisioned. Throughout this text, it will be clearly shown that thermodynamics simplifies once entropy is removed from the science. This being the case then what exactly is entropy? Is it something or is it nothing at all? If the importance of entropy is to be downgraded then what does this mean?

Consider computers; as amazing accomplishments they remain fraught with issues, all of which were unforeseeable to earlier computer's engineers and/or programmers. So with each and every new version of any program e.g. Windows, one has to continually update fixes in order to begin the circumnavigation of the computer galaxy. Ideally current technology could be redesigned with new simpler code enabling all computers to run smoother. Although bits and bytes remain simple concepts, after a gazillion transitions, the over complication is now too much to comprehend, yet alone fix. Moreover, doing so would render millions of computers obsolete hence is undoable. So we are stuck with these computers but who really cares since these habitually cursed machines generally satisfy our perceived needs.

Similarly thermodynamics remains fraught with problems that have festered for too long. Indoctrinated into the elite few, its lucidity remains intolerable to the masses. Rather than being based upon simple constructive logic it is based upon the big fix i.e. entropy; A.K.A. the parameter without clarity, A.K.A. statistical thermodynamic's cornerstone. As a mathematical contrivance[2], entropy enables the second law to become postulate supreme, all reinforced by equating its mathematical constructs to known empirical findings. The equating has unwittingly rendered its fundamentals into circular logic i.e. constructive logic be damned.

This author's understanding that useful expanding systems lose energy (lost work[1]) into the surroundings e.g. atmospheric potential energy increases, allows for the abolishment of the obtrusive second law! So rather than claiming that entropy increases (randomness?) explains why perpetual motion is unrealistic, a constructive logic based explanation for irreversible processes will be presented here.

Sadly, the current status of thermodynamics remains a mature science that few dare challenge. Hence we are continually told to accept this overly complicated science! Must future generations endear the exercises in madness that we endeared such as the illogical relentless unnecessary shuffling of all those partial derivatives? Unlike computers, the thermodynamic fix is actually doable. Accordingly, this book stands as a complete revolution by rewriting the science without any reliance upon entropy, or its accomplices e.g. the second law!

Humans naturally are comforted by surrounding themselves with people who were believe in the same fundamentals. As endearing as it may be, the echoing of similar ideas not only reinforces one's bravado, it equally closes our minds. Whether right or wrong, such reinforcing echoing closed bubbles of ideas equally occur in religion, politics and the sciences! Accordingly, any monumental change is never easily rendered, especially when the unrelenting fully indoctrinated remain in control.

Are any scientists humble enough to accept that they too, are part of this over complication? Certainly, few would willing relinquish their educational blinders. Even fewer will admit to their perfectly written facades. This book has been humbly written seeking just one!

References:

1) Mayhew, K.W. "Second law and lost work", Phys. Essays **28**, vol 1, 152 (2015)
2) Mayhew, K.W. "Entropy: An ill-conceived mathematical contrivance", Phys. Essays **28**, vol 3, 352 (2015)
3) Uffink, Jos "Bluff your way in the second law of thermodynamics" Utrecht University, 2001; Arxiv.org/abs/cond-mat/000532
4) Clausius R "Verber die bewegende Kraft der Warme und die Gestze die sich daraus fur die Warmelehre selbst ableiton lassen"(1850) English translation by W.F. Magie (Mendoza 1960)
5) Clausius R "Abhadlungungen uber de Mechalische Warmetheorie Vol 1" (1964)
6) Kelvin manuscript: "on the dynamical theory of heat" (1851) reprinted Archives of: The history of exact sciences **16**, 281-282
7) Kelvin "Mathematical and physical papers Vol 1 (1882) Cambridge University Press.

The True Beauty of Complexity

Have you ever had someone arguing with you use a complex assertion that you do not fully understand. There is a certain humiliation that one experiences whenever this occurs. The meek will walk away, while brave will seek a better understanding. When the brave determines that the someone does not grasp all the ramifications of their own claims i.e. they have been baffling you with BS, there is an undeniable immeasureable perverse joy that one feels.

Chapter 1: **Traditional Basics and Problematic Concerns**

In his *"Treatise on Thermodynamics"*[1] Planck acknowledges that there are two ways to formulate thermodynamics. 1): *"We may take for granted the correctness of the mechanical view of nature, and assume that all changes in nature can be reduced to motions of materials points between which there act forces which have a potential. Then the principle of energy is simply the well-known mechanical theorem of kinetic theory, generalized to include all natural processes."* Or 2): As is traditionally done; *"leave open the question concerning the possibility of reducing all natural processes to those of motion, and start from the fact which has been tested by centuries of human experience and repeatedly verified"*…"no way possible to have perpetual motion"[1pg40], thus crafting the second law into the supreme postulate!

In order to prerve the integrity of the second law numerous concerns are ignored,thus at times enshrining the science with conviently ignored problematic concerns. This chapter accomplishes two things. Firstly, it ensures that we are on the same page concerning terminology and concepts. Secondly it brings to light some of these concerns, opening your eyes to the prospect that the science needs to be rewritten based upon constructive logic i.e. the other way that being Planck's previously stated 1).

In this Book

The rewriting of the science without any reliance upon entropy nor the second law in the ensuing chapters will clearly demonstrate a simplification of the sciences. This will consider kinetic versus potential energy, along with associated mechanical aspects all under the guise that molecular interaction are rarely elastic. Moreover the need for the second law becomes non-existent once you realize that the main reason processes are irreversible is that work is done onto the atmosphere by expanding systems!

System Parameters

If a volume of space contains matter and/or energy, the state of that space can be thermodynamically defined using the following five fundamental parameters, and their corresponding Systeme International d'Unites (S.I. units):

1) Entropy (S), S.I. units J/K
2) Absolute Temperature (T), S.I. units K
3) Pressure (P), S.I. units N/m^2
4) Volume (V), S.I. units m^3
5) Internal Energy (ε), S.I. units J

Entropy (S) and absolute temperature (T) are referred to as the thermal parameters, while pressure (P) and volume (V) are mechanical parameters. A system containing energy and/or matter can have a real or an imaginary boundary with the *surroundings* encompassing everything that envelops that boundary. Accordingly, a system can be arbitrarily or realistically drawn, such that it encloses all energy and matter of interest. Generally, a system should be drawn on a scale, such that all parameters of relevance can be construed as being homogeneous throughout the system.

How one defines a system depends upon the problem at hand. For example, an engineer may want to know the power requirements of a refrigerator with the refrigerator as a whole being considered as a single system, while the power being the rate at which energy is supplied. If the primary interest is the refrigerator's compressor, then the compressor can be considered as being a system and the remainder of the fridge becomes either a separate system or its surroundings.

Intensive parameters are independent of the system's size, while *extensive* parameters are proportional to the system's size. Generally, temperature and pressure are homogeneous throughout a system, i.e. intensive parameters. Conversely, volume, entropy and internal energy, all depend upon the system's size and hence are extensive parameters. Intensive versus extensive is not limited to the above five parameters, e.g. density and specific heat (per gram) are intensive parameters. Extensive parameters are additive, while intensive are not.

Intrinsic properties are parameters that define a system. Capital letters are used for intrinsic properties that represent the whole system. For intrinsic properties that are expressed on a per molecule basis, convention dictates that small letters are used. For example "V" represents the total volume of a system, while "v" represents the molecular volume within that system. Properties deemed *extrinsic* are only written in their capital letter form.

Defining a System

Any system in equilibrium has a pressure, volume and internal energy, which correlates to entropy multiplied by temperature via the following *parameter relation*[7,8]:

$$TS = \varepsilon + PV \qquad 1.1$$

The concept that temperature multiplied by entropy really has it basis with Clausius' mid-19th century assertion. Both TS, and PV, are defined in terms of units of energy, i.e. the joule:

1) Units for TS: $K(J/K) = J = $ joule
2) Units for PV: $(N/m^2)(m^3) = Nm = J = $ joule

Traditional: Eqn 1.1 defines the relationship among these parameters in terms of energy.

Internal Energy, Pressure & Volume

Volume is readily understood in terms of its three dimensional construct, as taught in grade school. Pressure in atmospheres is based upon the force that a 76 cm column of mercury exerts upon 1 square cm cross-section at 45 degrees latitude on the Earth's surface.

The internal energy (ε) is taken to be the energy associated with the microscopic random disordered motions of the atoms and/or molecules within a system. This traditional perspective is sometimes referred to as the "invisible microscopic energy".

Problematic concern A: *Seemingly eqn 1.1 implies that the energy of a system is defined in terms of the microscopic energy plus any macroscopic work as defined by PV change. Which sounds great until you ask the following: Should the macroscropic energy of a system not simply be a result of the summation of the system's microscopic energies?*

Temperature & Entropy: Traditional vs Our new Perspective

Everyone has felt hot and cold materials thus providing us with a qualitative understanding of temperature, which can be quantitatively measured by using a thermometer. When a thermometer reaches *thermal equilibrium* with the system, then it is the thermometer's *thermometric property* that has changed allowing the measurement of the system's temperature. Quantitative temperature will be revisited.

In the 19th century, Rudolf Clausius realized that something when multiplied by temperature represented energy. Since then, *entropy* has taken on an array of various meanings. To many its definition (wrongly?) revolves around the 20th century consideration that entropy signifies a system's disorder; essentially entropy represents the "randomness of matter in incessant motion"[2]. An early 21st century but equally suspect definition is that entropy is "the dispersal of a system's molecular energy"[3]. A more recent yet still suspect definition belongs to Atkins[4] "S is a measure of the quality of that energy; low entropy means high quality, high entropy means low quality".

Problematic concern B: *Entropy remains the poorest understood parameter, specifically "no one knows what entropy really is"[5], which is part of Von Neumann's statement to Shannon when Shannon was trying to figure out what to call a variable in his information theory.*

Types of Systems

Three fundamental types of systems exist, each dependent upon the nature of our system's boundary and how energy and matter both flow across its boundaries:

1) An open system occupies a particular region of space from or into which both mass and/or energy may cross the system's boundaries;
2) An isolated system contains a fixed quantity of both energy and matter;
3) A closed system contains a fixed quantity of matter. Closed systems have two further breakdowns:

   a) Adiabatic boundary whereupon there is no heat exchange.
   b) Rigid boundary through which no mechanical work can be exchanged.

Thermodynamic Change

Thermodynamics concerns the correlation of the changes (Δ) to the five parameters that define a system's state, during some process. Therefore, a process can be defined in terms of some combination of: $\Delta S, \Delta T, \Delta P, \Delta V$, and $\Delta \varepsilon$. We accept that momentarily transition states may exist, which may not be readily defined. Our concern becomes the transition between *equilibrium states* where eqn 1.1 is valid.

Changes to a system's state can be written as:

$$T_f S_f - T_i S_i = \varepsilon_f - \varepsilon_i + P_f V_f - P_i V_i \qquad 1.2$$

where the subscripts "f" and "i" respectively represent the system's final and initial state.

Eqn 1.2 can be rewritten using the mathematical symbol delta (Δ), as follows:

$$(T_i + \Delta T)(S_i + \Delta S) - T_i S_i = \Delta \varepsilon + (P_i + \Delta P)(V_i + \Delta V) - P_i V_i \qquad 1.3$$

Multiplying through and collecting the terms, gives:

$$S_i \Delta T + T_i \Delta S + \Delta T \Delta S = \Delta \varepsilon + V_i \Delta P + P_i \Delta V + \Delta P \Delta V \qquad 1.4$$

Fig 1.1 graphically illustrates an increase to all the system's parameters. If all parameters are increasing then all deltas (Δ) in eqn 1.4 are positive.

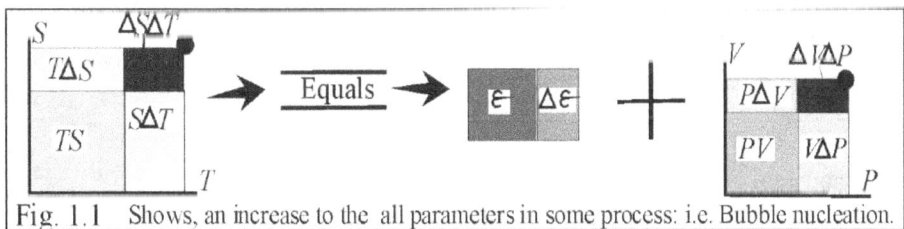

Fig. 1.1   Shows, an increase to the all parameters in some process: i.e. Bubble nucleation.

In the limit of thermodynamic change being infinitesimally small, then: $\Delta S \rightarrow 0$, $\Delta T \rightarrow 0$, $\Delta \varepsilon \rightarrow 0$, $\Delta P \rightarrow 0$ and $\Delta V \rightarrow 0$. Therefore, $\Delta S \Delta T <<< S \Delta T$, and/or $\Delta S \Delta T <<< T \Delta S$. Similarly, $\Delta V \Delta P <<< P \Delta V$, and/or $\Delta V \Delta P <<< V \Delta P$. Using these approximations, eqn 1.4 becomes:

$$S_i \Delta T + T_i \Delta S \approx \Delta \varepsilon + V_i \Delta P + P_i \Delta V \qquad 1.5$$

Eqn 1.2 can also be written in differential form:

$$d(TS) = d\varepsilon + d(PV) \qquad 1.6$$

For infinitesimally small change, eqn 1.6 can be approximated by:

$$TdS + SdT \approx d\varepsilon + PdV + VdP \quad 1.7$$

It is important to recognize that eqn 1.7, as the differential equation of eqn 1.2, remains valid if, and only if, the system's parameters changes are infinitesimally small, i.e.: $dSdT <<< SdT$, $dSdT <<< TdS$, $dVdP <<< VdP$ and/or $dVdP <<< PdV$. Situations exist where parameter changes are not infinitesimally small hence a more accurate result is obtained using eqn 1.2, rather than eqn 1.7. Consequentially, eqn 1.2 remains the general relation and the approximation is eqn 1.7!

Enthalpy

Enthalpy (H) is based upon PV space, and is defined as[7,8]:

$$H = \varepsilon + PV \qquad 1.8$$

Changes to enthalpy can be written:

$$\Delta H = \Delta \varepsilon + \Delta(PV) \qquad 1.9$$

Rewriting eqn 1.9 in differential form:

$$dH = d\varepsilon + d(PV) \qquad 1.10$$

For an *isenthalpic* system (constant enthalpy $\Delta H = 0$) eqn 1.9 becomes:

$$\Delta \varepsilon = -\Delta(PV) \qquad 1.11$$

For isenthalpic systems any changes to internal energy ($\Delta \varepsilon$) is countered by changes to PV space (ΔPV). The differential form of eqn 1.11 is:

$$d\varepsilon = -d(PV) \approx -(PdV + VdP) \quad 1.12$$

Problematic concern C: *One cannot help feel that the real benefit of enthalpy is that it allows the science to bypass the poorly understood parameter known as entropy*

Simplified Systems

It is easiest to deal with *simplified systems* wherein certain parameters remain constant, while the others are system variables. Such systems are known as one of the following[7,8]:

1) *Isobaric system*: A constant pressure system ($\Delta P = 0$).
2) *Isothermal system*: A constant temperature system ($\Delta T = 0$).
3) *Isometric (Isochoric) system*: A constant volume system ($\Delta V = 0$).
4) *Isentropic system*: A constant entropy system ($\Delta S = 0$).

For an isobaric process ($dP = 0$), eqn 1.7, becomes:

$$TdS + SdT = d\varepsilon + PdV \qquad 1.13$$

If the process is also isothermal ($dT = 0$), eqn 1.13, becomes:

$$TdS = d\varepsilon + PdV \qquad 1.14$$

Obviously, eqn 1.14 is a subset of the general relation, for isobaric and isothermal processes. Note: Traditionally, a convoluted approach is used by starting off with eqn 1.14 and then deriving all thermodynamic relations via various transformations (See Chapter 15).

To calculate isobaric & isothermal changes to internal energy ($d\varepsilon$) eqn 1.14 becomes[7,8]:

$$d\varepsilon = TdS - PdV \qquad 1.15$$

For the case of an isobaric & isothermal process wherein the system's internal energy remains constant ($d\varepsilon = 0$), eqn 1.14 simplifies to:

$$TdS = PdV \qquad 1.16$$

For an isentropic process ($dS = 0$), eqn 1.7 becomes[7,8]:

$$SdT \approx d\varepsilon \mid PdV \mid VdP \qquad 1.17$$

If the above process was also isobaric then:

$$SdT \approx d\varepsilon \mid PdV \qquad 1.18$$

Keeping certain parameters constant and others as variables can be continued resulting in an array of simplistic differential equations, as will be discussed in Chapter 16.

In this Book

It will be clearly shown that the work done (PdV) by the system is external to the system i.e. often the system's surroundings. Hence the change to internal energy is really the change to system's energy, while *PdV* actually describes work done to the surroundings <u>through</u> an expanding system's wall. This will alleviate many of the above described problematic concerns. Note: Many texts[6] actually wrongly consider that work is done into a system's walls irrelevant of the walls being real or imaginary. Although this wrong consideration is a mathematically plausible the reality is that such math actually show that the work is through the walls[8,9].

Ideal Gas

An ideal gas is one wherein the gas molecules have no intermolecular bonding hence the bonding energy is zero: $U=0$. Note: Traditionally the bonding energy is taken to be part of the internal energy. Therefore, changes to an ideal gas's internal energy are also considered zero, i.e. $d\varepsilon = dU=0$. Real gases tend not to be ideal due to electromagnetic attraction, or repulsion between molecules in the gaseous state. Two examples being:

1) Polar gas molecules behave like magnets floating in space; hence they have an attraction to their neighbor's dipole moments, e.g. vaporous water molecules.
2) Similarly charged ionized gaseous molecules have an electromagnetic repulsion rather than attraction.

The mathematics of the energy associated with bonding of polar molecules is discussed in Appendix A.1 and are often dealt with using van der Waals' and/or Clausius' equation. The bonding potential (U) is attributed to the electromagnetic attraction between molecules, which decreases as the intermolecular distance increases. <u>Sufficiently dilute</u> gases at most temperature regimes sufficiently above their boiling points often can be approximated as ideal gases. Note: Hard to condense gases i.e. CO, H_2, N_2, O_2, tend to best approximate ideal gases[1].

If $dU \approx 0$, then simple compression or expansion of the gas by an external force does not alter the energy associated with that gas. Hence, for such an ideal gas under compression or expansion:

$$PV = \text{constant} \qquad 1.19$$

Robert Boyle (1627-1691) was the first to envision the fundamental principle that PV equates to a constant for an ideal gas being compressed or expanded by an external force; therefore, eqn 1.18 is known as *Boyle's law* (1660) and is sometimes also called *Boyle-Marriotte law*. Since PV remains constant during either compression or expansion, it follows that for infinitesimal change:

$$PdV = -VdP \qquad 1.20$$

The ideal gas law is commonly written in various forms, one being[7,8]:

$$PV = nRT \qquad 1.21$$

where n is number of moles, and R is the ideal gas constant or $R = 8.31$ J/mol·K

The *equation of state* for an ideal gas considers the microscopic energy of each molecule, and as such, is written[7,8]:

$$PV = NkT \qquad 1.22$$

where N is the number of gaseous molecules in the volume, (V) and k is Boltzmann's constant, $k = 1.38 \times 10^{-23}$ J/K.

The impact of infinitesimal change to pressure and/or volume within a closed system containing an ideal gas can be calculated by differentiating eqn 1.22:

$$d(PV) \approx PdV + VdP \approx NkdT \qquad 1.23$$

Both N and k are constants for an ideal gas system, therefore for isothermal expansion or compression: $dT = 0$ and $PdV = -VdP$, satisfying equations 1.19 through 1.22. Conversely, for non-isothermal processes then $dT \neq 0$ and $PdV \neq -VdP$. Accordingly, Boyle's law is actually limited to isothermal processes!

In this Book

It will be shown that the ideal gas law is a law with limitations that previously were misunderstood!

Quantitative Temperature

The quantitative nature of temperature is measured with a thermometer. Constant pressure and/or volume thermometers, both of which use a gas as the *thermometric medium*, are what all other thermometers are compared to. This includes modern electronic devices with digital readouts. Understandably temperature can be readily defined for temperatures ranges at which gases obey the ideal gas law. Specifically, for ideal gases their volume changes at a rate of 1/273 per degree Celsius temperature change.

Accordingly, the concept of temperature can be accurately measured at all temperatures except those approaching absolute zero. Absolute zero cannot be readily measured rather it is extrapolated to be negative 273 degrees Celsius. Hence absolute zero is somewhat arbitrary, and for many it is taken to be an entropy/second law based construct. Simply put absolute zero is the temperature at which an ideal gas of finite volume has no pressure, or if you prefer the temperature at which vibrational energies within condensed matter ceases to exist..

The measurement of temperature requires a scale so that comparisons can be made. In 1745, Carolus Linnaeus decided that 0°C, and 100°C would respectively represent the freezing, and boiling points of water, thus creating the Centigrade scale. In 1948, the Centigrade scale was dropped in favor of using degrees Celsius.

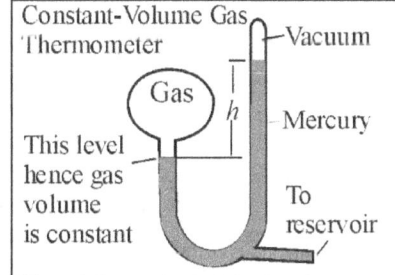

Fig. 1.2 The height h of mercury allows us to calculate the gas's pressure hence temperature based upon ideal gas law

In 1887, P. Chappuis studied constant volume thermometers, where a gas residing in a glass bulb, acts as the thermometric medium, meaning it is placed in thermal contact with the system whose temperature is being measured. As shown in Fig. 1.2, the gas's volume is held constant by either adding, or subtracting, mercury through the tube labeled "to reservoir".

By knowing the density of the mercury in the tube, the gas's pressure is readily calculated by measuring the height (h) of mercury inside of the tube. Knowing the gas's pressure and constant volume, the temperature is then calculated using the ideal gas law, i.e.:

$$T = PV / Nk \qquad 1.24$$

One does not necessarily need to know the number of molecules (N) in order to utilize a constant volume thermometer. You could compare the ratio of temperatures for two systems in terms of their pressure ratio, i.e. $T_1/T_2 = P_1/P_2$, and knowing one of the system's temperatures then enables one to measure the other system's temperatures.

Quantitative Temperature is really a comparative

When measuring the temperature of a system, the thermometer is placed in thermal contact with it. Thermal equilibrium is obtained when the influx equals the efflux of thermal energy between the thermometer and the system whose temperature is being measured.

If two systems in thermal contact are at the same temperature then the net exchange of thermal energy between them would be zero. This is the basis of what is known as the *zeroth law* of thermodyanmics, which treats *thermal equilibrium* as a transitive property. Two systems are considered to be in thermal equilibrium with each other, if the following two conditions, hold true[7,8]:

a) both systems are in an equilibrium state; and
b) both systems remain in equilibrium when they are brought into thermal contact.

As a transitive property the zeroth law states that *thermal equilibrium* exists[7,8]: "*If two systems/bodies are in thermal equilibrium with a third, then they must be in equilibrium with each other.*" Intuitively, the third system may be considered as being a thermometer.

The zeroth law of thermodynamics can be considered as similar to, but slightly different than the first law of thermodynamicthat being the fundamental principle that energy is conserved, i.e. energy can be converted from one form to another, but it cannot be created nor destroyed. Accordingly, the total influx of energy into a system must equal the energy change within that system minus the magnitude of the efflux of energy out of that system.

For a given substance or system, its thermal energy density is directly proportional to its temperature for most temperature regimes experienced here on Earth. This does not mean that a thermometer compares the thermal energy densities because different substances in different states all have different thermal energy densities. Although differing systems have differing thermal energy densities at a given temperature, their temperature always defines the net direction of flow of heat that being from hot to cold.

Temperature can also be taken as a comparative between systems and/or as to what is felt when standing outside on Earth, where the thermal energy density is primarily derived from our Sun. Interestingly, when a system is in thermal contact with our atmosphere, then that system generally exchanges thermal energy with our atmosphere whose thermal energy density was primarily attained from the sun. This is fundamental to so many phenomena in part because our atmosphere often acts as the mother of all heat baths/sink/reservoir. Remember, thermal equilibrium means that the systems are at the same temperature in which case the systems are exchanging equal amounts of thermal energy.

Thermal Energy

In condensed matter, the molecules are so close that electromagnetic (EM) intermolecular bonds exist between all the molecules. When considering systems of condensed matter, the *thermal energy* is contained within the vibrations associated with both the intermolecular and intramolecular bonds. Specifically, *intermolecular vibrations* are between molecules, while *intramolecular vibrations* are between the various atoms that constitute the molecules.

Phonons are packets of energy related to the random lattice vibrations in solids, which are a function of the crystalline substance's temperature. Phonons are theoretical equivalent to photons. Specifically, a phonon is an electromagnetic (EM) particle within a crystalline substance, while a *photon* is an EM particle in *freespace* that being a volume without matter. Since crystalline substances have a lattice structure that prefers specific phonons, our expectation is that crystalline substances preferentially interact with specific frequencies of photons.

Liquids and amorphous solids lack the crystalline structure to which phonons are mathematically related. Even so, it is accepted that the thermal energy contained within such substances can be considered as phonons (packets of energy) that are treated in the same manner. Liquid molecules also have freedom of movement (e.g. convection), and accordingly they can possess both translational and rotational energies. However, both of these energies are generally considered to be minor in comparison to vibrational energy. Therefore, most of the thermal energy within all condensed matter can be attributed to vibrational energy.

A more exacting deliberation would include the fact that changes to a system's energy results in changes to the vibrational energy, which in its simplest terms can be thought of as the motions of intermolecular and intramolecular EM bonds. Such motions transform vibrational energy into photons. Equally, photons that interact with matter as thermal energy can be transformed into the movement of molecular EM bonds.

Any analysis should consider that most condensed matter absorbs and emits thermal energy, which for the most part consists of photons whose frequencies are less than that of light. Moreover, the hotter the matter is the more thermal radiation it emits, which is the basis of thermal imagery devices such as infrared military night goggles.

Visible light

Visible light is EM radiation generally at slightly higher frequencies than thermal energy. Certainly color in matter implies the absorption of certain frequencies and the reflection of others. Note: Many adsorbed frequencies may contribute as heat. An interesting consideration; can light actually be seen with our eyes? Light in space remains invisible to our eyes, it is only when it interacts with matter that we actually see it! For example a light ray through dust/smoke is visible but that same ray through a vacuum is not so discernable. Similarly, it is only when light reflects off of, or refracts through matter that its presence is actually revealed by our eyes. For an interesting and controversial take see/google Goethe's theory. This does not necessarily mean that darkness is anything but the absence of the light, as Goethe's would believe. Rather I believe that this has more to do with how our eyes evolved. Imagine that all rays of light were seen then we would be blinded by those rays. Thermal (infra-red) imagining is really a just a wavelength shift by a device, enabling our eyes to see the energy that has interacted with, and is now being emitted by warm/hot bodies of matter. Food for thought.

Measurement of Temperature

For condensed matter, temperature is considered as being a measurement of the vibrational energy associated with the kinetic motions of the system's molecules, i.e. the system's kinematics. Such an interpretation becomes problematic when considering gases.

Consider the measurement of the temperature of a gas, as is illustrated in Fig. 1.4. Obviously, thermal energy is transferred between the thermometer and the volume of gas, via a combination of:

1) The gas molecules physically exchanging their kinetic energy with the molecules within the thermometer through collisions with the thermometer; and

2) Thermal energy being absorbed from, and emitted into the surrounding freespace results in the exchange of thermal energy between the system and thermometer.

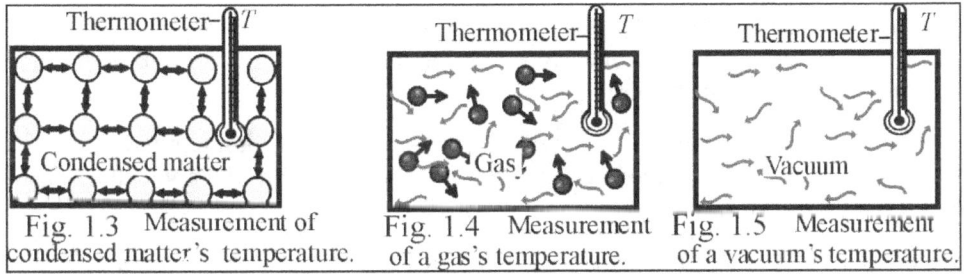

Fig. 1.3 Measurement of condensed matter's temperature.   Fig. 1.4 Measurement of a gas's temperature.   Fig. 1.5 Measurement of a vacuum's temperature.

The net result is that the molecules within the thermometer attain thermal equilibrium with the system. Again, thermal equilibrium occurs when both the thermometer and gaseous system have an equal influx vs efflux of thermal energy.

Temperature Consideration

Problematic concern D: *Strangely, traditional thermodynamics only considers temperature in terms of a system's kinematic, which is fine for condensed matter Fig 1.3. Can the above described thermal radiation [2]] simply be ignored when contemplating temperature? Certainly the total energy associated with thermal radiation often is minute when compared to other thermal energies within a given system.*

The problem becomes most obvious when one considers a vacuum, as is illustrated in Fig. 1.5. The traditional interpretation is that a matter-less vacuum possesses no molecular motion, and hence has no temperature i.e. zero kinematics of matter thus has no temperature. Strangely however, if a thermometer is placed into such a vacuum containing thermal radiation, then the thermometer obtains a temperature reading, solely due to the exchange of thermal radiation. Specifically, the molecules within the thermometer eventually attain thermal equilibrium with the surrounding thermal radiation, although no kinetic energy actually existed within the vacuum until the thermometer was placed inside.

17

A metaphysical argument arises. Traditionalists argue that by putting a thermometer into the vacuum, there is now a temperature associated with the thermometer but not with the surrounding vacuu but is the thermometer not actually measuring the vacuum's temperature? E.g, consider that a thermometer is put into an immense vacuumtghat is full of thermal radiation. Although the energy associated with thermal radiation is often minute when compared to the energy of molecular kinematics, the fact that the vacuum's volume is immense means the eventual thermometer's temperature reading will be that a of the vacuum. Furthermore, because the speed of light is vast, a significant quantity of heat can be exchanged within a vacuum even when the thermal radiation density in that vacuum remains diminutive in comparison to thermal energy contained within most condensed matter.

Consider the dark side of the moon being much colder than the bright side! It seems farcical that the word "cold" is used if it no longer applies to relative temperatures. Certainly, the moon's condensed matter involves kinematics. What about a few millimeters above the moon's surface? Does the term temperature no longer apply? Are we to believe that there is no thermal equilibrium between the matter on the moon and the space that surrounds it?

Arguably clarity could be obtained by saying: If a thermometer makes a measurement in a system, then that system has a temperature! However this too may be problematic because a better understanding of thermal radiation is needed. This should make more sense after reading the ensuing chapters.

Certainly at a given temperature, systems containing matter:

1) Will tend to exchange thermal energy faster than vacuums; and
2) Will have a higher thermal energy density than freespace i.e. matter tends to concentrate thermal energy and hence increase the thermal energy density within a given volume.

The general exception occurs when we are dealing with thermal radiation at high temperatures, i.e. "radiation heat transfer", for which significant heat exchange can occur even through a vacuum, e.g. systems at blast furnace type temperatures.[11]

Conclusion at this point traditional thermodynamics should reconsider its stance concerning temperature. Of course their argument goes beyond the energy associated with kinematics often being significantly greater than that associated with radiation. Specifically, it is based upon the traditional insistence that probability/mathematical based statistical thermodynamics is more relevance than common sense, e.g. another case of maintaining the second law as some supreme postulate.

In this Book

It will be demonstrated that the accepted limiting temperature to the kinematics of matters is wrong!

Joule's Gas Expansion Experiment

The experiment illustrated in Fig 1.6 and 1.7 is known as Joule's experiment for gases. James Prescott Joule concluded that since no temperature change was found in the heat bath that his experiment shows that the gas's internal energy is a function of temperature but not volume. Joule's experiment is far from perfect, e.g. if energy were extracted from the surrounding heat bath then would it be measurable? Even so, it has been verified by others performing more exacting experiments e.,g. Lord Kelvin's version of Joules' experiment. Obviously the isothermal expansion of an ideal gas implies that $d\varepsilon = 0$.

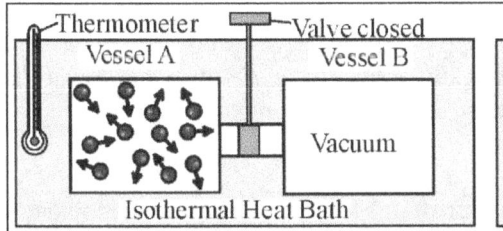
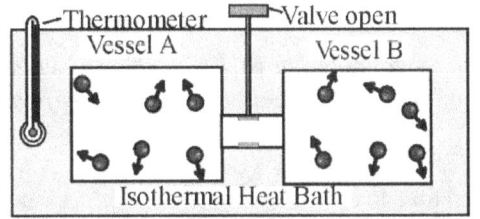

Fig 1.6   Shows a dozen moving molecules in an isothermal system separated from an equal volume isothermal vacuum, by a closed valve. A.K.A.: Joules experiment.

Fig 1.7   Shows the same dozen molecules of Fig 1.6, with the valve opened allowing for the molecules to disperse evenly between both Vessel A and Vessel B.

Ideal gas Paradox

Problematic concern E: *Bearing in mind the previously stated definitions of entropy, consider the ideal gas in Vessel A, as shown in Fig 1.6. A valve is opened and the gas is allowed to isothermally disperse into Vessel B, as is illustrated in Fig 1.7. We expect that $PdV = -VdP$. In other words, as the gas's volume doubles its pressure decreases by half.*

As this ideal gas's volume increases, the molecules' randomness, and/or the dispersal of energy, must increase. Therefore by certain definitions, its entropy should increase ($S\uparrow$). If a system's entropy is increasing, and there is no total energy change within the system, then shouldn't we expect that the ideal gas's temperature will decrease ($T\downarrow$), such that: $TdS = -SdT$? But that makes no sense because the process is isothermal ($dT = 0$), allowing Boyle's law (Boyle-Mariotte law) to remain valid, i.e.: $PdV = -VdP$. Certainly, if the internal energy is related to the potential associated with intermolecular bonding, then the ideal gas's internal energy does not change.

Seemingly, there is something wrong with Joules' understanding. What could it be? Perhaps the ideal gas law is only an approximation! One could rightfully argue that the heat bath kept everything isothermal, but the above experiment should remain isothermal without the heat bath. Other possibilities:

1) Perhaps, we must reconsider eqn 1.7: $TdS + SdT = d\varepsilon + PdV + VdP$

If $PdV = -VdP$ and $dT = 0$, then 1.7 implies:

$$TdS = d\varepsilon \qquad\qquad 1.25$$

If eqn 1.25 defines changes to our isothermally expanding ideal gas, then one cannot isothermally expand a gas and maintain constant internal energy within that system of gas. In which case the internal energy of our isothermally expanding ideal gas has seemingly increased: ($d\varepsilon\uparrow$). Does this mean that the internal energy changes, while the intermolecular bonding energy remains constant, as expected for an ideal gas? It all seems convoluted.

2) Perhaps we must reconsider what entropy is!

Since: $PdV = -VdP$ and $dT = 0$, and if $d\varepsilon = 0$, then $dS = 0$. Consider our previously given two definitions of entropy. During the isothermal expansion of the ideal gas, both the randomness of molecules in incessant motion and/or the dispersal of the gas molecules energy have increased, yet there is no predicted entropy change? Seemingly, the virtues of entropy should be queried! Has Atkin's consideration of the quality of the energy changed? Perhaps but even that remains weak!

Ultimately the ideal gas law has suffered a paradox. One might argue that our analysis is overly-simplistic. But to do so implies that the ideal gas is complex, which it is not. Or that eqn 1.1 cannot be obtained by the integration of eqn 1.14. And herein resides the issue.

In order to circumnavigate the above logic, thermodynamics may have unwittingly complicated the simple, in part by shuffling the differential equations around, all in order to protect the false postulate.

In this Book

It will be discussed that Joule's experiment really only demonstrates that no work can be done onto a vacuum! I.e. it shows that the expansion into a vacuum does not change the energy of gas. In context of entropy; if this had anything to do with randomness of molecules, then randomness has nothing to do with energy of gas. It will be shown that much of our 20th century teachings has been fool's gold all because the science remained postulate blind

Heat Transfer

Thermodynamics concerns the transfer of thermal energy (heat) both into, and out of, a system. Such heat transfer is could be dealt with in terms of *TS* space, where entropy (S) is considered in its simplest guise: Entropy is something that when multiplied by temperature defines thermal energy, hence isothermal entropy change is often traditionally defined the Clausius equation, which is commonly written in the following form[6,7]:

$$\Delta S = \Delta Q / T \qquad 1.26$$

Where ΔQ is the thermal energy (heat) change. Eqn 1.26 can be rewritten:

$$\Delta Q = T \Delta S \qquad 1.27$$

From a purely mathematical perspective, eqn 1.27 implies that the thermal energy change (ΔQ) is directly proportional to the entropy change (ΔS), and that the proportionality constant is temperature (T).

Problematic concern F: *There is a profound drawback to the Clausius equation; it does not consider the thermal energy change (ΔQ) in terms of temperature change (ΔT), but only entropy change (ΔS) i.e. it is an isothermal relationship. This bodes the question; what exactly is entropy? Answer remains the same no one really knows. Great!*

Moreover at first glance, it befuddles the mind concerning how does one have a thermal energy change within a system and no temperature change. To some the answer is; this applies to systems where influx (energy entering) equals the efflux (extracted energy). Certainly this limits the applicability of eqn 1.27.

To others the answer is somewhat more complex, in that infinitesimal changes are contemplated, i.e. energy exchanges that are too small to be measured by a thermometer as a temperature change (ΔT). The reality remains that ΔT is only not noticed because the thermometer is not accurate enough to read such a minuscule ΔT. This is NOT exactly the same as $\Delta T = 0$.

The above absurdity is further hidden by traditional claim that these concepts are developed based upon heat reservoirs i.e. the definition of a heat reservoir/bath/sink being a system wherein exchanges of thermal energy do not alter its temperature! Again just because temperature change was infinitesimal, does not mean it did not occur!

Think back to previously discussed Joule's experiment. Joule rendered his conclusion based upon the fact that in his experiment the heat bath's temperature did not change. Certainly any energy associated with the expanding gas should be infinitesimally small compared to the heat bath's thermal energy. It is hard to fathom why, after Joule's experiment that none jumped all over the dangers of developing a science based upon such consideration of infinitesimals.

Ultimately, eqns 1.26 and 1.27 are based upon the rather poor conceptualization of an isothermal system's thermal energy change. Or, does the exchange between heat reservoirs simply mean that the equations do not apply to real systems, where thermal energy changes are noticeable? Any way you look at it, this is another consequence of formulating a science around the second law as a postulate.

Ideal Gas Constant and Specific Heat/ Heat Capacity

Problems associated with Clausius equation can be avoided by thinking in terms of specific heats and/or heat capacities, which allows us to consider a system's thermal energy change in terms of temperature change. Why would anyone prefer Clausius's consideration over specific heat/heat capacity?

A problem with heat capacity being that it does not remain constant through all temperature regimes. Planck[1] discussed that a zero calorie was taken as the energy required to raise a gram of water from 0 to 1° C. And that this differs from what is taken to be an actual/mean calorie that being the energy required to increase one gram of water's temperature from 14.5 to 15.5° C, which equals 1/1.008 of a zero calorie. Seemingly this is a small difference, but it is also over a relatively narrow temperature range. Even so, considering heat capacity as a constant over most (not all) temperature regimes generally provides us with a useable approximation.

Overlooking the above problem, the preference lay in the fact that as traditionally written, isothermal entropy change fits well with statistical probability based conscripts which are used to defend the second law as a postulate. And are to this author are part of the traditional self-serving over complication of the science that is really embedded with circular logic.

Consider the specific heat per unit mass, where the subscript "y" describes the variable that is considered constant. The isobaric specific heat per gram (or per kilogram) (unit mass in SI system) for a given substance is[7]:

$$c_{p'} = (1/m)(dQ/dT)_p \qquad 1.28$$

Similarly, the isometric specific heat per gram for a given substance, becomes[7]:

$$c_{v'} = (1/m)(dQ/dT)_v \qquad 1.29$$

Similarly, the isobaric molar capacity for a given substance is[7]:

$$C_p = (1/n)(dQ/dT)_p \qquad 1.30$$

Similarly, the isometric molar heat capacity for a given substance is[7]:

$$C_v = (1/n)(dQ/dT)_v \qquad 1.31$$

The isobaric heat capacity is greater than the isometric heat capacity for a gas ($c_p > c_v$ or $c_{p'} > c_{v'}$). Specifically, the correlation between the molar isobaric heat capacity (C_p), and the molar isometric heat capacity (C_v), for an ideal gas is given by Mayer's relation[7]:

$$R = C_p - C_v \qquad 1.32$$

For a monatomic ideal gas, the result of eqn 1.32 fits our empirical findings. Specifically, for such a gas $c_v \approx 3R/2$, and $c_p \approx 5R/2$. Eqn 1.32 is often rewritten in different formats, e.g. [7]:

$$R = C_v(C_p/C_v - 1) \qquad 1.33$$

which leads to[7]:

$$C_v = R/(C_p/C_v - 1) \qquad 1.34$$

which can be rewritten as[7]:

$$C_v = R/(\gamma - 1) \qquad 1.35$$

where γ is the *adiabatic index*[7]: $\gamma = C_p/C_v$, that being the ratio of heat capacities

In this Book

The explanation for the difference between the two heat capacities will become apparent in the ensuing chapters, namely Chapter 5, where it is association with work is discussed. Specifically, this difference is due to work done onto the surrounding atmosphere.

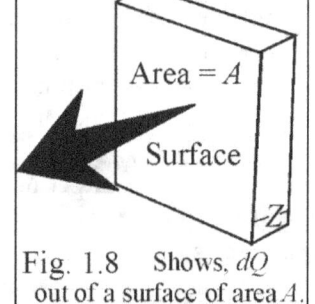

Fig. 1.8   Shows, dQ out of a surface of area A.

Thermal Conductivity

Thermal conductivity represents the ability of a system to transfer thermal energy. In order to better understand thermal conductivity, consider a surface with area A, as is shown in Fig. 1.8. Assuming it is not in temperature equilibrium therefore, heat is transferred through the surface at the rate given by[7]:

$$dQ/dt = -\kappa dT/dZ \qquad 1.36$$

where T= temperature, t = time, κ = coefficient of thermal conductivity and dZ= thickness

Reversibility

Reversibility is an idealistic concept concerning a system's state, wherein a system in some process can readily return to its initial state. Processes are generally irreversible i.e. cannot return to their original energy state without an input of external resources, e.g. an input of energy. The fact that most processes are irreversible has consequences such as prevention of perpetual motion, which is traditionally wrongly explained in terms of entropy and its accompanying postulate, that being the second law. Furthermore, reversibility requires that all thermodynamic change be infinitesimal[2] that forms a basis of what is current dogma, which often removes the science from reality.

In this Book

In the ensuing chapters it will be shown that irreversibility of all processes can be explained without the requirement of the second law. Moreover this demonstrates that the second law is a false postulate.

First Law of Thermodynamic:

In terms of infinitesimals the first law can be written in terms of a system's internal energy change (du), the energy input (dq_{in}), and work done (dw) by the system, that being the combined law:

$$du = dq_{in} - dw = dq_{in} - Pdv \qquad 1.37$$

du is positive when $dq_{in} > Pdv$. Conversely du is negative when $dq_{in} < Pdv$. Note: Writing $dw = Pdv$, means eqn 1.37 is limited to reversible work. Similarly, reversible heat is accepted as $dq_{in} = Tds$. For some reversible thermal energy input:

$$du = dq_{in} - dw = Tds - dw \qquad 1.38$$

For isobaric processes the first law is traditionally accepted as:

$$du = dq_{in} - dw = Tds - Pdv \qquad 1.39$$

Problematic concern G: *The inherent problem remains that neither eqn 1.37 nor 1.38 were for any real process because reversibility (mechanical or thermal) remains an idealistic rather than a realistic concept. Seemingly tradition asks us to accept that two wrongs make a right. Perhaps!*

Certainly one could argue that these are minor wrongs; however a wrong is never absolutely right! Moreover, the term law requires absolute! Another way of viewing this is reversibility is limited to quasi-static processes, which are not necessarily realistic processes.

In this book

It will be shown that irreversibility of all processes can be explained in simple terms with no reliance upon either entropy or the second law. Remember a basis of traditional thermodynamics is the use of second law to explain irreversibility, which will be shown to be a complication of reality all based upon a misunderstanding of lost work.

Problematic concern H: *Equating $dq_{in} = TdS$ is not exactly based upon any constructive logic rather it was defined that way by our 19th century greats. And through the 20th century we were told to embrace it. Yet any notion that this proves anything is nothing short of circular logic because it was defined as equality, rather than formulated upon constructive logic.*

Continuing eqn 1.39 gives:

$$Tds = du + Pdv \qquad 1.40$$

Substituting in for isometric specific heat (c_v), $du = c_v dT$ one obtains:

$$Tds = c_v dT + Pdv \qquad 1.41$$

Problematic concern I: *Consider a process in which no work is done ($dw = 0$). This leads to the following trivial result: $dq_{in} = du$. This implies that when no work is done, then: $Tds = du = C_v dT$ which is not particularly comforting because isothermal entropy change is equated to isometric heat capacity (C_v) times temperature change. It is doable because nobody knows what entropy is therefore entropy change can be anything or perhaps even strange remain something for everything!*

Integrating when $dw = Pdv = 0$ gives: $dS = c_v dT / T$ which leads to natural logarithmic functionality.

Problematic concern J: *The inherent logic remains illusionary because, a constant temperature system is on one side of the equality, while a temperature change exists on the other side of the same equality.* Similarly, *constant volume on one side of the equality and entropy change on the other.*

Okay, logic be damned! Continuing, divide eqn 1.41 by T gives:

$$ds = c_v dT / T + Pdv / T \qquad 1.42$$

For a mole of ideal gas molecules the ideal gas law leads to $P/T = R/V$. Hence:

$$ds = c_v dT / T + Rdv / v \qquad 1.43$$

Integrating results in:

$$\Delta S = C_v In(T_2 / T_1) + RIn(V_2 / V_1) \qquad 1.44$$

One can see why mathematical entropy has been embraced. Overlooking the previously discussed fundamental problems, you may become beholden to this world of natural logarithmic functions, which fits so well with probability based statistical thermodynamics. As to what is entropy's true guise? Based upon eqn 1.44 no one really knows!

Problematic concern K: *Consider a process where work is done while the energy input is zero i.e. $Q_{in} = 0$. Since $dq_{in} = TdS$ therefore there is no entropy change, but how can that be? Is not entropy related to work by eqn 1.40? Are we to believe that this correlation only exists when the system's internal energy is constant? Talk about weak constructive logic!*

Problematic concern L: *Eqn 1.44 concerns entropy change and entropy is traditionally accepted as being valid over all temperature regimes. To many this has specific special meaning especially for temperatures approaching absolute zero where as previously stated our understanding of temperature becomes arbitrary. If temperature is arbitrary then based upon 1.44 so too is entropy. Moreover heat capacity, which is not constant over all temperature regimes, in all likelihood loses all meaning near 0 K.*

Reconsider the enthalpy relation ($H = U + PV$). The combined first and second law are sometimes traditionally rewritten like the enthalpy relation as:

$$dh = du + d(PV) \approx du + Pdv + vdP \qquad 1.45$$

Solving for du in eqn 1.40 and substituting into eqn 1.45 gives:

$$dh \approx Tds - Pdv + Pdv + vdP = Tds + vdP \qquad 1.46$$

Eqn 1.46 is often referred to as the enthalpy relation or combined first & second law (b), or even Gibbs equation (b). It all looks grand except at certain levels what was conceived simply ignores fundamental issues, preventing the science from adhering to constructive logic.

In this Book

The above series of problematic concerns will be addressed. Furthermore the differences between a system's energy and its ability to do work will be addressed. I.e. putting energy and work in the same equation and then to claim that they have the same functionality to temperature may be irrational. It is like adding apples and oranges and not realizing that the answer isnow in fruit. To the indoctrinated this may seem absurd but in the ensuing chapters the differences between a system's ability doing work, and its thermal energy, will be investigated.

Joules Weight Experiment

Another of Joules experiments involves weights on a rope that drive a series of paddles rotating on shaft in a liquid. Herein it was determined that work and energy are one and the same based upon the liquid's temperature increase as the weights dropped thus rotating the paddles in the liquid. This equality of work and energy is further backed by various mathematical analyses in most textbooks. And it is true that work and energy are often one and the same, but this is not always the case, as previously stated!

In this book

Joules claims that his experiment shows that all the work that is put into the system in his experiment was readily turned into thermal energy. Interestingly it will be determined that the converse is not necessarily true. Specifically all the energy of a system cannot be extracted as work. Furthermore, Joules experiment also shows is that increased motions in a liquid leads to heating of that liquid, which implies that intermolecular collisions are not elastic, as is traditionally taught. This too will be dealt with herein.

Joules Gas Expansion Experiment: Revisited

Reconsider Joules expanding gas into a vacuum experiment. Planck[1pg51] discusses Lord Kelvin's version of this experiment, which is really a version of what is known today as a throttling process, where what he calls the external work done is defined by:

$$W = P_1V_1 - P_2V_2 \qquad 1.47$$

Problematic concern M: Planck was willing to consider external work in terms of: $W = -d(PV)$.

In this book

The implication must be that work done can be defined in terms of PdV will be examined e.g. the ideal gas law relates to work. Again this simple relation is traditionally avoided to protect the postulate (second law), and this too will be discussed throughout this book as simpler understandings are presented!

Entropy Change in Heat Transfer

For reversible isothermal heat transfer, the following is traditionally used:

$$ds - dq_{rov}/T \qquad 1\ 48$$

For path independent processes eqn 1.48 can be rewritten as:

$$\Delta S = Q_{rev}/T \qquad 1.49$$

Problematic concern N: *The fact is that heat transfer is always from high to low temperature, and that this is fundamentally not reversible, unless the temperature difference is so infinitesimally small that it approximates zero. This seemingly separates such analysis from reality.* Continuing with traditional analysis; for two heat reservoirs in thermal contact hence exchanging thermal energy:

$$\Delta S = Q_{rev}/T_c - Q_{rev}/T_h = Q_{rev}(T_h - T_c)/T_h T_c \qquad 1.50$$

Based upon eqn 150, the traditional claims are that heat must flow from hot to cold and that the netentropy change is always greater than zero. How constructive was the logic in getting to this point? There is no disagreement that the net flow of heat is always from hotter to colder but the traditional reasoning is disagreeable at so many levels.

Problematic concern O: *Eqn 1.50 requires reversibility between heat reservoirs, which limits its usefulness to systems whose temperature does not change during the given process. To further exasperate the situation the temperature difference must be real yet approximate zero!* *As a realistic applicable equation, eqn 150 remains suspicious at best*

In this Book

The reason two heat reservoirs were traditionally considered is simply because heat reservoirs contain so much heat that the extraction of some thermal energy can go unnoticed i.e. no measurable temperature change. The real reason that it is unnoticed is that thermometers tend not to be accurate enough to read minuscule temperatures changes associated with such thermal energy changes to heat reservoir, which is fundamental for a system to be considered a heat reservoir/bath. Just because temperature change is infinitesimally small, does not mean it did not occur! This book does not white-wash reality.

Traditional Free Expansion

Consider an expanding system or specifically the traditional writing for entropy change in the free expansion of a system (ΔS_{sys}) from state 1 to state 2:

$$\Delta S_{sys} = \int_1^2 ds = \int_1^2 dU'/T + \int_1^2 PdV/T \qquad 1.51$$

Since the process is unrealistically deemed isothermal, therefore: $dU' = 0$ and eqn 1.51 for a freely expanding system containing a mole of molecules becomes:

$$\Delta S_{sys} = \int_1^2 PdV/T = RIn(V_2/V_1) \qquad 1.52$$

The entropy increase as described by eqn 152 is to the expanding system itself. Sounds great, in part because it fits with the 20th century assertion that increases in randomness is associated with energy change.

Problematic concern P: *If the expanding gas is ideal and isothermal then based upon Boyle's law there is no change to the gas's energy. Therefore just because a gas has expanded does not mean that its energy has changed although one might argue that the gas's randomness has changed.*

It remains interesting that Ben-Naim[5] rightfully points out that randomness is not a particularly scientific term because when describing randomness of various systems, the answer remains in the eyes of the beholder. Furthermore Planck[1] realized that work is often done onto the surrounding atmosphere. It just seems strange that Planck did not make/state the following connection.

Problematic concern Q: *Concerning eqn 1.52 traditionalists failed to recognize that in free expansion that there is an exchange of energy between the expanding system and its surroundings.*

In this Book

It will be discussed that expanding systems tend to do work onto their surrounding atmosphere and that this is generally lost work[8,9]. In order to understand how free expansion has fooled thousands, all you need to realize is that it was conceived for quasi-static expansion. Thus for a system that does work, instead of cooling the expanding system's temperature, the expanding system remained isothermal because thermal energy is allowed to pass from the isothermal surroundings through the walls and into the expanding system. Moreover when discussing free expansion the expanding force is not aklways clearly defined, which has ramifications, as will be discussed throughout this book. This enables us to now understand the essence of this accepted gross misunderstanding!

Entropy and the Second Law

Herein, the conceptualization of entropy has already been challenged. What about the second law? The second law states that for any process, the isothermal entropy change of any isolated system is always equal to or greater than zero. And as previously stated this is traditionally used to explain why real life processes tend to be irreversibility!

Problematic concern R: *The second law loses it universal appeal because it is limited to ISOLATED systems. Few systems here on Earth are truly isolated. Specifically, any system that experiences a volume increase must displacement its surroundings atmosphere, in which case it is NOT an isolated system. This clearly dethrones the second law as some universal supreme law i.e.* **Second law is a false postulate***!*

Our reality; whether we are considering an expanding system powering a device, or a chemical reaction where the volume of the products surpasses the volume of the reactants, then we are considering system's that must displace our atmosphere's mass. To thinkof any other outcome is to claim that our atmosphere has no mass, or that its mass is not contained within Earth's gravitational field[8,9].

Accepting the equation written on Boltzmann's tombstone, then entropy can be defined by:

$$S = kln\Omega \qquad 1.53$$

where Ω is number of microstates, k is Boltzmann's constant

In this Book

As previously stated the second law's absolute validity will be dethroned mainly by showing that simpler explanations exist for all that it is wrongly claimed to explain.

Is this not simply logical, that molecules experiencing continuous intermolecular collisions with tend to disperse? Of course constraints generally prevent complete dispersal. Furthermore, thermal energy (heat) also tends to disperse often resulting in the heating of cooler systems at the expense of the hotter systems. This is irrelevant of whether it is heat from a fire, hot plate, exothermic reaction, or friction!

What Happens to Entropy?

What exactly is thermodynamic entropy remains for the world to decide. Possible definitions for entropy include:

a) A measure of how much effort would be required for a system to return to its original state. Since eqn 1.50 is only an approximation for entropy change generated by thermal energy changes, it is hard to gauge how scientific this understanding would be.

b) A heat capacity for non-homogeneous systems i.e. Eqn 1.49. Heat capacity could remain for homogenous systems of single state of a single type of matter.

c) Associated with work, hence $PdV = TdS$. Of course this needs some thought as it is really based upon the entropy change within a freely expanding system, which is a concept based upon the illogical association between randomness and energy.

d) As Boltzmann's guise i.e. eqn 1.53. If Boltzmann's entropy remains then Ω should only be a function of the system's energy. Hence the traditional concept that entropy relates to the randomness within an expanding system, is fool's gold at best. In other words the relation between entropy to volume becomes suspect.

e) Atkin's guise that being the quality of energy. As will be seen in this book; the higher a system's pressure is in relation to the surrounding atmosphere, the more work per unit volume that the system can do. Ditto the higher a system's temperature is, the more energy per unit volume that can be extracted from that system.

Entropy can only be any one of the above, or something else or even expunged. It cannot remain something for everything, because without some exacting clarity, it remains meaningless.

Closing Remarks

The traditional insistence of writing thermodynamic around a false postulate (second law) has led to the mathematical contrivance[10] known as entropy, being habitually used although nobody knows what it means. This creates many problematic concerns that few have dared to address with any conviction.Never forget; although empirical data can disprove a theory, it cannot necessarily prove any one theory i.e. more than one given theory can explain given empirical findings. Hopefully even those indoctrinated in the science may actually open their minds to other plausible simpler explanations for our various empirically verified findings. Concepts like Helmholtz free energy will be shown to be right but for the wrong reasons!

In the ensuing chapters, new considerations of the science's fundamentals will be presented. It will be up to the reader to determine whether our new perspective is more palatable what is accepted. It is hoped that you will be open minded, void of indoctrination and abide by the principles of Occam's razor (Ockham's razor), which is paraphrased: *"All things being equal, the simplest solution remains the best"*.

Hopefully you will find that the solutions described in this book are simpler than what is currently accepted traditional mainstream. After clearly demonstrating that a simpler constructive logic based explanation exists, some of the problems with entropy and second law based thermodynamics will be revisited in Chapter 16.

References:

1. Planck, Max "Treatise on Thermodynamics" Third edition, London, Logmans, Green and co., 1917
2. Tilley D.E. "Contemporary College Physics" Benjamin Cummings Publishing Don Mills Ont Canada 1979
3. Lambert F.L. "Entropy is simple, qualitatively". J Chem. Educ 79 187 (2002)
4. Atkins , P. "Four laws that drive the universe" Oxford University Press Oxford England 2007
5. Ben-Naim "A Farewell to Entropy: Statistical thermodynamics Based on Information." World Scientific Publishing Co Hackensack NJ 2011
6. Reif, F."Fundamentals of Statistical and Thermal Physics", McGraw-Hill, New York, 1965
7. Reif F."Statistical Physics", McGraw-Hill, New York, 1967
8. Mayhew, K.W. "Second law and lost work", Phys. Essays **28**, vol 1, 152 (2015)
9. Mayhew, K.W. "A new perspective for kinetic theory and heat capacity", *Progress in Physics* Vol. 13 (**4**) 2017 pg 166-173
10. Mayhew, K.W. "Entropy an ill-conceived mathematical contrivance" Phys. Essays **28** vol 3, 352 (2015)

Chapter 2: Thermal Energy and Kinetic theory

Thermodynamics' main concern the interactions of thermal energy with matter, or if you prefer, how systems of matter behave as a function of temperature. Herein will be discussed the fundamentals of thermal energy and kinetic theory. It should be stated that traditional kinetic theory does not match empirical findings for heat capacity particularly well. This will be discussed, while an improved kinetic theory will be presented. A theory that better matches known empirical findings as envisioned by this author and published in on-line Journal "Progress in Physics" in July 2017, followed by April 2018. Note a variation giving the same results but using a slightly different analysis was first published in this book's first edition.

Thermal Energy

As was discussed in Chapter 1, thermal energy resides within condensed matter as both *intermolecular vibrations* are between molecules, and *intramolecular vibrations* between the various atoms, all of which can be considered asenergetic vibrations between bonds whose packets of energy are *phonons*. Accordingly, the photons associated with the thermal radiation from our Sun are fundamental to any consideration of the temperatures that is witnessed here on Earth. Certainly other sources of thermal energy contribute, such as heat associated with pressure, heat from irreversible processes etc etc.

Liquids and amorphous solids lack the crystalline structure to which phonons are mathematically related. Even so, it is accepted that the thermal energy contained within such substances can be considered as phonons (packets of energy) that are treated in the same manner. Liquid molecules also have freedom of movement (e.g. convection), and accordingly they possess both translational and rotational energies. However, both of these energies are generally considered as being minor in comparison to vibrational energy. Therefore, most of the thermal energy within all condensed matter can be attributed to vibrational energy within that matter.

It is understood that condensed matter adsorb thermal energy by the absorption of thermal photons, turning them into phonons. Conversely condensed matter release previously absorbed thermal radiation by emitting photons. Importantly from a perspective of volume, all matter tends to concentrate thermal energy i.e. increase the thermal energy density within a given volume. As complicated as explanations can be, all that is needed is to understand that thermal energy is somehow held within condensed matter!

Simply defined thermal energy is heat. A better definition is that thermal energy consists of a spectrum of thermal photons and/or phonons, i.e. those whose wavelengths are sufficiently long that they are readily absorbed by condensed matter, becoming either intramolecular or intermolecular vibrations. For the most part thermal energy consists of a spectrum of infrared wavelengths, however depending upon the substance and temperature, thermal energy may also include microwave and/or visible & UV light.

Energetics within Condensed Matter

When dealing with the energy of molecules in condensed matter there are two energy terms. The first term represents the energy associated with the momentum of the molecule i.e. the molecule's kinetic energy. The second term represents the potential energy, which depends upon the location of the molecule and/or its elements. This is similar to a harmonic oscillator, e.g. two masses attached via a spring, wherein we associate both a kinetic and a potential energy (See Appendix B.5). If x = position coordinate along the x-axis, $\vec{p}_x$ is the momentum along the x-axis and α = spring constant, then the energy of a one-dimensional harmonic oscillator can be written as[1,2]:

$$\overline{E} = |\vec{p}_x|^2 / 2m + \alpha x^2 / 2 \qquad 2.1$$

When applied to the energy within condensed matter, both the terms on the R.H.S. of eqn 2.1 are quadratics, thus the equipartition theorem gives[1,2]:

Mean kinetic energy $= \overline{E}_k = |\vec{p}_x|^2 / 2m = kT/2$ 2.2

Mean potential energy $= \overline{E}_p = \alpha x^2 / 2 = kT/2$ 2.3

The total mean vibrational energy ($\overline{E}_v$) of a one-dimensional harmonic oscillator is the summation of the kinetic and potential energies, that being[1,2]:

$$\overline{E}_v = \overline{E}_k + \overline{E}_p = kT/2 + kT/2 = kT \qquad 2.4$$

For a crystalline solid, each molecule is considered as a simple harmonic oscillator about its lattice site with the molecular motions being along all three orthogonal axes (x,y,z). Moreover, the mean energy along each axis is taken to be equivalent, with a mean magnitude defined by eqn 2.4. Accordingly, for N molecules the total thermal energy (E_T) along the three axes is:

$$E_T = 3NkT \qquad 2.5$$

For a mole of molecules ($N = 6.02 \times 10^{23}$ molecules) $Nk = R$, where: R is the universal gas constant = 8.31 J/mol/K. Then eqn 2.5 becomes:

$$E_T = 3RT \qquad 2.6$$

Note that eqn 2.5 and eqn 2.6 consider the total thermal energy as being purely vibrational, therefore for condensed matter: $E_T = E_v$.

Note: Those familiar with traditional thermodynamics will realize that eqn 2.5 and 2.6 are what one would obtain based upon traditional analysis using equipartition and degree of freedom arguments e.g. each molecule has 3 degrees of freedom with each degree of freedom having a mean energy defined by kT. In so much as the results for condensed matter agree with empirical findings we do not agree with the implications to gases, hence herein the merits of such arguments are diminished, beyond the fact that it is a mathematical construct rather than a logic based result. Basically condensed matter adsorbs thermal radiation, and then equally distributes it in all directions throughout that matter via molecular vibrations.

Equipartition & Crystalline Solids

In order to calculate the thermal energy (dQ) that can be extracted from a crystalline system resulting in a temperature change (dT), eqn 1.2.5 is used: $dQ = C_v dT$. The subscript "v" indicates that the molar heat capacity is taken at a constant volume[7]:

$$C_V = dQ/dT = 3R \qquad 2.7$$
$$= 6 \text{ cal/mole/K} = 25.10 \text{ joules/mole/K}$$

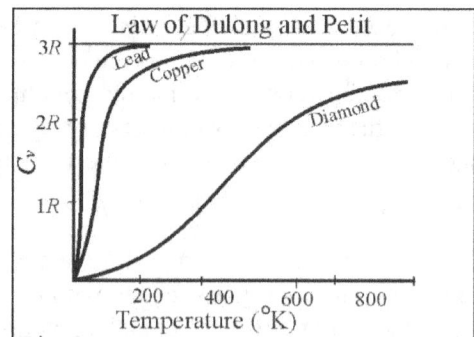

Fig 2.1   Shows the temperature dependence for the molar heat capacity (C_v) at constant volume. All metals and nonmetals have a molar heat capacity between lead: $C_V \rightarrow 3R$ at about 200°K, and diamond: $C_V \rightarrow 3R$ at about 2000°K. [9]

Eqn 2.7 has been shown to apply to most condensed crystalline matter and is known as the *Law of Dulong and Petit*. Fig. 2.1 shows the molar heat capacity as a function of temperature for three substances, namely: lead, copper, and diamond.

For most crystalline substances, the law of Dulong and Petit is not valid when at extremely low temperatures. Moreover, the heat capacity of solids tends to zero, as the temperature tends toward absolute zero, at which point the Debye's theory is often used (See Appendix B.4).

For most crystalline substances at room temperature (395 K) is sufficiently high that the molar heat capacity can be equated to $3R$. Rather than on a molar basis, eqn 2.7 can be rewritten in terms of its N molecules, for most crystalline substances i.e.:

$$C_V = dQ/dT = 3Nk \qquad 2.8$$

Eqn 2.8 states that the thermal energy stored per degree Kelvin (or Celsius) is $3Nk$, for most crystalline matter, i.e. the total thermal energy (E_T) stored in an N molecule crystalline substance, at temp T, can be approximated by eqn 2.5: $E_T = E_v = 3NkT$.

As seen in Fig 2.1, diamond is an exception to the above in that its molar heat capacity does not approach $3R$ until its temperature reaches 2000 K. As explained to this author by an acquaintance: Diamond has a perfect infinite tetrahedral lattice thus there are no/few intermolecular vibrations in diamond. In fact a one-carat diamond is a one-carat molecule making diamond the largest solitary molecule on Earth!

System of Gas

There are four types of energy associated with a volume of gas:

1) Rotational energy
2) The translational energy
3) Vibrational energy within any polyatomic gas molecules.
4) Radiation energy residing in freespace, i.e. thermal radiation.

Monatomic gases have no vibrational energy, while polyatomic gases will possess vibrational energies between the molecule's atoms along their atomic bonds, i.e. they absorb and emit thermal photons. The energy associated with the thermal radiation[3] (in freespace) generally is infinitesimally small when compared to the energies of motion of the molecules. Thus, when calculating a system's total energy, such thermal radiation can often be omitted. Even so, it is prudent to incorporate the radiation energy as part of our logic.

A New Perspective for Kinetic Theory

The interaction of gases with condensed matter is the basis of kinetic theory. The theory has its origins in the 18th century with Bernoulli; however the current traditionally accepted understanding is more based upon 19th century conceptualizations by the likes of Maxwell, Clausius and Boltzmann. That being a combination of the law of equipartition (theory) and the degrees of freedom argument, which may be more a mathematical conjecture than some logically, construed theory. Interestingly the following is Maxwell describes some attributes of the theory in his 1875 paper.[4]

"The kinetic energy of the molecule may be regarded as made up of two parts--that of the mass of the molecule supposed to be concentrated at its centre of mass, and that of the motions of the parts relative to the centre of mass. The first is called the energy of translation, the second that of rotation and vibration. The sum of these is the whole energy of motion of the molecule.

The pressure of the gas depends, as we have seen, on the energy of translation alone. The specific heat depends on the rate at which the whole energy, kinetic and potential, increases as the temperature rises.

Clausius had long ago pointed out that the ratio of the increment of the whole energy to that of the energy of translation may be determined if we know by experiment the ratio of the specific heat at constant pressure to that at constant volume.

He did not, however, attempt to determine à priori the ratio of the two parts of the energy, though he suggested, as an extremely probable hypothesis, that the average values of the two parts of the energy in a given substance always adjust themselves to the same ratio. He left the numerical value of this ratio to be determined by experiment.

In 1860 I investigated the ratio of the two parts of the energy on the hypothesis that the molecules are elastic bodies of invariable form. I found, to my great surprise, that whatever be the shape of the molecules, provided they are not perfectly smooth and spherical, the ratio of the two parts of the energy must be always the same, the two parts being in fact equal. This result is confirmed by the researches of Boltzmann, who has worked out the general case of a molecule having n variables."

It is interesting to note Maxwell's surprise at the ratio of energies (the translational motion the rotational and vibrational) being equal, as not even he expected this result! This, and the ensuing work of Boltzmann and others concerning statistical ensembles, led to the 20th century understanding of the determination of molecular energy using statistical analysis, the law of equipartition and degrees of freedom based argument. The basics of equipartition theory and degrees of freedom are further discussed in Appendix B.3 of this text. For those interested in more encompassing analysis, other texts do give a more thorough discussion e.g. Reif[1], Carey[5].

Problematic traditional kinetic theory

Due to inconsistencies between theory and empirical findings for heat capacities of gases, traditionally accepted kinetic theory often illogically claim various exceptions. In literature you may find any combination of the following exceptions:

1) Monatomic gases have no rotational energy.
2) Diatomic gases have no vibrational energy or that they too have no rotational energy.

Amazingly the above exceptions are blindly accepted in ordr that the purely mathematical based equipartition theory better matches known empirical findings. Limiting monatomic and/or diatomic gases to translational energy is like saying a curve ball in baseball has no rotational energy! The accepted argument being that small radius means small to no angular momentum. This is ridiculous because given the same impulse/force passing on a certain angular momentum, simply means that the smaller the radius the higher the angular velocity should be. In other words the expectation is that for a given collision, the smaller the radius, the larger the angular velocity is. Accordingly smaller radius molecules can have the similar angular momentum, as larger radii molecule have.

It is somewhat irrational to claim that triatomic molecules all of a sudden have full rotational energy ($kT/2$ per degree of freedom), while both monatomic and/or diatomic molecules have none. Furthermore, the contention that diatomic gases have no vibrational energy, is like saying the bonds are different for diatomic gases than triatomic and/or polyatomic gases!

Interestingly, Einstein[8] acknowledged that reason quantum theory was developed, was in part to explain why traditional kinetic theory and known empirical finding for gas's heat capacities do not match more closely than they do. The completeness of what is said will become more apparent when heat capacities are discussed in later chapters of this book. As a side-point this authors wonders to what extend quantum theory would have been fully embraced if the great minds of over a century ago had adhered to kinetic theory as presented herein.Seemingly quantum physics may require a similar rethink.

Our new perspective

Herein an alternate perspective than the traditionally accepted theory, will be given. By applying some of the conscripts of equipartition to condensed matter and then making other logical deductions for the kinetic theory of gases, this author's theory is presented, that being an improved theory that is a superior fit with known empirical findings[6.7]. This is simply not a case of arbitrarily choosing some aspects of traditional/statistical thermodynamics over other ones; rather it is a consequence of logical deduction, backed by empirical data. The irrefutable evidence confirming our new understanding will be given in Chapter 5 and then other aspects will be discussed in both Chapters 6 &7.

Visualizing the Translational, Rotational Energies of a Monatomic Gas

In Fig.2.2, a sufficiently dilute monatomic gas is illustrated where the collisions are primarily between the gas molecules and their surrounding walls.

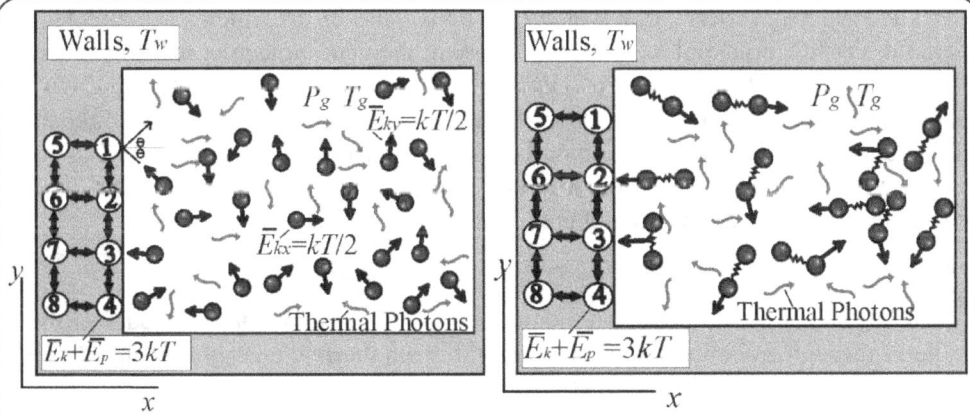

Fig. 2.2 Shows an ideal monatomic gas, at pressure: P_g and temperature: T_g, surrounded by walls at temperature T_w. The gas molecules have no vibrational energy.

Fig. 2.3 Shows an ideal diatomic gas, at pressure: P_g and temperature: T_g, surrounded by walls at temperature T_w. The gas molecules all have vibrational energy.

Traditional equipartition and degrees of freedom: The mean kinetic energy per gas molecule along the x-axis is: $\overline{E}_{kx} = kT/2$. Similarly along the y and z-axis, the mean kinetic energies respectively are: $\overline{E}_{ky} = kT/2$, and $\overline{E}_{kz} = kT/2$. The energy of a wall molecule along the x-axis is due to both kinetic and potential energy: $\overline{E}_{kx}+\overline{E}_{px}=kT$. And similarly for the y and z axis: $\overline{E}_{ky}+\overline{E}_{py}=kT$ and $\overline{E}_{kz}+\overline{E}_{pz}=kT$. The mean total energy of each wall molecule is: $\overline{E}_{k}+\overline{E}_{p} = 3kT$

Our new perspective has the gas's mean translational kinetic energy along the three axis plus its mean rotational energy equal to the above described wall molecule's kinetic energy.

Note: Both the gases are in thermal equilibrium when the wall's temperature equals the gas's temperature: i.e. $T_w=T_g$, which also equals the temperature associated with the surrounding radiation (blackbody/thermal).

Case 1): Start by imagining that a monatomic gas molecule collides dead on with a wall molecule, e.g. the gas molecule hitting wall molecule #3 in Fig 2.2. Assume that the gas molecule only obtains translational energy from the vibrating wall molecule. Accordingly, the gas molecule's resulting mean kinetic energy ($\overline{E}_k$) (due to impact) would be purely translational. It is like hitting a golf ball square on and having the majority of the golf ball's energy being purely translational, thus attaining a long distance down towards the golf green.

Case 2): Next imagine that the monatomic gas molecule hits the wall molecule at some angle other than $\theta = 90^\circ$. As expected, the gas molecule would obtain both rotational and translational energy from the vibrating wall molecule such that the total mean resultant energy (due to impact) of the gas molecule would be similar to that in Case 1). It is like hitting a golf ball with the same force/impulse of impact but slicing the ball, hence giving the ball a significant amount rotational energy in comparison to its translational energy, thus watching your golf ball plummet into the woods, shortly after contact.

Case 3): Finally imagine a rotating and translating monatomic gas molecule striking the wall transferring both its rotational and translational energies onto the wall molecule. The wall molecule cannot rotate therefore both energies will only result in the wall molecule attaining vibrational energies from the impact, where the wall molecule's vibrational energy is along some combination of its three orthogonal axes.

Note in the above: The total mean energy of a wall molecule is still $3kT$, of which kT is directed along the x-axis, which includes both the potential and kinetic energy of the wall molecule, with the mean kinetic energy of the wall molecule along the x-axis being $kT/2$. This energy along positive x-axis is then passed onto any gas molecules that collide with the wall molecule, resulting in both translational and rotational energy of that gas molecule. This is different than traditional accepted theory, which is based upon the mathematical degree of freedom argument and $kT/2$ along each degree as claimed by equipartition, hence the gas molecule's translational energy along the x-axis is $kT/2$.

Mathematics for the Translational, Rotational Energies of a Monatomic Gas

Since what is going on has been visualized, then the proper equations can be determined. For the group of wall molecules 1 through 8 shown in Fig 2.2, the total mean thermal energy associated with a vibrating wall molecule, along the x-axis, is defined by:

$$\overline{E}_x = kT \qquad\qquad 2.9$$

As was previously discussed half of a wall molecule's mean thermal energy would be kinetic energy, and half would be potential energy, Accordingly, the mean kinetic energy ($\overline{E}_{kx}$) of a wall molecule along the x-axis, is based upon eqn 2.2 that being:

$$\overline{E}_{kx} = kT/2 \qquad\qquad 2.10$$

Due to the massive size of the wall in comparison to gas molecules, the wall' s continuous vibrations, and the continuous collisions, the gas molecule's mean energy along the x-axis is defined by 2.10. This equally applies to gas molecule-wall collisions along both the y and z-axis.

This all can be viewed another way; the wall behaves like a massive machine pumping its kinetic energy onto the smaller gaseous molecules, along all three axes. Since each orthogonal wall will impart the same mean kinetic energy onto each gas molecule, then the mean translational plus rotational energy of the gas molecule [$\overline{E}_{k(t,r)}$] equals three times the wall's mean kinetic energy along any one axis, i.e.:

$$\overline{E}_{k(t,r)} = 3kT/2 \qquad\qquad 2.11$$

For N gaseous molecules, the total kinetic energy (translational plus rotational) [$E_{T(t,r)}$] is[6]:

$$E_{Tk(t,r)} = 3NkT/2 \qquad\qquad 2.12$$

The magnitude for total energy as defined by eqn 2.12 equals what is traditionally accepted for the translational energy. However our analysis and nomenclature for the energy of gas molecule emphasizes that it is both translational and rotational energy of the gas that is obtained from the wall molecule's kinetic energy. It should be emphasized that eqn 2.12 is only valid for systems wherein gas-wall collisions dominate when compared to gas-gas collisions.

Translational, Rotational, Vibrational Energies of a Polyatomic Gas

Fig 2.3 illustrates a system of dilute diatomic gas molecules in a container. The vibrating wall molecules still pass the same amount of kinetic energy onto the diatomic gas molecule's center of mass with each collision, as was the case for the monatomic gas. It really does not matter how the diatomic molecule strikes the wall molecule, e.g. the same principles apply whether it be a diatomic gas molecule that is striking wall molecule #2, or the one striking wall molecule #3 in Fig 2.3. Note: For a more enlightened analysis of diatomic collisions, see Appendix B.5.

So as was the case for the monatomic gas, the diatomic gas molecule's total translational plus rotational energy remains defined by eqn 2.11. What about the vibration energy of the diatomic gas molecules? As is the case for condensed matter, this is related to the absorption and emission of the blackbody/thermal radiation surrounding the diatomic gas molecules. Therefore the mean vibrational energy within a diatomic gas molecule is also defined by eqn 2.4: $\overline{E}_v = kT$.

Accordingly, the mean total energy ($\overline{E}_{tot}$) for the sufficiently dilute diatomic gas molecule must be the addition of its mean translational energy and rotational energy plus its mean vibrational energy. That being:

$$\overline{E}_{tot} = \overline{E}_{T(t,r)} + \overline{E}_v = 3kT/2 + kT = 5kT/2 \qquad 2.13$$

Now consider a triatomic gas molecule. The mean vibrational energy should be twice that of a diatomic, therefore for a triatomic gas molecule the mean vibrational energy ($\overline{E}_v$) is:

$$\overline{E}_v = kT + kT = 2kT \qquad 2.14$$

Similarly, the total mean energy of the triatomic gas molecule must be the addition of its mean kinetic (translational and rotational) energy, plus its mean vibrational energy, that being:

$$\overline{E}_{tot} = \overline{E}_{Tk(t,r)} + \overline{E}_v = 7kT/2 \qquad 2.15$$

If the polyatomic gas molecule absorbs/emits thermal photons in a manner similar to diatomic molecules, then an n"-molecule polyatomic gas molecule should have a mean vibrational energy of[6]:

$$\overline{E}_v \cong (n''-1)kT \qquad 2.16$$

where n'' signifies the number of atoms in each gas molecule, which shall be called *polyatomic number*. The mean total energy ($\overline{E}_{tot}$) for a polyatomic gas molecule becomes the addition of its mean kinetic energy plus its mean vibrational energy. In other words equations 2.12 and 2.16 are added together, thus obtaining for the mean total energy ($\overline{E}_{tot}$):

$$\overline{E}_{tot} \cong (n''-1)kT + 3kT/2 \qquad 2.17$$

Collecting the terms, gives:

$$\overline{E}_{tot} \cong (n''+1/2)kT \qquad 2.18$$

The total energy [E_T] for N gas molecules, becomes[6]:

$$E_T \cong NkT(n''+1/2) \qquad 2.19$$

Consider n moles of gas then eqn 2.19 becomes[6]:

$$E_T \cong nRT(n''+1/2) \qquad 2.20$$

Eqn 2.19 is a theoretical approximation, which matches exceptionally well with empirical findings for heat capacities of gases for all values of the number of atoms in each gas molecule, where $n''< 4$. At $n''= 4$, there is a marked change between empirical findings and this theory however there is a small marked decrease in heat capacity for the empirical findings, when compared to theoretical. Importantly the difference between the two never changes i.e. the difference between empirical findings and this theory remains constant for $4 < n''< 20$. It must be emphasized that the fit of this theory with empirical finding is far superior when compared to the fit with the traditional kinetic theory. Note this will all make more sense after reading Chapter 5.

At first this author thought that the difference was perhaps due to variations in the molecule's shape and how they may interact with the surrounding thermal radiation. Having second thoughts this author then realized that it was long linear gaseous molecules whose empirical findings did not match theory. At which point this author started to think it was because of *flatlining*.

Flatlining

Imagine a linear molecule ($n''\geq 4$) flatline's against a wall as illustrated in Fig 2.4. With one wall molecule is moving outward from the wall, while its neighbor is moving inward into the wall.

The long flatlining molecule still bounces off of the wall but not as cleanly as a smaller molecule would. Accordingly, the wall molecules mean kinetic energy is not cleanly pumped onto any flatlining molecules. There may be other explanations but this one certainly fits and will be discussed further in Chapter 5.

A new perspective for thermal equilibrium & the role of walls

Fig. 2.4   Shows a $n''=4$ gas, at pressure: P_g and temperature: T_g, surrounded by walls at temperature T. The gas molecule hitting wall molecule 2 adheres to kinetic theory but the gas molecule hitting wall molecules 5 thru 8 flatlines thus do not receive as much energy from the wall.

Consider that both a given volume of dilute gas and surrounding walls are in thermal equilibrium at temperature T, as is illustrated in Fig. 2.3. This means[6]:

1) The walls are in thermal equilibrium with the enclosed blackbody radiation e.g. both are related to the same temperature (T).
2) The gas molecule's translational plus rotational energy is energy equilibrium with the molecular vibrations of the wall molecules at T.
3) The gas molecule's vibrational energies are in thermal equilibrium with the enclosed blackbody radiation e.g. both are related to the same temperature (T).

The combination of these three states of equilibrium causes the energies of the gas molecules and wall molecules to correspond with each other as well as their surrounding thermal radiation, i.e. if isothermal walls did not surround the gas then true equilibrium between the gas molecules and the surrounding thermal radiation may not exist. A fact overlooked by traditional analysis, one that will be revisited in Chapters 5 thru 7. Note for radiation equilibrium does not necessarily mean that same spectrum is emitted as is adsorbed, rather it means that the energy flux (rate) is the same for absorption as emission.

Gas's Kinetic Energy

The kinetic energy of the walls from three orthogonal axes is passed/pumped/imposed onto an enclosed gas molecule, resulting in both the gas's kinetic and rotational energies. However in the future chapters of this book when discussing issues like work, the gas's kinetic energy means both the translational and rotational energy because it is both of those energies from the gas that is passed on, enabling systems to perform work. Also, this is because most often work done simply results in a kinetic and/or potential energy increase of something external to the system i.e. surroundings.

Also herein it was never determined to what degree such kinetic energy would be rotational versus translational. The implication being that the translational velocities of gases may actually be lower than Maxwell's velocity distribution implies, although the energy of the gas as implied by the distribution functions should remain approximately correct especially for smaller (i.e. monatomic, diatomic,triatomic) sufficiently dilute gases, expectation remains for small molecules translational > rotational energy . Another case of the sciences getting answers right for the wrong reasons. To what exact extend a sufficiently dilute gas's energy is rotational vs translational is now up for debate.

Traditional: Loschmidt's Paradox (1876)

It is of interest that that Loschmidt paradox (A.K.A. irreversibility paradox) which puts the time reversal of fundamental processes at odds with the second law that was used to describes macroscopic systems. Namely that Joseph Loschmidt's challenged Boltzmann's H-theorem, which used traditional kinetic theory to explain the entropy increase of a non-equilibrium state ideal gas, when the gas molecules are allowed to collide. Specifics are left to others, but our new understanding does impose challenges to such accepted doctrine.

Closing Remarks

We discussed the kinematics of matter, presenting a simple explanation as to how the energy of a system relates to Boltzmann's constant (k). Basically the orthogonal walls of a closed system act as massive pumps, pumping/imparting/imposing a mean energy of $kT/2$ onto each of the significantly smaller gas molecules. The important revelations being:

1) That the kinetic energy of the wall molecules equals the dilute gas' translational plus rotational energy. This says nothing about the relative values of rotational and translational energies of a gas, except that they are to be added and equated to the summation of the wall molecule's mean kinetic energies along all three axes

2) The total energy of the dilute gas being equated to the translational plus rotational energy, plus any vibrational energy. In Chapter 5 it will be shown that this insight better explains the accepted empirical findings for the heat capacity of gases, than the traditional understanding.

3) The thermal radiation of the blackbody radiation that exists in interior volumes of freespace is what allows for thermal equilibrium to exist in many systems.

4) The above stated thermal radiation is responsible for the vibrational energies of the dilute polyatomic gases within a system i.e. polyatomic gases and condensed matter are to be treated equally.

5) The energy associated with the above stated radiation are generally infinitesimally small when compared to the kinematic energies associated with matter. Exceptions being vacuums and extremely high temperature systems e.g. blast furnaces.

At first glance the above defies the traditionally accepted mathematically based equipartition theory. However, one might consider this as a more logical altered equipartition wherein the rotational plus translational energy of sufficiently dilute gases are part of (equated to) one and the same degree of freedom. Of course degree of freedom is a mathematical consequence whose definition and/or understanding will require significant changes.

It should be further stated that most empirical data is obtained using closed systems contained within walls. The ramifications of such walls will become readily apparent in Chapters 5 through 7, as to will the proof be given concerning what was discussed herein. Note: See Graph 5.1, where empirically obtained heat capacities of gases clearly prove that our new perspective concerning kinetic theory is superior to the traditionally accepted theory.

 Perhaps the most important ramification that will become apparent is that molecular collisions are for the most part NOT elastic i.e. inelastic. Understand that traditional kinetic theory requires elastic collisions and therefore CANNOT be right.

Reconsider Planck's Statement in Chapter 1

If we are to adhere to Planck's previously stated (Chapter 1) words concerning the first way to formulate thermodynamics. *"We may take for granted .... reduced to motions of materials points between which there act forces which have a potential. Then the principle of energy is simply the well-known mechanical theorem of kinetic theory, generalized to include all natural processes."*

Remember traditional thermodynamics is based upon Planck's stated second way, that being the one the renders the second law as the postulate supreme. Obviously, if we are rewrite thermodynamics not only will we need to use our new improved kinetic theory but we will have to come to a better understanding of the mechanical nature, which will be dealt with in the ensuing chapters.

References:

1. Reif, F., "Fundamentals of Statistical and Thermal Physics", McGraw-Hill, New York, 1965
2. Reif, F., "Statistical Physics", McGraw-Hill, New York, 1967
3. Eisberg,R., Resnick, R. "Quantum Physics", John Wiley & Sons   Toronto 1974
4.   J.C. Maxwell, J. Chem. Soc (London), **28**, 493-508, (1875) [facsimile published in Mary Jo Nye, The Question of the Atom (Los Angeles: Tomash 1984)]
5. Carey, V. "Statistical Thermodynamics and Microscale Thermophysics", V. Carey, Cambridge U 1999
6. Mayhew, K.W. "A new perspective for kinetic theory and heat capacity", *Progress in Physics* Vol. 13 (**4**) 2017 pg 166-173
7. Mayhew, K.W. "Kinetic theory: Flatlining of Polyatomic Gases", *Progress in Physics* Vol. 14 (**2**) 2018
8. Einstein A. and Stern O. "Einige Argumente Fur die Annahme einer molekularen Agitation beim absolute Nullpunk" (Some Arguments for the Assumption of Molecular Agitation at Absolute Zero). {Ann. Phys.} 40 551: 551-560 (1913)

Chapter 3: Work: The Basics

In this chapter the basic concepts pertaining to work are discussed. Again there will be similarities with traditional thermodynamics; however there will also be some profound differences. Appendix B.6 provides an additional enhanced understanding of work.

What is Work?

The fundamental principle of mechanical work (W) can be expressed in terms of the dot product of force ($\vec{F}$) and distance (dx), i.e.[1,2]:

$$W = \vec{F} \bullet dx \qquad 3.1$$

Our first experience in defining work generally involves the displacement of a weight, that being the movement of a mass against a gravitational field. For example: A man lifting a mass, then the work done (W) is[1]:

$$W = M\vec{g}dh \qquad 3.2$$

where $\vec{g}$ is the gravitational constant, M is mass and dh is the height the mass is raised. Note herein the angle between $\vec{g}$ and dh is $\theta = 0$, hence $\cos\theta = 1$. And $\vec{F} = M\vec{g}$ and $dx = dh$.

The work is done onto this mass, resulting in its increase in potential energy (dE_p) is defined by:

$$dE_p = M\vec{g}dh \qquad 3.3$$

Although, all mechanical work does not necessarily involve the displacement of weight (mass against gravity), the correlation between mechanical work, and the displacement of weight, are fundamental to our new perspective. Furthermore, such work is path-independent, i.e. only depends upon endpoints!

Compression/Expansion of Ideal Gases in Closed Systems

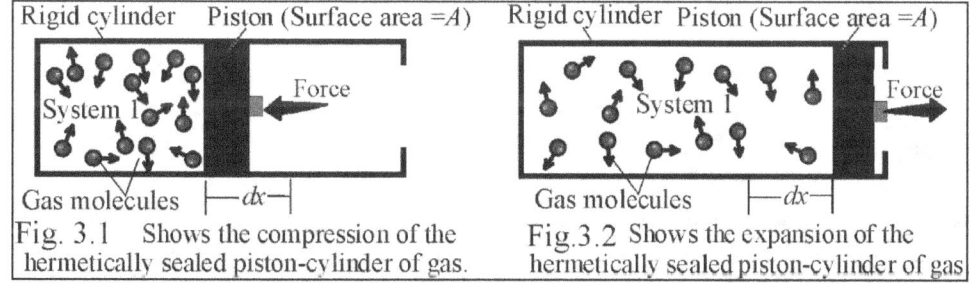

Fig. 3.1   Shows the compression of the hermetically sealed piston-cylinder of gas.

Fig.3.2 Shows the expansion of the hermetically sealed piston-cylinder of gas

Consider a hermetically sealed piston-cylinder apparatus filled with an ideal gas at 1 atm pressure as System 1. The ideal gas can be compressed by applying a force upon the piston, thus displacing it inwards a distance dx, as illustrated in Fig 3.1. Conversely, the ideal gas can be expanded, therefore displacing the piston outwards a distance, dx, as shown in Fig 3.2.

For both the expansion and compression, the work required to move the piston can calculated in terms of a force ($\vec{F}$) over a distance (dx). This can be transformed into a pressure-volume relation by dividing the force by the piston's surface area (A) and then multiplying the distance, dx, by the piston's surface area:

$$W = (\vec{F}/A)Adx \qquad 3.4$$

Eqn 3.4 assumes path independence. Experience tells us that a gas's compression often results in a temperature increase. Since the ideal gas law ($PV = NkT$) tells us that a temperature increase leads to $PV \uparrow$ therefore the simply assumption of path independence is not precisely correct. However, if the compression is quasi-static then any temperature increase could radiate out through the cylinder's walls, in which case the quasi-static compression could be considered as being isothermal.

Moreover, even for isothermal compression, the work is not exactly path independent. Since, $F/A=P$, and $Adx = dV$, then for path dependent infinitesimal work (w), in terms of pressure, and infinitesimal volume change (dv) we can write:

$$w = Pdv \qquad\qquad 3.5$$

The reason the work is path dependent is that the instantaneous force required to move the massless, frictionless piston is directly proportional to the pressure difference between the System 1, and its surroundings at P =1 atm. Therefore, the farther away from the neutral position (P=1 atm), that the piston is pushed, or pulled, then the greater the pressure change with becomes, hence the greater the magnitude of force required to move the piston another incremental distance, dx, must be.

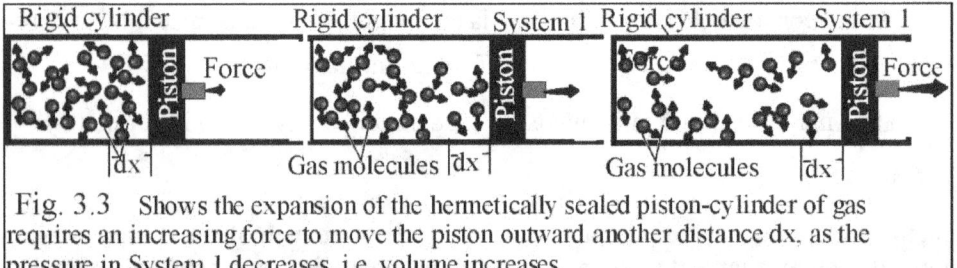

Fig. 3.3   Shows the expansion of the hermetically sealed piston-cylinder of gas requires an increasing force to move the piston outward another distance dx, as the pressure in System 1 decreases, i.e. volume increases.

The magnitude of the required force increases as the piston moves a further incremental distance dx, as illustrated by the arrow's increasing size in Fig. 3.3. The Force vs Distance: Graph 3.1 illustrates the increasing required force, as the distance, dx increases.

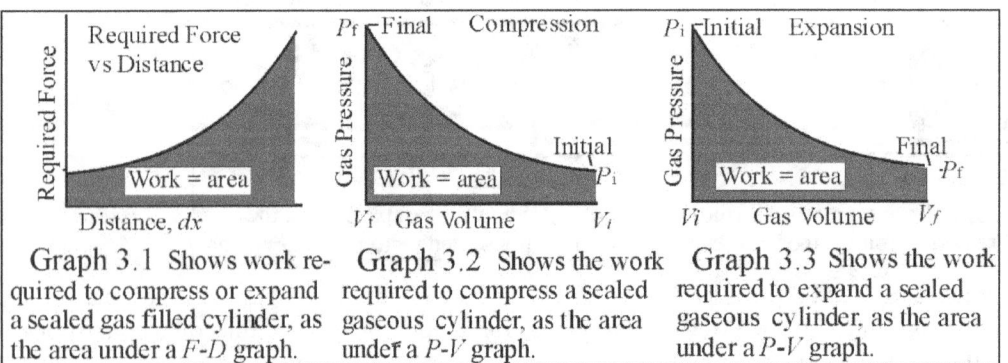

Graph 3.1 Shows work required to compress or expand a sealed gas filled cylinder, as the area under a F-D graph.

Graph 3.2 Shows the work required to compress a sealed gaseous cylinder, as the area under a P-V graph.

Graph 3.3 Shows the work required to expand a sealed gaseous cylinder, as the area under a P-V graph.

Similarly, the P-V Graph 3.2 illustrates the work required for our apparatus. On the L.H.S., is shown the compression of the system, while on the R.H.S. is shown the expansion of the system and once more, the area under the curve equates to the work required.

Ideal Gas and Isothermal Work

A mathematical analogy for this work is obtained by defining the process as isothermal ($dT = 0$), thus the ideal gas law [$PV = NkT$] tells us that the pressure multiplied by volume equals a constant (C'), i.e.:

$$PV = C' = NkT \qquad 3.6$$

Rewriting eqn 3.6:

$$P = C'/V \qquad 3.7$$

When the work required varies with the location at which the force is applied, the line integral is often used to calculate the work involved, in which case work becomes the integration of infinitesimal changes (see Appendix Fig B.6.6). Therefore, the total isothermal work (W_{iso}) is obtained by the integration of infinitesimal isothermal work (dw), as follows:

$$W_{iso} = \int dw = \int_{V_{1i}}^{V_{1f}} P_1 dv_1 \qquad 3.8$$

where subscripts "1", "f" and "i" respectively represent System 1, final, and initial, states.

Notice that infinitesimal work (dw) was transformed into infinitesimal volume change (dv). Work now takes the more general form that being the summation of all the work associated with each infinitesimal volume change. Substituting eqn 3.7 into eqn 3.8, gives:

$$W_{iso} = \int dw = C' \int_{V_{1i}}^{V_{1f}} dv_1 / V_1 \qquad 3.9$$

Performing the integration gives:

$$W_{iso} = C' \ln(V_{1f}/V_{1i}) \qquad 3.10$$

Based upon eqn 3.6, eqn 3.10 can be rewritten for the isothermal work as:

$$W_{iso} = (NkT)\ln(V_{1f}/V_{1i}) \qquad 3.11$$

Eqn 3.11 adheres to the traditional conceptualization for work. However, it possesses the following inherent ambiguity that is not traditionally acknowledged.

"PdV" Versus "VdP"

Ponder, what was the reason for choosing to integrate: Pdv instead of: Vdp, in eqn 3.8? Certainly, as the System 1's volume increases, its pressure decreases. Obviously, eqn 3.8 could have equally been written in term of infinitesimal pressure change (dp) rather than dv. Does this mean that the choice was arbitrary? Investigating: Let us rewrite eqn 3.8, in terms of path dependent dp, obtaining:

$$W_{iso} = \int dw = \int_{P_{1f}}^{P_{1i}} V_1 dp_1 \qquad 3.12$$

Performing the same substitutions and calculations, gives the following for our expanding System 1:

$$W_{iso} = (NkT)\ln(P_{1i}/P_{1f}) \qquad 3.13$$

For an ideal gas (one that obeys kinetic theory), the empirical value obtained by eqn 3.13 would be identical to that obtained by eqn 3.11. Therefore, a more general equation for an ideal gas becomes:

$$W_{iso} = (NkT)\ln(P_{1i}/P_{1f}) = (NkT)\ln(V_{1f}/V_{1i}) \qquad 3.14$$

Obviously, the isothermal work required can be determined in terms of either isobaric volume change, or isometric pressure change, within System 1. Did System 1 experience any significant energy change during either its compression, or expansion? No! The total energy of a monatomic gas would still be defined by its total kinetic energy, i.e. eqn 2.12: $E_{Tk(t,\,r)} = 3NkT/2$. Of course if the process was not isothermal then this would not be the case.

Since both the expansion and compression were considered as being isothermal processes, and System 1 experienced no significant energy change. Now ask: Onto what was the work done? Furthermore, the final explanation must explain why eqn 3.14 remains empirically correct, although theoretically ambiguous.

Work onto a Vacuum

Instead of a gas consider the expansion of a vacuum by an external force. E.g. take a syringe, and place the plunger all the way to the bottom and then hermetically seal the syringe's opening with your finger as illustrated in Fig 3.4. Finally apply a volume expanding force. Since the internal pressure is not a function of the system's volume, then the required mechanical work should be:

$$W = P_{atm}dV \qquad\qquad 3.15$$

where, the subscript "*atm*" signifies the Earth's atmosphere.

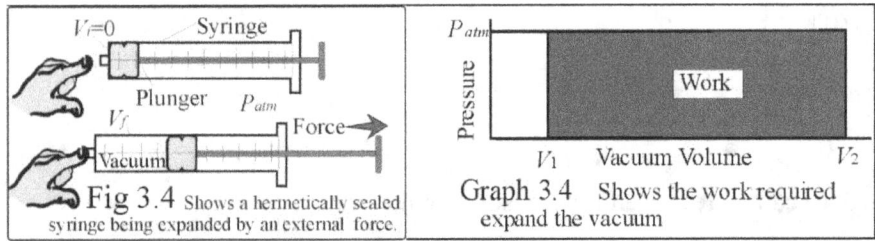

Fig 3.4 Shows a hermetically sealed syringe being expanded by an external force.

Graph 3.4 Shows the work required expand the vacuum

For the isothermal expansion of a closed system by an external force as illustrated by both Fig. 3.3 and Fig. 3.4 not only does the system's energy remained constant, but the processes is also mechanically reversible. That is to say if the applied force was removed then system would return to its original state.

Man on the Moon

Consider that the man on the moon is studying the Earth, as illustrated in Fig 3.5. When the piston was pulled outwardly from the cylinder (system expanded), the man on the moon noticed that the Earth's atmosphere has expanded. So although he has no idea as to why this expansion took place, he writes in his book that the Earth's atmosphere had experienced an isobaric volume increase.

The man on the moon then concludes that work must have been done onto the Earth's atmosphere. He then decides to calculate this work done and writes that the work is defined by:

$$W_{atm} = P_{atm}(V_{atm(f)} - V_{atm(i)}) \qquad 3.16$$

where, the subscripts "*f*" and "*i*" respectively signify

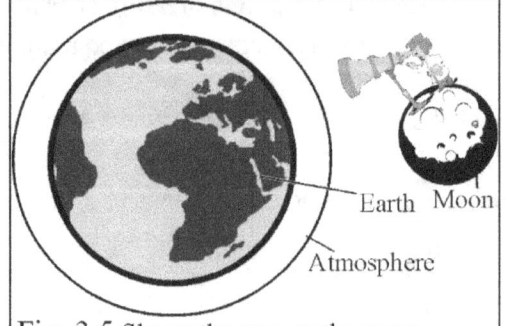

Fig. 3.5 Shows the man on the moon studying the Earth and its atmosphere

the final and initial volume. Certainly eqn 3.16 is based upon eqn 3.15 but the man on the moon has wrongly assumed that this signifies an energy increase to the Earth and/or its surrounding atmosphere.

If only the man on the moon realized that this was a practical joke by the humans because although it appears that the Earth and its atmosphere had an increase in energy, the reality is that the Earth's energy remained constant because the witnessed expansion was a result of negative work onto the expanding volume.

Negative work

Although "negative work" conceptually feels counterintuitive, it is mathematically expressed by placing a negative sign in front of the work term. The inside of the expanded syringe in Fig 3.4, represents negative work due to the volume of space that is being created the inside of the syringe.

Since the initial volume (V_i) of gas within the expanding piston-cylinder is zero, then our expansion is the creation of a vacuum. Therefore, ignoring any blackbody/thermal radiation that enters through the cylinder's walls, then system's total final energy is the final kinematic energy within that created volume, which is zero. Accordingly, the negative work done (W_{neg}) onto the volume inside of the syringe is the work required to create the vacuum:

$$W_{neg} = -P_{atm}(V_f - V_i) = -P_{atm}V_f = -W_{atm} \qquad 3.17$$

Has the total energy of the atmosphere changed? No it has not! Mathematically it is the addition of eqn 3.16 plus eqn 3.17, whose sum is zero work. Another way of viewing this is; the atmosphere's upward displacement was due to the introduction of a *volume of nothingness* (VON), then the total energy of the atmosphere has not been altered. Even if one considered any blackbody/thermal radiation within the syringe that entered through the walls, it was from the atmosphere, thus the total energy change of the atmosphere plus system's/syringe's contents remains zero.

Importantly negative work tends to create unstable volumes. In the case of the above expanded syringe, once the plunger is released, then the plunger will come crashing down into the cylinder eradicating the recently created volume (VON) within the syringe. One could rightfully argue that since the pressure inside of the syringe was less then atmospheric, then the atmosphere's pressure drove the plunger into the syringe, i.e. the mechanical understanding. From an energy perspective, the plunger went crashing down into the syringe because the interior of the syringe signified negative work, when compared to its surroundings. Ultimately, when a VON displaces a mass in a gravitational field, then this is negative work onto that volume, even though the volume's expansion may appear to be positive work onto the atmosphere.

Again how one considers work maybe a matter of perspective. From a thermodynamics perspective, the expansion of the syringe does not mean much. The syringe full of nothing could be brought it into outer space, then if the plunger is released, nothing really changes. However, claiming that nothing happens is from the perspective of inside the syringe. The reality is that that a volume of something replaced the VON within the syringe, when the syringe is brought into outer space i.e. Earth's atmosphere's surroundings. Okay we are verging upon metaphysics but you get the idea!

Consider what man on the moon sees. Although it went unnoticed here on Earth, the man on the moon clearly saw the Earth's atmosphere's volume increase, as the syringe was pulled. He then looks at his extraordinary accurate instrumentation and notes the Earth's temperature remained constant. Did the actual energy of the Earth and its atmosphere isothermally increase? He ponders; perhaps some cosmic rays added energy into the system, Earth. So the man on the moon checks and no cosmic rays struck the Earth's atmosphere at that instant of time. He then recalibrates his equipment, and no errors are found. Facing a conundrum, the man on the moon turns blue.

If only the man on the moon knew all the intricacies going on in the system Earth, he would then understand that there was no real energy change, due to the syringe's expansion. It must be emphasized that adding or subtracting a VON into a system does not change the system's total energy. Note: At this point many readers may be wondering; "when does the atmosphere's energy increase in the above expansion of hermetically sealed piston cylinder". This will be explained shortly.

Next consider that the syringe's volume starts at a specified non-zero value, and then negative work is done onto the volume of gas within the syringe. Now the pressure inside of the syringe decreases as its volume increases. Considering the gas inside of the syringe is ideal then the isothermal ideal gas relation ($PV = C' = NkT$), applies. Therefore based upon eqn 3.14 the negative work (W_{neg}) done onto the volume of gas within the syringe becomes:

$$W_{neg} = -(NkT)\ln(P_{1i}/P_{1f}) = -(NkT)\ln(V_{1f}/V_{1i}) \qquad 3.18$$

Graph 30.2 illustrates the magnitude of work required, as defined by eqn 3.18. From the perspective of Earth and its surrounding atmosphere, once more the expansion of the syringe did not represent a change to the system's energy. Although Earth's atmosphere was upwardly displaced, the magnitude of the potential energy increase equals the magnitude for the negative work associated with the expanded volume.

Potential Work

What about the compression of a gas, whose work was illustrated in Graph 3.2? The gas's compression causes a pressure increase within System 1, and a decrease to the atmosphere's volume. If done quasi-statically then such compression can be isothermal otherwise compression will result in a temperature increase.

When the compressing force is removed from the piston, then the increased pressure within the compressed gas drives the piston back to the mechanical equilibrium position at 1 atm pressure. Herein, the compression of a gas can be considered in terms of *potential work*. For example, the compression of the System 1 as was illustrated in Fig 3.1 is really an increase in the System 1's potential to displace the Earth's atmosphere or to move man and/or machinery.

Moreover, when the force of compression is removed then the potential work returns System 1 back to its original position. Herein, once more work involves the displacement of Earth's atmosphere against gravity but in this case it is at the expense of the energy within the compressed system.

For an isothermal compressed gas whose volume decreased while its pressure increased, then the change to the potential of this gas to do work (W_{pot}) is:

$$W_{pot} = (NkT)\ln(P_{1f}/P_{1i}) = (NkT)\ln(V_{1i}/V_{1f}) \qquad 3.19$$

We will return to various forms of compression later in this Chapter. Before we do some basic understandings will help.

Natural logarithmic function

The natural logarithmic functions/plots are found throughout thermodynamics. Interestingly, many believe it is because entropy and statistical thermodynamics, when perhaps our reality is that it is the other way around i.e. statistical thermodynamics thrives because this functionality exists.

Consider the mechanical filling of a vessel with gas. As the pressure within the vessel increases, it will require increasingly more work to fill the vessel with a given pressure increase. This is because as the vessel's pressure increases, the gas within the vessel pushes into the filling nozzle with an increasing outward force. Thus an increasing amount of energy is required to push gas into the vessel at a constant rate, or the gas will enter the vessel at a slower rate if the driving force remains constant.

Similar consideration is needed when filling a system with heat. Heat flows both ways i.e. from both cold to hot, as well as from hot to cold, with the net flow being from hot to cold. Accordingly, as the cold system's temperature increases the net rate of heat transfer decreases or an increasing hot system temperature would be required in order for the net heat flow to remain constant.

Furthermore, as will be discussed in Chapter 8, our Sun's power density per unit wavelength increases logarithmically with temperature. And in Chapters 4 and 14, rates and logarithmic functionality will be further discussed.

Discussion: Work onto the Atmosphere

Elaborating upon the concept of work being done onto the atmosphere, reconsider the hermetically sealed piston-cylinder as was illustrated in Fig 3.3. When the piston is pulled outwards a distance, *dx*, then the expansion of the piston-cylinder apparatus displaces a column of our atmosphere, as illustrated in Fig 3.6.

Consider that the piston-cylinder apparatus remains in an expanded position. If one were to travel upwards in the atmosphere, eventually an altitude will be reached wherein the pressure surrounding our piston-cylinder apparatus equals the lowered pressure within our expanded apparatus. The act of bringing both the piston-cylinder and its gaseous contents to this higher elevation requires work. The energy required to elevate the piston-cylinder's gaseous contents equates to the work that was required to expand the piston-cylinder.

Expansion of cylinder is equivalent to:Upward displacement of a column of atmosphere

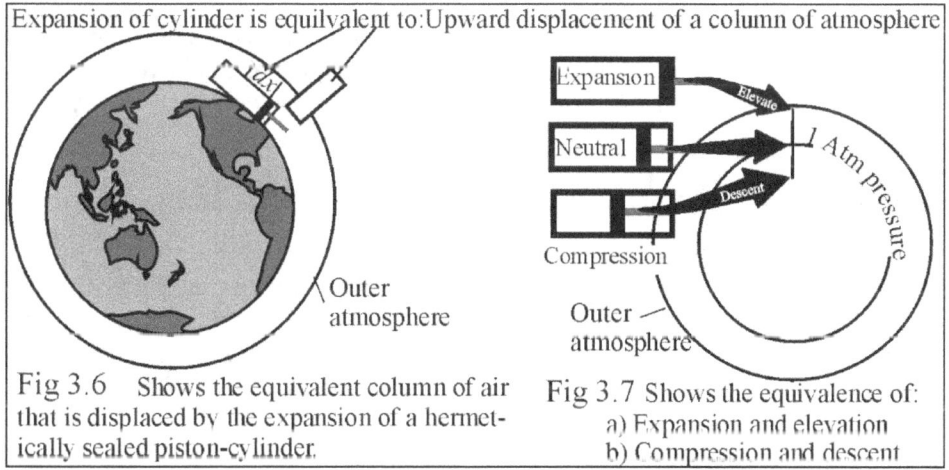

Fig 3.6 Shows the equivalent column of air that is displaced by the expansion of a hermetically sealed piston-cylinder.

Fig 3.7 Shows the equivalence of:
a) Expansion and elevation
b) Compression and descent

This can be envisioned as follows; the work done is the change in potential energy that would be associated with the mass of gas within the piston cylinder, between the two elevations. Let "*dh*" represent the change in elevation and "*M*" be the total mass of the ideal gas inside of the cylinder, then:

$$W = M\bar{g}dh \qquad 3.20$$

The atmosphere's pressure does not decrease linearly with its height above the Earth's surface. Rather, the correlation between the height above Earth's surface and atmospheric pressure is obtained from the

Law of Atmospheres (A.K.A.: *Barometric law*)[3]:

$$P(h) = P_0 e^{-m'gh/(kT)} \qquad 3.21$$

where $P(h)$ is the function of how the pressure changes with height, P_0 is the constant, taken to be the pressure at height $h=0$, which in general is 1 atm, and m' is the molecular mass (g/mole) that being the mass of one mole of molecules. Note: Eqn 3.21 treats the atmosphere as being isothermal, which is not precisely correct.

Dividing both sides by the constant, P_0, and then taking the natural logarithm gives:

$$In[P(h)/P_0] = -m'gh/kT \qquad 3.22$$

In terms of height (h), eqn 3.22 becomes:

$$h = -(kT/m'g)\ln[P(h)/P_0] \qquad 3.23$$

Based upon logarithmic rules, eqn 3.23 can be rewritten as:

$$h = (kT/m'g)\ln[P_0/P(h)] \qquad 3.24$$

Defining " h " as the distance above the Earth's surface. Then substituting eqn 3.24 into eqn 3.20, gives the work required to expand our piston-cylinder apparatus:

$$W = (MgkT/m'g)In[P_0/P(h)] \quad 3.25$$

Canceling out the gravitational constant (g), and realizing that mass (M) divided by molecular mass (m') gives number of molecules (N), the work (W) becomes:

$$W = (NkT)\ln[P_0/P(h)] \qquad 3.26$$

Realizing that $P(h)$ is the piston-cylinder's internal pressure when the apparatus is in the expanded position, and that P_0 is the original pressure inside of our apparatus, then eqn 3.26 can be rewritten in terms of volume. For the expansion of an ideal gas, we now have: $P_0/P(h) = P_{1i}/P_{1f} = V_{1f}/V_{1i}$. Therefore, eqn 3.26 becomes:

$$W = (NkT)\ln[P_0/P(h)] = (NkT)\ln(P_{1i}/P_{1f}) = (NkT)\ln(V_{1f}/V_{1i}) \qquad 3.27$$

We are now enlightened to why eqn 3.14 was empirically valid, yet ambiguous. Specifically, it equates to eqn 3.27, which provides a theoretical explanation that actually makes sense! Moreover, we begin to understand why work done by an ideal gas can be thought of equally in terms of either: PdV, or VdP, although it has little to do with the total energy of that gas! Obviously, the work done onto gases, often simply equates to the changes in the elevation of their mass within a gravitational field.

Discussion of Traditional Work

Traditional thermodynamics faltered because it limited ideal work; $W = PdV$, i.e. mechanical work is force over a distance. Yet the same result can be obtained by defining work as $W = VdP$!

Consider the traditional logic; "*Work upon a system represents a boundary effect to that system, rather than signifying a change to the system's internal energy*"[1,4]. A traditional implication being that work is done onto the system's walls. Although mathematically based, such traditional logic falters. To understand just ask: Why is the same amount of work required for imaginary walls the same as for real walls, i.e. work is the same irrespective of what the walls are made of and irrespective of what the surroundings are?

Furthermore, another traditional assertion; an isothermal entropy increase ($TdS \uparrow$) cannot then be reversed ($TdS \downarrow$) without the input of energy. For constant internal energy; $W = PdV = TdS$. Ask what does an increase in randomness have to do with work into walls or for that matter energy? Moreover, is not the concept of randomness subject to one's individual interpretation [5]. There remains only one logical conclusion. Rather than onto the walls, this author contended that work is generally done <u>through the walls</u>[4] onto the system's surroundings, which is more often than not, the Earth's atmosphere.

Interestingly, if one now googles this they may find work is done onto the walls, as was often traditionally claimed, or they may also find that some now realize that the work is done through the wall onto the surroundings. Even so clarity that the surroundings is the atmosphere is not always given. Interestingly; if you read Planck's treatise on thermodynamics (1917), you will find that he discusses work onto the atmosphere but his discussion is more flippant than something providing clear logic.

A New Perspective of Work

Up to this point various interpretations of work have been considered, with emphasis upon reversible mechanical work and considerations of why work is often contemplated in terms of logarithmic functions.

Clarifying our new inchoation; reconsider the relation for change, i.e. eqn 1.6: $d(TS) = d\varepsilon + d(PV)$. It was discussed that work can often be equally expressed in terms of either: PdV or VdP. Realizing that for infinitesimal change: $d(PV) \approx PdV + VdP$. Can it now be boldly stated that?

$$W = d(PV) \qquad 3.28$$

In Chapter 1 it was discussed that Planck considered a version of eqn 3.28, i.e. eqn 1.47: $W = P_1V_1 - P_2V_2$, which was used for describing a throttling process. Even so simply claiming eqn 3.28 universality remains tricky, as more clarity is needed. Any broad statement that $d(PV) \neq 0$ means that a system is actually doing work cannot be made. Nor can it be said that a system where $d(PV) = 0$, is not doing work.

To understand this, just reconsider the expanding piston cylinder apparatus of Fig. 3.3. For the ideal gas within the apparatus $d(PV) = 0$, and yet work was being done, e.g. an applied force pulled the piston outwards from the cylinder, which required work. However, this work was done onto the surrounding atmosphere while negative work was done onto the volume occupied by expanding gas in the cylinder! Accordingly, eqn 3.25 lacks the clarity of onto what is the work done. Herein, the preference would be to write:

$$W_{done} = W_{atm} = d(PV)_{atm} \qquad 3.29$$

Although, eqn 3.29 now provides clarity concerning onto what the work is done. One has to be careful in so far as how it is applied, in part because the work may be path dependent.

Now ask what did the man on the moon believe he saw? He considered the atmosphere as being isobaric and wrote:

$$W_{done} = W_{atm} \approx P_{atm}dV_{atm} = P_{atm}dV_{system} = P_{atm}dV \qquad 3.30$$

Is there any validity to what the man on the moon wrote? Yes there is, rather than an expanding force doing work upon the system, if the system/atmosphere is expanding due to its own internal energy/energy change then that system does work onto the surround atmosphere! And this work is defined by eqn 3.30, that being path independent work done onto the surrounding atmosphere or if you prefer work as an exact differential! Furthermore, our atmosphere is not homogeneous in terms of either pressure, or molecular volume, therefore P_{atm} must be the atmosphere's pressure at the elevation that the process occurs.

The above path independent work can be viewed in terms of changes to the Earth's atmosphere's potential energy, which is to say it represents an upward displacement of the atmosphere's mass against Earth's gravitational field. Specifically, accepting that the Earth's atmosphere has mass then its upward displacement requires work, which is fundamentally no different than the lifting of any mass. Think about a system inside of our atmosphere expands, then that system's volume increase must try to increase the atmosphere's volume, and the atmosphere only has one way to move, that being away from Earth's surface i.e. upwards.

Why was "try" used in the above sentence? The reason is that the system's expansion may also result in a sudden localized pressure increase to the surrounding atmosphere. Since the atmosphere is an open system, an instantaneous pressure increase may revert into a volume increase due to mechanical equilibrium between molecules. Equally the localized pressure increase might result might result in an increase to intermolecular friction (molecular dissipation), which in turn results in an increase in heat i.e. $T\uparrow$. Note, as was discussed in Chapter 1; Joules weight experiment saw an increase in molecular friction, resulting in a temperature increase.

Obviously, there is no real way to know exactly what the man on the moon would see even so it was still a useful analogy. Whatever the end result is, whether it be a volume, pressure and/or temperature increase of the atmosphere, the amount of work done onto our atmosphere by the expanding system is quantified by eqn 3.30.

Lost Work

The next question is; can the above discussed work be reversed? Irrelevant of its form, the energy associated with such work will eventually become evenly dispersed throughout the random motions of the atmosphere's molecules. In other words such work will be forever lost by the expanding system into our atmosphere! Hence shall be referred to as *lost work*! [6,7,8]

Accordingly, lost work is energy that is forever lost into our atmosphere by an expanding system. This can be equally viewed as an increase in the atmosphere's potential energy. The criteria being that the work is done by the expanding system, rather than being done onto the system itself or even the erroneous traditional notion that work is done onto the expanding system's walls.

Volume Decreases

What about systems whose volume decreases? Reconsider the previously discussed expansion of a hermetically sealed piston-cylinder, wherein the atmosphere's total energy did not change by the creation of a volume of nothingness (VON). After expansion, when the piston is released, it went crashing down into the cylinder, then and only then does the total energy of the atmosphere actually increase. This occurs as potential energy within the atmosphere is transformed into kinetic energy within the atmosphere, i.e. as the atmosphere's gas molecules come crashing downwards thus filling the VON. And more often than not the net result is the infinitesimal/small heating of the atmosphere e.g. infinitesimal in comparison to the atmosphere's total thermal energy, it is not necessarily inconsequential!

Accordingly shrinking system's do no work onto their surroundings, although the net result may be heating of the atmosphere i.e. atmosphere's gases are again disturbed from equilibrium except now the atmospheric gas's potential energy changes into atmospheric gas kinetic energy, which generally results in heat e.g. a form of viscous dissipation. Note: This all will make more sense after reading the ensuing chapters!

The Ability to do work

So what is $d(PV)$ in eqn 3.29? It is the change to a gaseous system's ability to do work. This change in ability does not take into consideration, onto what the work is being done. Certainly one could argue that the ability of the expanding piston cylinder apparatus of Fig. 3.3 to do work did not change and they would be right, because the gaseous system's energy did not change! Another way of saying this is that the system's gas remained isothermal when the external force was applied.

However, when the external force was applied, the potential to do work decreased because the gas's pressure decreased. What is the real difference between the ability of to do work and potential to do work as described in this text? The ability is idealistic concept based upon how much work can be done in relation to some surroundings whose pressure is approaching zero. The potential to do work is how much work can be done in relation to the current surrounding's pressure, which most often is 1 atm pressure.

What does the ideal gas law now represent? The ideal gas law, $PV = NkT$, now represents the *ability of a system of ideal gas to do work, as a function of temperature*.[6,7] And this further alienates our new perspective from traditional thermodynamics.

Gas Compression Reconsidered

In order to better understand let us now reconsider compression of a gas. This time an external mass compresses the gas as illustrated in Fig 3.8. Obviously, the external mass applies an extra external downward force onto the gas and this causes a pressure increase within System 1.

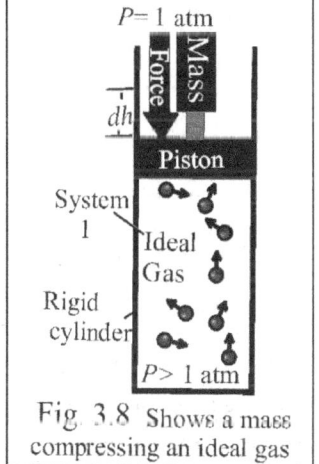

Fig. 3.8 Shows a mass compressing an ideal gas

Now consider that the cylinder is not insulated so that any additional heat created in this process can radiate out from System 1. So ask, has the energy of the gas changed. Assuming that the gas is sufficiently dilute and ideal, i.e. obeys kinetic theory, then as its pressure increases, it mean molecular volume decreases. Therefore the isothermal compression of the isothermal ideal gas did not change its kinetic energy! Thus the gas's total energy remained constant.

In terms of work: $d(PV) = 0$. Therefore the ability of the gas to do work also does not change! What about the potential of the gas to do work. Since the potential to do work is in comparison to the surrounding atmosphere then eqn 3.19:
$W_{pot} = (NkT)\ln(P_{1f}/P_{1i}) = (NkT)\ln(V_{1i}/V_{1f})$ defines its increase!

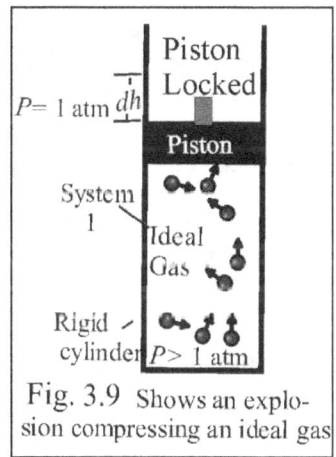

Fig. 3.9 Shows an explosion compressing an ideal gas

Next consider that isometric closed rigid system of gas experiences an internal explosion i.e. piston is locked in position (dh=0) as in Fig 3.9. If the explosion's energy is released quasi-statically and the walls are not insulated then any heat generated by the explosion will be allowed to leave System 1, hence no real change will occur.

However if the explosion is realistic then the ideal gas within System 1 will heat up at which point its the ability to do work increases. Does the total energy of System 1 change? That is tricky but if we assume that prior to the explosion there was a stored potential energy, and that the explosion simply represents a change of this stored energy's state, then the answer becomes no. An example would be an exothermic chemical reaction as will be discussed in Chapter 14.

If System 1 is fully insulated then any generated heat will remain within System 1. What then would be the potential energy increase to insulated System 1? It would be defined in terms of its isometric pressure increase within System 1 by:

$$W = VdP_{(System1)} \qquad 3.31$$

Note: One has to be careful here because the change in either the potential or ability to do work does not necessarily equate to the change in system's energy. This will become obvious in the ensuing chapters.

Temperature & Work

What happens to a gas's temperature, if it does work onto its surroundings? I.e. the piston in Fig 3.9 is unlocked hence $dh>0$.

Begin by reiterating that the kinetic energy of a sufficiently dilute gas in a closed system is defined by eqn 3.12: $E_{T(k,r)} = 3NkT/2$. If a thermally isolated gas uses some of its internal kinetic energy in order to perform work, then that gas must experience a temperature decrease. Accordingly, an expanding system of gas in an insulated closed system that does work will tend to cool down.

Temperature, Work, & Natural P-T Systems

When a system's pressure is increased, its temperature tends to increase. Obviously, pressure increases results in an increase in molecular friction, thus creating heat, which often leads to a temperature increase! Certainly intermolecular friction explains *viscous dissipation*, which is based upon the concept of frictional heat being produced by inelastic intermolecular collisions whenever a real gas or liquid, experiences forced net motion. As previously stated this was seemingly confirmed, yet conveniently ignored, by Joule's weigh experiment.

It has to be emphasized that this all means that elastic collisions is an idealized traditional misconception, where an increase in intermolecular elastic collisions should simply increase the rate of any heat transfer, rather than increase the system's temperature i.e. traditional kinetic theory of gases.

Seemingly, a correlation exists between the weight felt by a system, and both the created pressure and temperature. We can define a *natural P-T system* as a system whose temperature is a direct result of the pressure that the system is under. By this definition, is the Earth's atmosphere a natural *P-T* system? The answer is probably not, because the temperature of Earth's atmosphere is so strongly influenced by other phenomena e.g. our Sun's radiation.

How about our Earth's interior? Considering that other sources of heat, e.g. radioactivity, are sufficiently small in comparison to heat associated with pressure, then the answer would be; yes the Earth's interior approximates a natural *P-T* system. This will be revisited when molecular collisions are discussed in more detail e.g. Chapter 7.

Limitations of our Analysis of Work

Potential energy (u) between two masses, M_1 and M_2 separated by distance r, is given by the relation:

$$U = -GM_1M_2/r \qquad 3.32$$

For a mass (M_1) located a height/distance h above the Earth's surface, in terms of the Earth's radius (r_e) and Earth's mass (M_e) eqn 3.32 can be rewritten as:

$$U = -GM_1M_e/(r_e + h) \qquad 3.33$$

Where G is the gravitational constant such that: $GM_e/r_e^2 = 9.81$ m/s^2

It can be shown that for the case of: $h \ll r_e$ then eqn 3.33 can be approximated by:

$$U \approx -GM_1M_e/r_e + M_1gh \qquad 3.34$$

Herein, we started with eqn 3.2: $W = M\bar{g}dh$. Apparently eqn 3.2 remains valid for: $h \lll r_e$ hence $M_1gh \gg GM_1M_e/r_e$ i.e. valid for expanding systems that upwardly displace our atmosphere. Thus for the purpose of comprehending thermodynamics as witnessed here on Earth, our analysis remains valid.

Traditional Demises

As previously stated, rather than explaining lost work in terms of energy given into the surrounding atmosphere, the traditional interpretation is that the expanding system experienced an increase in the idea of randomness, which was suposidly mathematically validated in terms of Boltzmann's isothermal entropy change, which ensured that lost work was wrongly explained ultimately leading to its indoctrination in the sciences. One versed in traditional thermodynamics may also realize that this has led to the dreadfully misunderstood, misapplied second law A.K.A. the supreme false postulate.

Second law and its implication to isothermal entropy has been wrongly used to explain why most systems are not reversible and/or why perpetual motion is generally impossible and/or why heat flows from hot to cold, as well as other peculiarities. The H-theorem is traditionally accepted as proof that intermolecular collisions increase entropy thus taken to be proof of the so-called second law. Accepting that intermolecular collisions are not elastic could lend itself to new interpretations for the H-theorem. Moreover, in the ensuing chapters we shall arrive at better explanations for what has been witnessed thus reinforce our questioning of the validity of traditional second law based interpretations.

Also too often traditional interpretations consider that the work done onto system always equal the system's energy change, which may or may not be the case.\

Closing Remarks

It has been shown that the traditional concept of work is structured around poorly perceived mathematical justifications. Firstly: PdV formulates a favored set of differential equations because it mathematically parallels work: Fdx. And this resulted in a bias against: VdP, and/or $d(PV)$, based solutions, although they can lend themselves to the same empirical. Note: Although not yet fully discussed herein, this helps explain why the ill-conceived entropy and second law of thermodynamics, endeared such persuasive arguments concerning their validity. As we are about to determine, their validity is void of any logical foundation.

Secondly, the traditional idea that work can be done onto/into walls is a most troubling mathematical result. Sure if the expansion of elastic membrane type walls is considered, then there would be some validity but such thoughts were universally applied to all walls, irrespective of their nature, including imaginary walls. Obviously, traditional conceptualization needs a rethink i.e. work is done through the walls onto the surroundings.

In that light, all this book can achieve is provide a starting point. In this chapter, the virtues of redefining work in terms of: $d(PV)$ was discussed. Furthermore, the realization that such work has to be done onto something, and that this work is always performed through a system's walls onto the surroundings. Moreover this work most often takes the form $W = P_{atm}dV$ i.e. work onto Earth's atmosphere, which can most often be viewed as an atmospheric potential energy increase.

Conversely contracting/shrinking systems do no work, however generally some of the potential energy stored in our atmosphere is converted into kinetic energy which most often results in heat.

Furthermore we begin the understanding that systems whose pressure is greater than that of their surroundings are systems that can do work onto those surroundings. In other words isometric pressure increase is an increase in the potential to do work but no work is actually done by the system until changes occur to the surroundings.

This all only makes sense; work through walls can also power what is outside of those walls, e.g. mechanical motion of devices. Moreover, it means that useful systems, especially those that power machines, tend to be systems that expand, irrelevant of them doing so either isobarically and/or isothermally. And as such, useful systems tend to upwardly displace our atmosphere, thus increasing its potential energy.

Reconsider Planck's Statement in Chapter 1

At this point we have attained a new undertsanding concerning the mechanical nature of expanding systems; i.e. here on Earth they all experience lost work in surrounding atmosphere! Now reconsider Planck's previously stated words from Chapter 1; *We may take for granted the correctness of the mechanical view of nature, and assume that all changes in nature can be reduced to motions of materials points between which there act forces which have a potential.* Remember traditional thermodynamics is based upon Planck's stated second way i.e. the second law as postulate supreme.

Certainly our understanding of the mechanical nature of thermodynamics was changed in Chapter 3, as was our comprehension of kinetic theory altered in Chapter 2. Our remaining goal shall be to attain a better understanding of how this all changes our fundamental perceptions in thermodynamics. This will occur in the ensuing chapters. Then and only then we can confidently expunge the false supreme postulate.

References:

1. Reif, F."Fundamentals of Statistical and Thermal Physics", McGraw-Hill, New York, 1965
2. Reif F."Statistical Physics", McGraw-Hill, New York, 1967
3. http://www. Scienceworld.Wolfram.com/physics/lawofatmospheres (2011)
4. Mayhew, K.W., "Improving our thermodynamic perspective" Phys. Essays ,24 vol 3, 338 (2011)
5. Ben-Naim "A Farewell to Entropy: Statistical thermodynamics Based on Information." World Scientific Publishing Co Hackensack NJ 2011
6. Mayhew, K.W. "Second law and lost work", Phys. Essays **28**, vol 1, 152 (2015)
7. Mayhew, K.W. "Entropy: An ill-conceived mathematical contrivance", Phys. Essays **28**, vol 3, 352 (2015)
8. Mayhew, K.W. "A new perspective for kinetic theory and heat capacity", Prog.in Phy. Vol 17 2017 pg 166-173

Chapter 4: **Systems Performing Work**

Work Done & Vacuum

Fig 4.1 illustrates a gaseous System 1, moving a massless, frictionless piston into a vacuum that being System 2. In this case zero work was done! The simplified reasoning is that work has to be done onto something, and the vacuum represents nothingness (VON), thus no work can be done[1]. Based upon eqn 3.28, the work done by System 1 onto System 2 is zero.

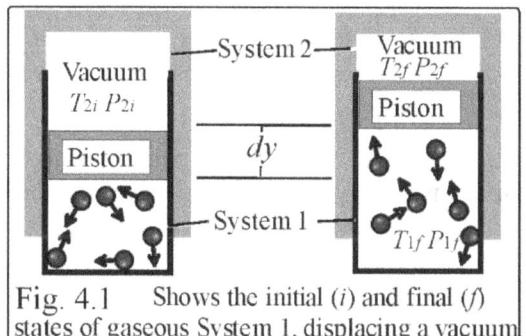

$$W = d(PV) = 0 \qquad 4.1$$

Fig. 4.1   Shows the initial (*i*) and final (*f*) states of gaseous System 1, displacing a vacuum

Furthermore, the <u>ability</u> of System 2 to do work did not change. This is analogous to a man lifting a weightless object, in which case the work done is zero. Of course an object is weightless if either, its mass is zero, or, it resides in zero gravity.

Remember that traditional thermodynamics had misconceptions that renders the erroneous result that work was done[1] onto the vacuum. Namely an association of work with randomness of expanding systems or that work is done onto a system's walls. We now realize that work is done onto the surroundings, through the walls. This is a gross oversight because after Joules experiment (see Chapter 1) this should have been understood! Such misunderstanding has also led to misconceptions concerning our expanding universe, as will be discussed in Chapter 6.

Work in Systems

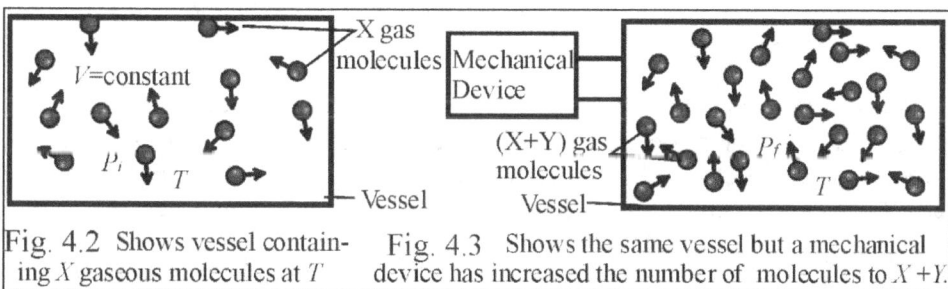

Fig. 4.2 Shows vessel containing *X* gaseous molecules at *T*   Fig. 4.3   Shows the same vessel but a mechanical device has increased the number of molecules to *X +Y*.

Now contemplate isothermal mass transfer that being the addition of gas molecules into an isometric system, e.g. a rigid vessel. Consider a rigid vessel of gas, containing *X* molecules of ideal gas, as is shown in Fig 4.2. Next, consider that a mechanical device e.g. a turbine increases the pressure in our rigid vessel by slowly pushing "*Y*" more molecules into the vessel, therefore the vessel now has: $N= X+Y$ molecules within it, as is illustrated in Fig 4.3.

One might say that the work required to increase the pressure is simply the pressure increase (dP) multiplied by the volume (V) over which the pressure increase occurs, i.e. $W = (VdP)_{vessel}$. However this fails to realize that as the pressure increases within the vessel, then an increasing force per molecule will be required, in order to add more gas molecules into the vessel. In such a situation we could use eqn 3.13 $W_{iso} = (NkT)\ln(P_{1i}/P_{1f})$ and write:

$$W_{vessel} = (NkT)\ln(P_{1i}/P_{1f}) \qquad 4.2$$

How accurate is path dependant eqn 4.2 for this process? In reality the act of increasing the pressure within the vessel will result in a temperature increase, so the hotter the vessel becomes the less accurate eqn 4.2 becomes. However if the vessel is not insulated, and the excess heat simply flows out into the surrounding atmosphere i.e. quasi-static $P\uparrow$, then eqn 4.2 approximates the isothermal work that was done.

The vessel can now perform work by simply opening a valve resulting in the displacement of the Earth's atmosphere: $W_{atm} = P_{atm}dV$. What is the increase in ability to do work onto the atmosphere? If the system remains isothermal then the increase in the ability to do work is due to the added Y molecules i.e.:

$$dW_{ability} = YkT = d(PV) \qquad 4.3$$

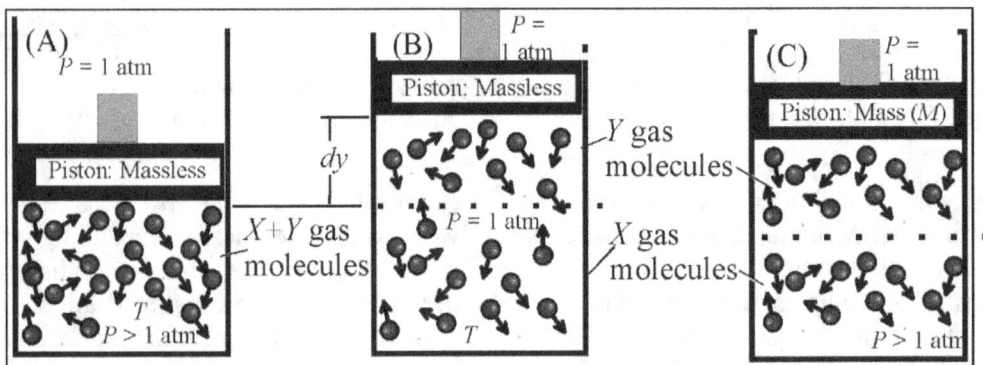

Fig. 4.4   Shows X+Y molecules in a piston-cylinder apparatus. The L.H.S.: (A) shows massless piston, whose pressure is higher than the surrounding atmosphere. The middle: (B) shows the volume occupied by the X+Y molecules at 1 atm, with a massless piston. The R.H.S.: (C) shows the volume occupied by the X+Y molecules when the piston has mass.

Now consider the piston-cylinder apparatus, in Fig 4.4 (A) containing $X+Y$ gaseous molecules, with X being the number of molecules required for the gas to be at 1 atm pressure, at its current total volume. If the process is quasi-static and the system is not insulated, then the work can be isothermal, in which case the path independent work done onto the surrounding atmosphere is:

$$W_{atm} = P_{atm}dV = YKT \qquad 4.4$$

Reconsider that the high-pressure gas shown in Fig. 4.4 (A) is in thermal equilibrium with its surroundings. Therefore, as defined by eqn 2.11: $\overline{E}_{k(t,r)} = 3kT/2$ the mean molecular kinetic energy would be the same for both the high-pressure gas, and its surroundings. Therefore, when the piston is driven upwards, it is because the flux of pressurized gas molecules that strike the inside of the piston is greater than the flux of atmosphere gas molecules that strike the outside of the piston. The net result is the piston moves upwards displacing the atmosphere thus doing work onto the atmosphere, at the expense of the potential work stored within the piston-cylinder.

Now consider that the frictionless piston has mass (M), as is illustrated in Fig. 4.4 (C). Since the gas molecules inside the piston-cylinder have to lift the weight of both, the piston, and the Earth's atmosphere, then the distance, dy, that the piston moves must be less than it was for the case of the massless piston in Fig. 4.4 (B). Accordingly, the work done by the gas must be the work required to displace the Earth's atmosphere, plus the work required to lift the piston's mass (M), that being:

$$W_{done} = P_{atm}dV + M\overline{g}dy = YkT + M\overline{g}dy \qquad 4.5$$

In the above, the process is assumed isothermal. In reality the expanding gas does work hence cools down. However if the work is considered quasi-static, then heat (infinitesimal or otherwise) can enter through the system walls from the surrounding atmosphere, thus keeping the process isothermal, thus keeping both eqn 4.4 and 4.5 valid.

Elaborating, the work done by an expanding system onto Earth's isobaric atmosphere is:

$$W_{atm} = \int dw = \int Pdv = P \int dv = P\Delta V \qquad 4.6$$

Conversely, the work done into an isometric system ($W_{isometric}$) generally represents changes to the potential to do work, which can be thought of as a change in the potential to displace the Earth's atmosphere [$W_{pot(atm)}$], where:

$$W_{pot(atm)} = \int dw = \int Vdp = V \int dp = V\Delta P \qquad 4.7$$

For both eqn 4.6 and eqn 4.7, the concept of work was expanded into infinitesimal work, and since, either V and T or P and T were approximated as constant then the integration of infinitesimal work is simplified.

Work & Balloon Inflation

Consider the work required to isothermally inflate a rubber balloon, as illustrated in Fig 4.5. We might think that the work required could be based upon eqn 3.28 and write:

$$W_{required} = d(PV)_{balloon} \qquad 4.8$$

Plotting the balloon's pressure versus its volume would give something similar to Graph 4.1, wherein the work required is the area shown under the slope where both the pressure and volume are increasing. In order to calculate the total work done in inflating the balloon one must realize that due to the balloon's rubber membrane, its internal pressure will vary with its volume. Assuming that the balloon's internal pressure increases linearly with volume, i.e. $P = C'V$ as is illustrated in P-V Graph 4.1.

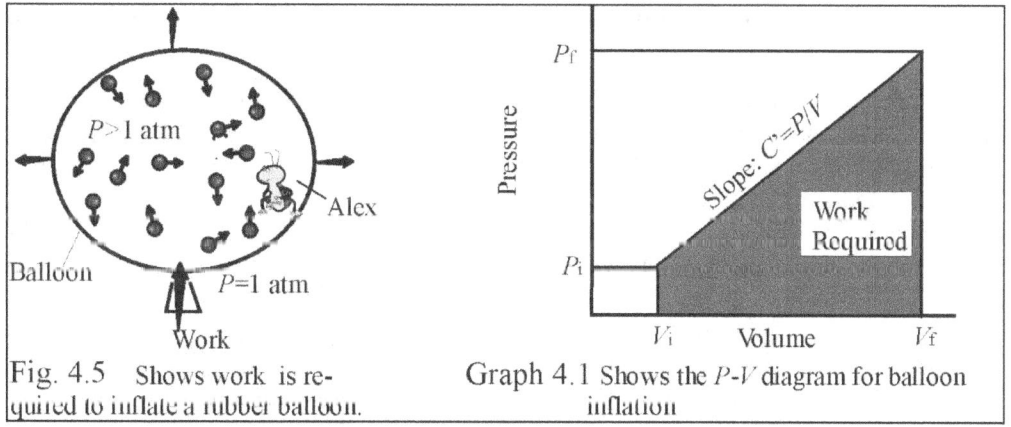

Fig. 4.5   Shows work is required to inflate a rubber balloon.

Graph 4.1 Shows the P-V diagram for balloon
inflation

Then the work is given by the area under the P-V Graph 4.1 can be calculated by:

Kent W. Mayhew

$$W_{required} = \int dw = \int_{V_{bi}}^{V_{bf}} P_{(balloon)} dv_{(balloon)} = \int_0^{V_{bf}} C'V_{(balloon)} dv \qquad 4.9$$

Choosing to define infinitesimal work in terms of Pdv is somewhat arbitrary. Integrating eqn 4.9, calculates the total work done in inflating the balloon as being:

$$W_{required} = C'V_{(balloon)}^2 / 2 \qquad 4.10$$

Does the above require that the balloon inflates isothermally? Certainly the reality must be as the balloon inflates, the pressure hence temperature within increases. Of course any temperature increase could result in a net flow of thermal radiation out through the balloon's walls into the surroundings thus maintaining thermal equilibrium: Quasi-static arguments again apply. Conversely rapid inflation of the balloon would not be quasi-static. This author can envision differing values for the constant (C') for the rapid inflation versus quai-static inflation.

In order to emphasize perspective, consider Alex, the bug inside of the balloon. As the balloon inflates, Alex may not realize that his world is expanding, and would have no idea that work was being done at all, especially onto the atmosphere. Oblivious, Alex would simple plod along the balloon's interior thinking it was a windy day.

Eqn 4.10 really on considers the work required to stretch the balloon. If we want to know the total work then we should add the lost work i.e. the work required to upwardly displace our atmosphere. Hence the total work done becomes:

$$W_{total} = C'V_{(balloon)}^2 / 2 + P_{atm}V_{ballon} \qquad 4.11$$

If the balloon punctures then the higher pressure from within, would disperse outwardly, i.e. the potential to do work from inside the balloon would become work done onto the earth's atmosphere. As it does work, the expanding gas will cool down, hence a net flow of thermal radiation (heat) will enter the expanding gas's volume, thus maintaining thermal equilibrium. Again quasi-static arguments apply.

Experiment for Temperature Pressure Relation

It has been discussed that the transfer of heat in quasi-static processes, often allows processes to be isothermal, which may have consequences upon our perception.

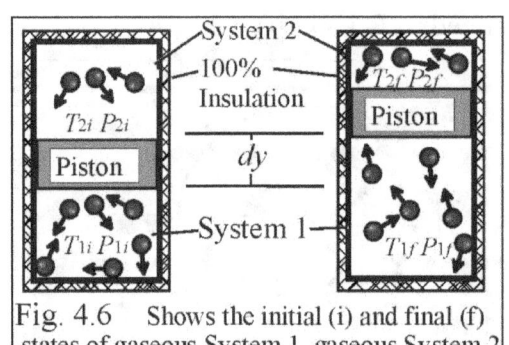

Fig. 4.6   Shows the initial (i) and final (f) states of gaseous System 1, gaseous System 2.

Consider the experiment shown in Fig 4.6. A frictionless piston separates two isothermal closed ideal gases, whose pressures are defined by: $P_1 > P_2$. Both systems are 100% insulated from their surroundings but heat can pass through the piston between systems thus keeping the quasi-static process isothermal. The net result is the isothermal expansion of System 1 equates to isothermal contraction of System 2.

Is any work done? If work is taken to be a displacement of mass against gravity, then one could argue that no real work is done, i.e. since isothermal the ability [$d(PV)$] of both systems to do work remained constant, as does their respective kinetic energies! However, the potential of System 1 to do work upon Earth's atmosphere was increased at the expense of System 2.

Start with the isothermal ideal gas relation ($PV = C' = NkT$). Therefore based upon the potential to do work here on Earth:

$$W_{Pot1} = -W_{Pot2} = (NkT)\ln(P_{1i}/P_{1f}) = (NkT)\ln(P_{2f}/P_{2i}) \qquad 4.12\ (a)$$

Or

$$W_{Pot1} = -W_{Pot2} = (NkT)\ln(V_{1f}/V_{2i}) = (NkT)\ln(V_{2i}/V_{2f}) \qquad 4.12\ (b)$$

Again eqn 4.12 (a) and (b) assume that the process is isothermal which would only be the case for quasi-static movements of the non-insulated piston.

Repeat the experiment with the piston 100% insulated so that no heat is transferred between systems. As System 1 expands its pressure will decrease, and so too will its temperature. Similarly, the temperature of System 2 would increase as its pressure increases. Thus the isothermal ideal gas law no longer applies! This seemingly simple experiment will now have a complex solution and maybe considered an adiabatic process whose solution more takes the form of a polytropic equation, which will be discussed in the Chapter 6, as well as Appendix B.10. An issue: Heat stored in the walls may not allow one to properly witness the real temperature changes.

Of course the surrounding/exterior insulation could be removed and any quasi-static change would appear isothermal, as both systems exchange heat with their surroundings. And for this isothermal process the ideal gas law would again be applicable.

Throttling (or Joule-Thomson) Process

As previously briefly discussed in Chapter 1, throttling is a rare process where traditional thermodynamics deals with work in much the same manner as our new perspective does. Throttling involves the flow of gaseous molecules within a tube, where the molecules must pass through a porous plug, as illustrated in Fig. 4.7. To calculate the work done by the ideal gas passing through the porous plug then eqn 3.28: $W = d(PV)$ is used i.e. a change to the gas's ability to perform work.

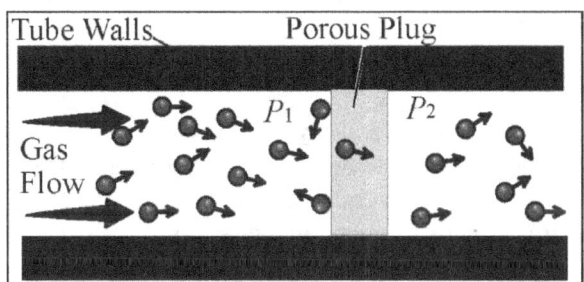

Fig. 4.7 Shows the throttling process, wherein a gas flowing in a tube must pass through a porous plug.

Next consider the gas in terms of its mean molecular volume (υ) and pressure (P). Let the subscripts "1" and "2", respectively represent the gas molecules before, and after, they pass through the plug. The work done onto the N molecules passing through the porous plug is:

$$W = N(P_2\upsilon_2 - P_1\upsilon_1) \qquad 4.13$$

The throttling process is considered isothermal i.e. the gas molecule's mean kinetic energy remains constant. Therefore, all the porous plug does is increase the gas' mean molecular volume (υ), resulting in a pressure decrease. Instead of a porous plug, consider a long tube wherein a gas flows. There will be a reduction in pressure as the gas molecules flow through the tube. Again, eqn 4.13 applies!

System within a System

When considering a system within a system, then both system's walls, and surroundings, must be defined. Fig 4.8 shows ideal gas System 1, surrounded by System 2, all surrounded by the Earth's atmosphere.

Consider that the walls of both systems are rigid and closed, with $P_1 > P_2$, while $T_1 = T_2$. Then puncturing a hole through the wall of System 1 would result in a pressure increase of System 2, hence increasing its ability to do work, at the expense of System 1. At least for infinitesimal isothermal change we could write:

$$V_2(P_{2f} - P_{2i}) = -V_1(P_{1f} - P_{1i}) \quad 4.14$$

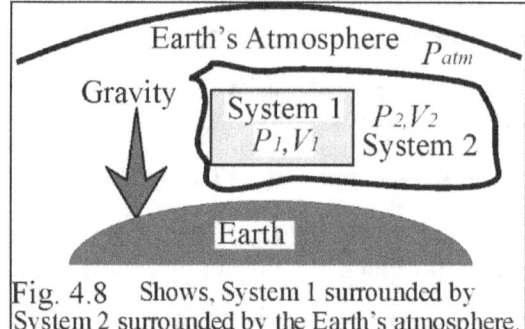

Fig. 4.8 Shows, System 1 surrounded by System 2 surrounded by the Earth's atmosphere.

The pressure in System 1 decreases as the gas molecules enter System 2. The mass transfer continues until both the pressure and mean molecular volumes (v) are identical, i.e.:

$$(Pv)_{1f} = (Pv)_{2f} \quad 4.15$$

Now ask was any work actually done? Answering this maybe a matter of perspective and what is known. Standing inside of System 2, then work was done onto System 2 by System 1.

If System 2 were not rigid but still closed, then it would expand, when the hole was put into System 1. Now System 1 does work onto System 2, which then does work in displacing our atmosphere. So long as heat can flow through all system walls then the quasi-static process can be isothermal, and the systems will end up with the pressure as the surroundings:

$$W_{done(1 \to 2)} = W_{done(2 \to atm)} = P_{atm}dV_{atm} \quad 4.16$$

Next consider a solitary system surrounded by the Earth's atmosphere. If this solitary system expands then what happens to the Earth's atmosphere? The Earth's atmosphere is a constrained open system, with gravity being the ever constraining force. Therefore the expanding system's volume increase equates to the atmosphere's volume increase, i.e. $dV_{sys} = dV_{atm}$. In other words the expanding system is really a subsystem of constrained open system Earth.

Accordingly, when considering work done onto the atmosphere one can express it in terms of either expanding system's volume increase or the surrounding atmosphere's volume increase i.e. $W_{atm} = P_{atm}dV_{sys} = P_{atm}dV_{atm}$. This all assumes that there are no real pressure and/or temperature increases experienced by the atmosphere due to the expanding system volume hence the atmosphere's potential increase is defined purely in terms of an isobaric volume increase.

Heating: Work Versus Energy

Can the work done be equated to change of the gas molecule's kinetic energy within any system? Let's investigate. The ability of a gaseous system to perform work, as a function of its temperature, is defined by the ideal gas law i.e.: $PV = NkT$.

Comparing the ideal gas law to eqn 2.12: $E_{Tk(t,r)} = 3NkT/2$, seemingly an increase in an ideal gas's ability to do work constitutes only 2/3 of an ideal gas's kinetic energy increase. Equally, when the same comparison is done as a function of temperature, the same ratio (2/3) is determined. This is readily proven by differentiating, both equations, with respect to temperature. I.e.:

$$d(E_{kT}) = (3/2)NkdT \quad 4.19$$

$$W = d(PV) = NkdT \quad 4.20$$

Remember: Eqn 2.12 hence eqn 4.19 applies to a sufficiently dilute gas in a closed/walled system.

It may seem odd that a closed system of gas cannot convert all of the energy into work. But its not! The 66.67% upper limit to the efficiency exists because not all of the system's gaseous molecules will be able to contribute all of their momentum to the system's expansion.

One must realize the following:

1)   Work involves the movement of mass in a unique direction i.e. along the positive z-axis or if you prefer the movement of a mass whose surface area in the x-y plane feels the force resulting in the movement being along the z-axis. The actual molecular flux that strikes the x-y plane is defined by eqn B.7.14: $\Phi_0 = (1/4)n\bar{v}$ (see Appendix B.7). This being the flux of molecules of ideal monatomic molecules that can actually contribute their energy to work.

2)   An enclosed sufficiently dilute gas's translational plus rotational energy is due to the energy obtained from interactions with the surrounding walls, and this energy is the summation of the energies from the three orthogonal walls. Accordingly, it as if the energy flux was a summation of energy from all six surrounding walls such that the flux of energy from each wall was proportional to eqn 1.4.5: $\Phi_0 = (1/6)nv$.

Accordingly, change to the gaseous system's energy does not equal its change in ability to do work. As discussed in Appendix B.7, the upper limit of a gas's ability to do work becomes: $(1/4)n\bar{v}/(1/6)n\bar{v} = 2/3$ i.e. 66.67%, of the gas's translational plus rotational energy. And remember that this is for sufficiently dilute ideal monatomic gases. The majority of gases have vibrational energy, hence may have varying efficiencies that are most likely lower.

Another way of considering the above being: When a gas is heated, all the molecules within that system will experience an increase in kinetic (translational plus rotational), as well as vibrational energies. The 2/3 only concerns itself with the kinetic energy of those gaseous molecules that can impart their increased momentum onto the system's walls during expansion, and this is for systems wherein eqn B.7.14: $\Phi_0 = (1/4)nv$ is valid that being relatively flat walls acting along some plane upon which the gas molecules exerts their momentum.

What if the system expands along all three axes? This would still makes little difference because the volume increase onto the atmosphere is still directed upwards hence the atmosphere's potential energy increase is approximated by the upward lifting of a column of air.

Difference between Energy and Work

The first step to solving any problem is asking the right question. Seemingly one of the first questions that should have been asked concerning thermodynamics being; what are the fundamental differences between energy and work? Certainly one could say that energy has no sense of direction while work is performed with a unique direction in mind. Or that energy exists along all axis, while work is performed along a given axis and is powered by energy impacting a plane perpendicular to that axis.

What about Vibrational energy?

It makes for interesting conjecture to ask; to what extend can the vibrational energy of polyatomic molecule contribute to work? Assuming that the vibrational energy of diatomic, triatomic and other polyatomic molecules is strictly due to the absorption and radiation of surrounding blackbody/thermal radiation then the answer seemingly is that vibrational energy does not necessarily contribute to work. Of course this assumes that all parts of the system is always exactly in thermal equilibrium. It also assumes that small polyatomic molecule receives as much vibrational energy from, as is gives onto all walls during all impacts. Of course large polyatomic molecule ($n">4$) are subject to flatlining and this may require other considerations. Note: Flatlining was discussed in Chapter 2 and will be discussed again in Chapter 5.

Heating: Isometric vs Isobaric

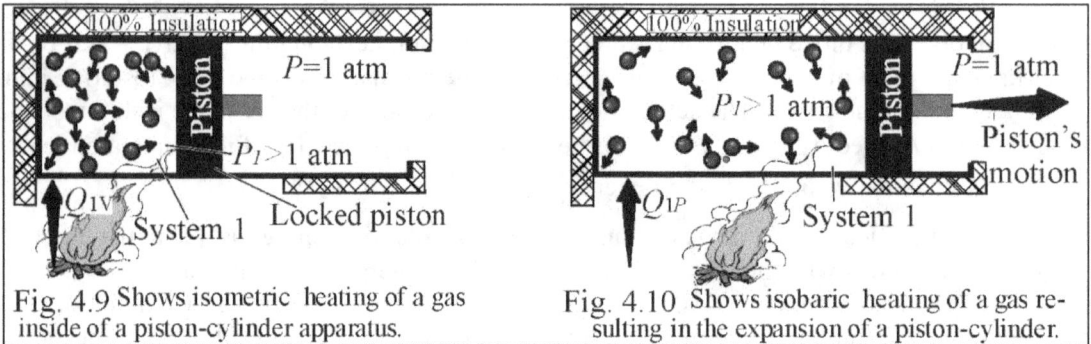

Fig. 4.9 Shows isometric heating of a gas inside of a piston-cylinder apparatus.

Fig. 4.10 Shows isobaric heating of a gas resulting in the expansion of a piston-cylinder.

Consider that a 100% insulated piston-cylinder containing an ideal gas as shown in Fig 4.9 that experiences isometric heating ($dV=0$) e.g. piston is locked in place. As the fire's thermal energy (Q_{1V}) heats the gas; its temperature increases ($T_1\uparrow$), hence the gas's kinetic energy increases. As the $T_1\uparrow$, the mean molecular velocity increases, hence the pressure exerted by the gas increases: $P_1\uparrow$. The heating process can be described in terms of kinetic theory. Based upon eqn 2:12, in terms of N molecules and Boltzmann's constant (k) the heat in is:

$$Q_{1V} = (3/2)Nk(T_{1f} - T_{1i}) \quad 4.23$$

Eqn 4.23 assumes that all of the energy change is associated with the gas's kinetic energy change. The change to the gas's potential to do work is 2/3 of the kinetic energy change that being:

$$dW_{pot} = Nk(T_{1f} - T_{1i}) \quad 4.24$$

Next consider the isobaric heating (Q_{1P}) as shown in Fig 4.10. Since the piston is no longer locked, the system can readily expand. If the piston is both massless and frictionless then in order for the piston to move P_1 only need be infinitesimally greater than 1 atm. And in order for the piston to keep moving to the right, the temperature of the gas must be constantly increasing. In terms of N molecules and Boltzmann's constant (k) and the work done onto the atmosphere:

$$Q_{1P} = (3/2)Nk(T_{1f} - T_{1i}) + P_{atm}(V_{1f} - V_{1i}) \quad 4.25$$

Since the ideal gas law applies then eqn 4.25 can be rewritten as:

$$Q_{1P} = (3/2)(P_{1f}V_{1f} - P_{atm}V_{1i}) + P_{atm}(V_{1f} - V_{1i}) \quad 4.26$$

In eqn 4.26: $P_1 > P_{atm}$. If the piston no longer moves, then: $P_1 \approx P_{atm}$. Now eqn 4.26 becomes for our system of monatomic gas:

$$Q_{1P} \approx (5/2)P_{atm}(V_f - V_i) \quad 4.27$$

Obviously for the same temperature change, the required thermal energy is greater for the isobaric case than the isometric case; $Q_{1P} > Q_{1V}$. This is because the isobaric case involves work onto the surrounding atmosphere, while the isometric case does not. Remember: In the above analysis the gas molecules were considered to be ideal and monatomic. For polyatomic molecules, one would need to also consider any changes to the gas's vibrational energy. See: Chapter 5, where isobaric vs isometric specific heats are considered.

Traditional Consideration

The correlation to the traditional consideration of the first law is readily discerned. Consider eqn 1.38, which can be rewritten as: $dq_{in} = du + dw$. When compared to our above analysis for isobaric heating as defined by eqn 4.25: $\int dq_{in} = Q_{1P}$, $\int du = (3/2)Nk(T_{1f} - T_{1i}) = dE_{tot}$ and $\int dw = -P_{atm}(V_{1f} - V_{1i})$. Although similarities exist, our analysis remains simpler because the analysis is based upon constructive logic rather than misconstrued entropy change; whatever entropy change actually means.

Heating: Potential Work

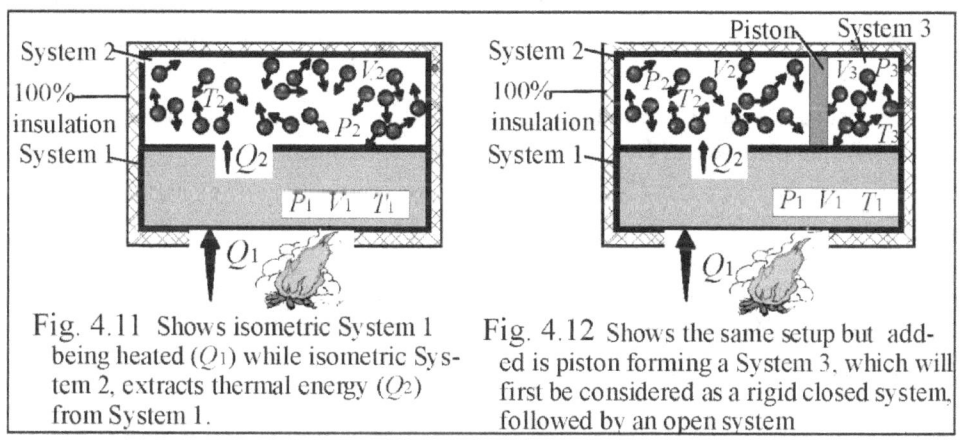

Fig. 4.11 Shows isometric System 1 being heated (Q_1) while isometric System 2, extracts thermal energy (Q_2) from System 1.

Fig. 4.12 Shows the same setup but added is piston forming a System 3, which will first be considered as a rigid closed system, followed by an open system

Consider that a closed isometric System 1 receives thermal energy (Q_1). Also thermal energy (heat) goes from closed System 1 into closed System 2, as shown in Fig 4.11. Accordingly:

$$Q_2 = Q_1 - (E_{T1f} - E_{T1i}) \qquad 4.28$$

The state of System 1 is not defined beyond it being isometric liquid i.e. condensed matter. Based upon kinetic theory for liquids, in terms of N molecules and Boltzmann's constant (k) the heat in is:

$$Q_1 = E_{T1f} - E_{T1i} = 3Nk(T_f - T_i) \qquad 4.29$$

Assuming that System 2 is a monatomic ideal gas then its energy increase could be expressed as:

$$Q_2 = (3/2)N_2k(T_{2f} - T_{2i}) \qquad 4.30$$

The increase in System 2's potential to do work is based upon eqn 4.2, or if one prefers is 2/3 of the gas's kinetic energy change. That being:

$$W_{pot2} = V_2(P_{2f} - P_{2i}) = N_2k(T_{2f} - T_{2i}) \qquad 4.31$$

Heating: Performs Work

Consider that a gas filled closed System 3 is added by having a friction-less piston separate closed System 2 from System 3, as shown in Fig 4.12. The fact that a piston resides between System 2 and System 3 does not really change anything from our analysis for Fig 4.11. Assuming that Q_2 is quasi-static heat and given sufficient time the temperature, mean molecular volume and pressure in System 3 will equal that in System 2.

Next assume System 3 is fully insulated e.g. both the piston and walls are fully insulated. As the temperature of System 2 increases, its pressure increases hence; $T_2 > T_3$ while $P_2 \geq P_3$. Furthermore, the temperature hence thermal energy of System 3 should also increase. But any temperature increase within fully insulated System 3 is solely due to the pressure increase. Thus System 3's ability to do work would increase i.e.

$$dW_{ability3} = d(PV)_3 > 0 \qquad 4.32$$

Based upon the ideal gas law eqn 4.32 can be rewritten:

$$\Delta W_{ability3} = Nk(T_{f3} - T_{i3}) > 0 \qquad 4.33$$

The problem remains; by how much the temperature increases is not known because it follows the natural P-T relationship for this gas, which is not known at this time. Obviously, System 3's potential to do work onto its surrounding would increase due to both its temperature and pressure increases.

If System 3's final temperature was at the same temperature as the surrounding atmosphere, then we might think in terms of isothermal work i.e. as System 3's volume isothermally decreased its pressure isothermally increased. Hence:

$$W_{pot} = (NkT)\ln(P_{3i}/P_{3f}) = (NkT)\ln(V_{3f}/V_{3i}) \qquad 4.34$$

Logic would dictate that if System 3's final temperature was greater than the surrounding atmosphere's temperature, then the total potential of hotter System 3 to do work:

$$W_{pot} = (NkT)\ln(P_{3i}/P_{3f}) + Nk(T_{f3} - T_{atm}) \qquad 4.35$$

Steam Power

Consider steam powering an engine, as shown in Fig 4.13. The fire's heat (Q_1) boils the water, which undergoes expansion (phase change), creating high-pressure steam, which then pushes the piston. This is similar to the analysis for Fig 4.12, except that herein System 1 is an open system of boiling water, while System 2 is the piston-cylinder whose expansion is approximately isothermal.

The reason that System 2 expands is the addition of energetic gaseous molecules $N_2 \uparrow$, i.e. both an energy and mass transfer from System 1 into System 2, as the water boils. Once the water is at its boiling point, then; $dT_1 = 0$, and the isobaric energy input into System 2 (Q_{2P}) can be defined by eqn 2.12: $E_{Tk(t,r)} = 3NkT/2$, but this time in terms of the boiling temperature (T_b), Boltzmann's constant (k) and change to N molecules, the energy input is:

$$Q_{2P} = (3/2)kT_b(N_{2f} - N_{2i}) \qquad 4.36$$

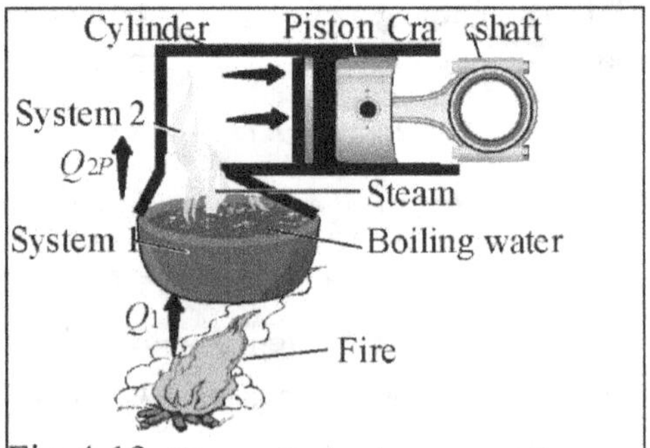

Fig 4.13 Shows the basic concept for a steam engine, i.e. boiling water creates steam that drives the piston.

The energy as defined by eqn 4.36 is really the work done onto the atmosphere eqn 3.30: $W_{atm} \approx P_{atm}dV$. And this is lost work!

The fire's thermal energy into System 1 is commonly called latent heat of vaporization [$L_{(l-g)}$]:

$$Q_1 = L_{(l \to g)} \qquad 4.37$$

As will be discussed in Chapter 10, the latent heat of vaporization includes the energy required to break the liquid's bonds (U) plus any work that is done. If the change to the liquid's bonding energy was zero ($\Delta U = 0$), then:

$$Q_1 = Q_{2P} = W_{atm} = P_{atm}dV \quad 4.38$$

In order to keep the piston moving, then $P_{(piston - cylinder)} > P_{atm}$. Note: Eqn 4.38 defines the energy required for $P_{(piston - cylinder)} = P_{atm}$. In other words reality (friction and piston's mass) requires that:

$$Q_1 \geq W_{atm} = P_{atm}dV \qquad 4.39$$

The goal of a steam-powered engine is not simply to displace the Earth's atmosphere rather its goal is to move man and machine (W_{out}), which also must overcome any friction ($W_{friction}$). Calculating the minimum energy required gives:

$$Q_1 = L_{(l \to g)} + W_{friction} + W_{out} \qquad 4.40$$

Eqn 4.40 assumes that all work elements are path independent. Again, all the heat (Q) that is put into our steam-powered engine cannot be readily transformed into work, i.e. all the energy that was put into a System cannot necessarily be extracted, controlled, nor harnessed as work. Traditionally such concepts are dealt with using relations such as Helmholtz free energy. This is not to say that the two are identical but the rational have certain similarities! Herein simpler perspectives are sought.

Reversibility

A reversible process is one that can equally proceed forwards or backwards. I.e. reversibility is an idealistic concept for a system's state, wherein it can change, and then readily returned to its original state. Conversely, an irreversible process is a process that cannot readily return to its original state without an input of resources, e.g. energy.

It should be stated that reversibility is a thought process traditionally embedded in thermodynamics as something profound, when in reality it should simply be part of common sense. Accordingly the concepts of reversibility will be re-evaluated based upon our new perspective and at the risk of seeming repetitive, what we have discussed will be applied to this concept.

Consider the system shown in Fig 4.14. Heat (thermal energy) warms System 1 (Q_1), which in turn increases the energy within System 2 (Q_2). However in this case, heat (Q_{lost}) is dissipated through the walls of System 2, hence is lost into our surrounding atmosphere therefore the process is irreversible. For such irreversible processes:

$$Q_1 = Q_2 + Q_{lost} \qquad 4.41$$

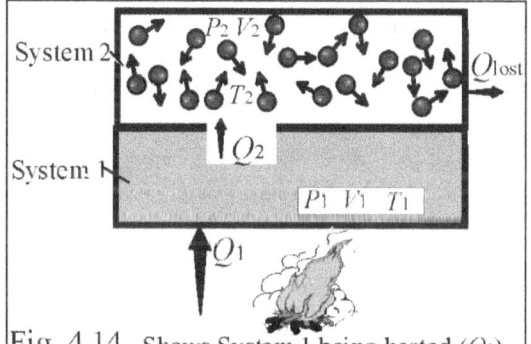

Fig. 4.14 Shows System 1 being heated (Q_1) while System 2, extracts thermal energy from System 1 (Q_2). Energy (Q_{lost}) is dissipated thru the walls into the surrounding atmosphere.

In Fig 4.14 there is an obvious direction for the flow of heat into the surrounding. Certainly when Q_{lost} enters the surrounding atmosphere, its heat will disperse throughout the atmosphere, therefore Q_{lost} cannot be magically collected and returned back into System 2, i.e. eqn 4.41 represents an irreversible process!

Obviously, for a process to be reversible then; $Q_{lost} = 0$. Moreover reversibility implies no real dispersal of thermal energy into some cooler surroundings, because reconcentrating energy back into its original state requires an input of resources. Reconsider Fig 4.14. Since the net flow of heat is always from hot to cold, then the best one could reasonably expect is that both systems reach thermal equilibrium, i.e. the external heat is turned off $Q_1 = 0$ and then, $Q_2 = 0$, thus thermal equilibrium will eventually be attained between all systems and their surroundings.

For two systems in thermal contact, thermal equilibrium is really a compromise, with the final temperature being somewhere between the two system's temperatures: $T_2 < T_{final} < T_1$. Even as a compromise thermal equilibrium means that some part (if not all) of the process is irreversible unless some imaginary processes occurs between two systems whose temperature difference is infinitely small i.e. $dT \rightarrow 0$ i.e. a theoretical process that accomplishes nothing.

Ultimately reversibility applies to quasi-static processes that involve no net energy loss and/or dispersal. At a fundamental level this renders thermal reversibility into idealism rather than realism. Certainly, real/useful processes that involve the transfer of a significant quantity of thermal energy (heat) generally are not reversible!

Arguably an exception to this is gravitational attraction, which often take exceptionally long durations before its relevance is realized. Moreover gravity does not influence heat per say, but it certainly influences matter, which adsorbs surrounding thermal energy thus concentrating it, hence increasing its density.

Equally, no real mechanical process can be reversible! Specifically, the motions of most all man-made devices involve friction resulting in dissipated heat, e.g. heat radiating into the surroundings. Okay, systems can be insulated from their surroundings but the gathering and then trying to directionally control the flow of dissipated heat, with any real efficiency remains an illusionary endeavor.

There are numerous other reasons that result in a given process being irreversible such as: Electrical resistance, shock waves in fluids (considered as part of viscous dissipation), inelastic deformation, magnetic hysteresis, mixing of substances, osmosis, flow of a viscous fluid along a solid surface, internal damping and mixing of similar substances at different temperature (perhaps considered as part of spontaneous heat transfer).

And the major reason as to why so many useful devices involve irreversible processes being: Often devices are powered by some useful process that involves system expansion (isobaric or otherwise), as part of some step in their cycle. Again this displaces our atmosphere hence lost work, hence irreversible process. Note: "Useful processes/systems" are those that can move man and/or machine. And remember lost work does not necessitate an atmospheric volume increase rather it quantifies and explains why the energy is lost and where the lost energy goes! That being into the surrounding atmosphere!

Illusionary Reversibility

However things are not always that discernable. Reconsider the compression of a hermetically sealed piston-cylinder apparatus, as was previously discussed. Omitting friction, is such work reversible? At first glance it may appear to be because when the compressing force is removed then the piston-cylinder expands back to its original state. Is both the atmosphere and system not back to their original states?

However in returning to its original state, it does work onto the atmosphere i.e. the potential to do work is transformed into work onto the atmosphere resulting in an increase to the atmosphere's potential energy. To fully understand one must realize that it was conveniently forgotten that when the gas was being compressed, then the atmosphere's gases experienced a kinetic energy increase, which resulted in an atmospheric temperature increase. What about the temperature of the expanding gas inside the syringe? Does it not decrease as it does work onto the surrounding atmosphere? This is also true but it omits the consideration that the temperature of the gas inside the syringe would have increased when compressed rapidly or if the syringe was fully insulated. Conversely if the compression was done quasi-statically and the syringe is not insulated, then the increased heat can escape through the syringe's walls making compression appear as an isothermal process.

Does this mean that the cooling during expansion (PdV) equate to heat gained (VdP) during compression? It may very well be the case as a non-insulated quasi-static expanding syringe can also appear isothermal i.e. heat flows back into the piston-cylinder through its walls.

Reconsider the expansion of the hermetically sealed piston-cylinder. Once expanded and the force is removed, the atmosphere's weight drives the piston inwardly until both the apparatus, and surrounding atmosphere, have seemingly returned to their original states. Again from the view of the piston-cylinder apparatus, the answer is yes, the process seems reversible. Yet again we conveniently forgot that as the piston comes crashing downwards that potential energy within the atmosphere is transformed into kinetic energy, hence resulting in the heating of the atmosphere although generally by an infinitesimal amount. And the term infinitesimal herein is in relation to the atmosphere's total energy.

Again the above can be reconsidered in terms of negative work, which involves the formation of VON (volume of nothingness)! Or if you prefer; think of it as the displacement of a volume of matter against a gravitational field. And this matter was previously in a state of equilibrium with gravity before its displacement. If you really think about it, you may realize that it was gravity that caused all this matter to coalesce in the first place. So, negative work is really a disturbance to the mechanical equilibrium that existed between that matter and the surrounding gravitational field.

Yet another way to view this is to say that during expansion negative work was done onto the volume occupied by the piston-cylinder apparatus, but not its contents. The concept of negative work applies to cavitation, which will hopefully be dealt with in a future book.

Our reality is that all real work done probably can never be recovered. This is a fact missed in traditional thermodynamics where theoretical considerations of work are too often considered in terms of idealistic reversible. Reversibility is actually taught so that we are become indoctrinated to the false postulate that being the second law. Talk about circular logic at its finest! Real work is done by real systems through their walls (real or imaginary) onto its surrounding, and this genie cannot be readily put back into the bottle without the expenditure of energy. Traditional fallacies need to stop!

Boiling: Pot vs Ocean

Boiling transfers mass into the atmosphere resulting in un-measureable infinitesimal pressure increase. Therefore, PdV is just an approximation for boiling processes when $VdP \approx 0$. Instead of a pot of water, consider that a large water body boils e.g. ocean. The mass transfer into the atmosphere certainly will significantly increase both its volume, and pressure. Since both P and V increase, then:

$$W_{atm} = d(PV)_{atm} \qquad 4.42$$

Rates

So far the total energy required in heating processes has been considered. It should be emphasized that natural logarithmic function does not necessarily dominate such thoughts. However, if one wanted to consider the rates of heat transfer, then as previously eluded to; heat flows both from cold to hot and from hot to cold, with the net flow being from hot to cold (Note: For similar materials this means high thermal energy density to low).

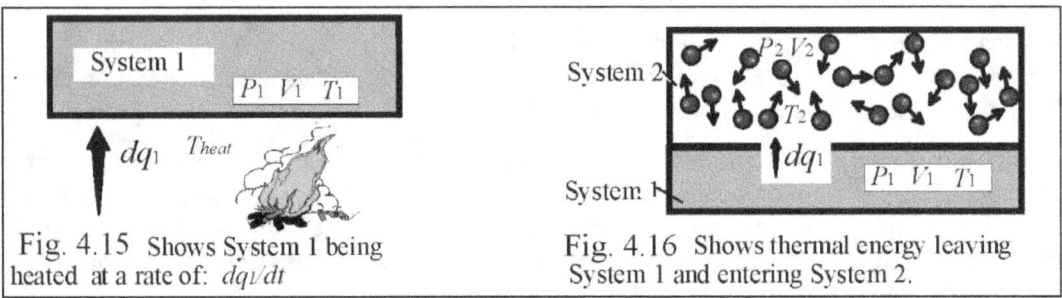

Fig. 4.15 Shows System 1 being heated at a rate of: dq_1/dt

Fig. 4.16 Shows thermal energy leaving System 1 and entering System 2.

Accordingly, the rate of heat exchange decreases as the cold system's temperature nears the hot sources temperature. Therefore, thermal energy exchange rates often should have logarithmic functionality $[In(T_f/T_i)]$ that the total energy change considerations do not necessarily convey. Consider the heating as illustrated in Fig 4.15. Now the rate of thermal energy entering System 1 can be written in terms of some constant (C) as:

$$dq_1/dt = CIn(T_1/T_{heat}) \qquad 4.43$$

Eqn 4.43 tells us that as System 1's temperature (T_1) approaches the temperature of the external heat source, then the rate of heat transfer will decrease, eventually becoming zero when $T_1=T_{heat}$.

Consider that System 1 is heating System 2 and that System 1 contains a given quantity of thermal energy (E_{tot}). Therefore as thermal energy (dq) is transferred from System 1 to System 2, in terms of relative temperatures: $T_1\downarrow, T_2\uparrow$. Obviously, as the temperatures of the two systems move towards thermal equilibrium, the net rate of heat transfer decreases towards zero. Again the rate of thermal energy leaving System 1 and entering System 2 can be written in terms of some constant (C''):

$$dq_1/dt = C'' In(T_2/T_1) \qquad 4.44$$

Fig. 4.17 Shows a studious man on a steam locomotive. From his perspective he can empirically define all mechanical work by: PdV, even if it is a theoretical mistake.

Conclusions & More Commentary on Traditional Interpretation of Work

Fig. 4.17 illustrates an individual, making measurements of the isobaric isothermal open Earth's surface. His perspective limits work to $W = PdV$, which correlates with $W = Fdx$, thus reinforces his belief. However, no clarity of onto what is the work done, wrongly leading him to the bizarre result of work done onto the walls, irrespective of whether the walls are real or imaginary. And this stated lack of clarity as to onto what is the work done has enabled thermodynamics to lose sight of reality.

Furthermore, limiting work to isobaric volume change limits us to processes where pressure increases and volume decrease, or visa versa. However, the mental tradition of limiting work to isobaric volume change has created havoc in the sciences. This is especially true when contemplating processes wherein both volume and pressure increase, e.g. bubble nucleation, whose work equation is[2]:

$$W = dE + d(PV) \approx dE + PdV + VdP \qquad 4.45$$

Where dE signifies energy associated with tensile layer formation in the ideal bubble.

Ripple effects of what is discussed herein will be felt everywhere.

Certainly, no machine can boast perpetual motion unless every aspect of it is absolutely frictionless i.e. no dissipative energy, involves no real heat flow, and neither it, nor its power source upwardly displaces any part of our atmosphere at any time. Again, this all has little to do with entropy or even the false postulate A.K.A. the second so-called law.

Furthermore, the formation of our atmosphere can be thought of in terms of work. And any disturbance to our atmosphere's equilibrium state can be expressed in terms of work.

References:

1. Mayhew, K.W., "Improving our thermodynamic perspective" Phys. Essays, 24 vol 3, 338 (2011)
2. Mayhew, K.W., "Energetics of Nucleation" Phys. Essays, 17 vol 4, 476 (2004)

Chapter 5: Heat Capacity and Conductivity of Gases

Thermal Energy Exchange defined by Heat Capacity

As was previously stated; thermal energy is herein considered as being the energy/radiation that when absorbed, results in intermolecular, and/or intramolecular vibrations within condensed matter.

Assume that two systems are in physical contact with each other, and that thermal energy from System 1 (Q_1), enters System 2. Then the change to the thermal energy (dE_{T1}) within System 1, in differential format would be:

$$dE_{T1} = Q_1 = C_{y1}dT_1 \qquad 5.1$$

In eqn 5.1, dE_{T1} is further calculated in terms of the heat capacity (C_{y1}), and the temperature change to System 1. The heat capacity (C_y) is generally expressed on either per mole or per unit mass basis, with the subscript "y" signifying the parameter (P,V) that is considered as being constant.

Dealing with condensed matter is straightforward. Fig. 5.1 shows System 1 consisting of condensed matter within a piston-cylinder, which is being heated by System 2. Consider a total thermal energy (Q_2) flows from System 2 into System 1 causing an isometric ($dx \approx 0$) temperature increase ($T_1 \uparrow$) within System 1, i.e. an increase in System 1's molecule's vibrational energies (both intermolecular and intramolecular).

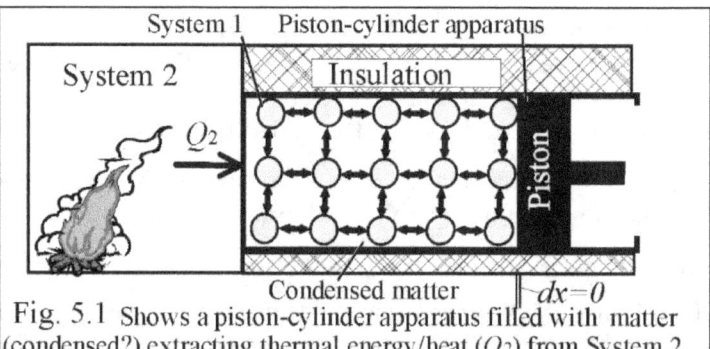

Fig. 5.1 Shows a piston-cylinder apparatus filled with matter (condensed?) extracting thermal energy/heat (Q_2) from System 2.

Based upon then eqn 1.28: $C_v' = (1/m)(dQ/dT)_v$ then in terms of the mass (m_1) within the piston-cylinder apparatus, the change in energy is:

$$dE_{T1} = dQ_1 = m_1 C_{v1'} dT_1 \qquad 5.2$$

If system 2 was a closed system (m_2) rather than a continual heat source i.e. fire, then in terms of the heat capacity (C_{y2}) of System 2, and the exchange of thermal energy:

$$dE_{T2} = -dQ_2 = m_2 C_{v2'} dT_2 \qquad 5.3$$

Conservation of energy dictates that in terms of flowing thermal energy:

$$dQ_1 = -dQ_2 \qquad 5.4$$

And in terms of total net thermal energy that has flowed between the two systems:

$$Q_1 = -Q_2 \qquad 5.5$$

The above was based upon the system's mass. Note: Engineers prefer to use heat capacity per unit mass[1], while chemists/scientists prefer per mole[2,3], and physicists often use per molecule.

Total Energy within a System

Could the total thermal energy contained within any system be approximated by using eqn 5.2, and considering that the temperature change starts at absolute zero (-273.15 °C = 0 K), and goes to the final temperature (T_{1f}) in question, e.g. room temperature (293 K). This assumes that the heat capacity is constant throughout all temperature regimesm which is not necessarily the case. As was pointed out by Planck[4] heat capacities can be approximated as constant through temperatures between 0°C and 100°C but certainly not for all temperatures. Actually our expectation should be as a system approaches absolute zero then the notion of constant heat capacity is invalid. Ditto for high temperature i.e. blast furnaces temperatures. In part the reasoning will make more sense after the Sun's input is discussed in Chapter 8.

Accordingly, obtaining the isometric heat capacity from tables, and knowing a substance's temperature, enables us to only roughly approximate how much thermal energy is contained in a given system. Hence, in terms of the isometric molar heat capacity (C_{v1}) and the final absolute temperature (T_{f1}), the total thermal energy (E_{T1}) in some System 1, consisting of n moles can be approximated by:

$$E_{T1} \approx nC_{v1}T_{f1} \qquad 5.6$$

Gases: Total Energy & Gas Constant

In Chapter 2, the total kinetic energy (E_{Tk}) of a system of N gas molecules was defined by eqn 2.12: $E_{Tk(t,r)} = 3NkT/2$. For a volume of monatomic gas with a sufficient number of molecules, and at a moderate temperature, then the total energy of that gas can be approximated as being purely kinetic (rotational plus translational). For such a monatomic gas, eqn 5.6 should equate to eqn 2.12:

$$E_T \approx nC_vT = 3NkT/2 = 3nRT/2 \qquad 5.7$$

Based upon eqn 5.7, obviously:

$$C_v = 3R/2 \qquad 5.8$$

Now reconsider a rough approximation for the energy of any gas, as defined by eqn 2.19: $E_T \cong NkT(n''+1/2)$. Therefore, for "n" moles of gas molecules, the rough approximation becomes:

$$E_T = nRT(n''+1/2) \qquad 5.9$$

Remember herein n represents the number of moles (a mole contains 6.02×10^{23} molecules) while n'' represents the number of atoms within each gas molecule.

Heat Capacity of Gases

The heat capacity of a gas is often empirically obtained by heating a wire surrounded by the gas in question, which is then surrounded by a coaxial cylinder, as illustrated in Fig 5.2. Reconsider eqn 5.9. Dividing through by temperature, then the total thermal energy for a mole of gas molecules per degree temperature is:

$$E_T/T = R(n''+1/2) \qquad 5.10$$

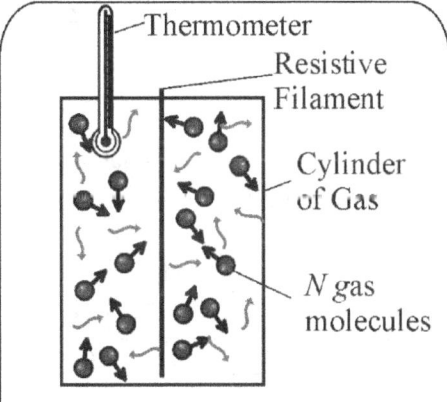

Fig. 5.2   Shows a hypothetical system for measuring a gas's specific heat. A gas in a cylinder, at pressure: P_g and temperature: T_g. is surrounded by walls at temperature: T. In thermal equilibrium: $T-T_g$. The gas is heated by electricity in the resistive wire and its temperature change is measured by the thermometer.

The total energy per degree temperature is the heat capacity, which is relatively constant for the temperature regimes that are generally witnessed here on Earth. In terms of the molar isometric heat capacity (c_v), Eqn 5.10 can be rewritten[5]:

$$C_v = R(n''+1/2) \qquad 5.11$$

The difference between isobaric heat capacity (C_p) and isometric heat capacity (C_v) for a mole of ideal gas molecules is the ideal gas constant (R) i.e. eqn 1.32: $R = C_p - C_v$. Based upon equations 1.32 & 5.11, the general equation for a gas's isobaric heat capacity becomes[5]:

$$C_p = R(n''+1/2) + R = R(n''+3/2) \quad 5.12$$

In Chapter 1, it was discussed that the adiabatic index (γ) is the ratio of heat capacities, which is given by: $\gamma = C_p/C_v$. Dividing eqn 5.12, by eqn 5.11, gives[5]:

$$\gamma = C_p/C_v = (n''+3/2)/(n''+1/2) \quad 5.13$$

Based upon the above equations, the following Table 5.1 is obtained, which is in very good agreement with the empirically proven values for the heat capacities of gases[1] for small n''. The empirically determined adiabatic indexes (γ) are shown in Table 5.2.[5]

Table 5.1

Type of gas	n''	C_v	C_p	γ
Monatomic	1	$3R/2$	$5R/2$	1.66
Diatomic	2	$5R/2$	$7RT/2$	1.40
Triatomic	3	$7RT/2$	$9RT/2$	1.29

Table 5.2

Monatomic	$n''=1: \gamma$	Diatomic	$n''=2: \gamma$	Triatomic	$n''=3: \gamma$
He	1.667	H_2	1.405	CO_2	1.289
Ne	1.667	N_2	1.4	NO_2	1.27
Ar	1.667	O_2	1.395	H_2S	1.32
Kr	1.665	OH	1.384	SO_2	1.29
Xe	1.65	CO	1.40		

Traditional Analysis and Heat capacity Discussion

As was shown in Chapter 2; our analysis given herein for thermal energy of gases does differ from the traditional degrees of freedom based analysis[5]. Since this author does not embrace the degree of freedom argument, it is left to the reader to decide i.e. apply Occam's razor! Importantly, our analys is a better fit with empirical findings. The fact that our interpretation of heat capacity fits so well with theory for triatomic, diatomic and monatomic gases should make traditionalists at least ponder. Note the traditionally accepted value equals our theoretical values for both monoatomic and diatomic. However for triatomic the traditionally accepted value is 24.9 [J/(mol*K)].

Even for $n''=4$. The fit is near perfect for many gases like hydrogen peroxide (H_2O_2: $C_v = 37.8$, $C_p=46.1$) & acetylene (C_2H_2: $C_v = 35.7$, $C_p=44.0$) the fit is rather good with our theoretical values ($C_v = 37.4$, $C_p=45.7$). However, ammonia (NH_3: $C_v = 27.34$, $C_p=35.7$) is not a particularly good fit. It raises questions; is there a difference between these molecules? Interestingly, both hydrogen peroxide and acetylene are bent elongated molecules while ammonia is pyramidal. The implication maybe molecular shape or size affects how it interacts with either the walls and/or surrounding blackbody/thermal radiation.

A Larger Data Set

It is hard to make conclusions based upon the above small data set. Table 5.4 provides a more encompassing set of values for the accepted isometric and isobaric molar heat capacities for various substances for; $0 > n'' > 27$. These accepted values were calculated using an engineering table from Rolle's book[1] as is shown in Table 5.5. Note engineer's specific heats are on a per mass basis, while scientists prefer heat capacities that are on a per mole basis. Table 5.4 and Table 5.5 are located at the end of this Chapter.

In Graph 5.1, both the isometric and isobaric theoretical (based upon equations 5.11 and 5.12) molar heat capacities are sketched against the number of atoms/elements (n'') in each molecule[5]. The accepted values for heat capacities vs number of atoms/elements are taken from Table 5.4 and are roughly plotted.

As previously stated, our theory and accepted results are a very good fit up for $n'' < 4$. What was not expected is the fact that the slope of the accepted values graph is visually very close to the slope of accepted value for $n'' > 9$. This can only be explained in terms of vibrational energy, as is done herein.

When compared to the traditional accepted theoretical curve it is obvious that

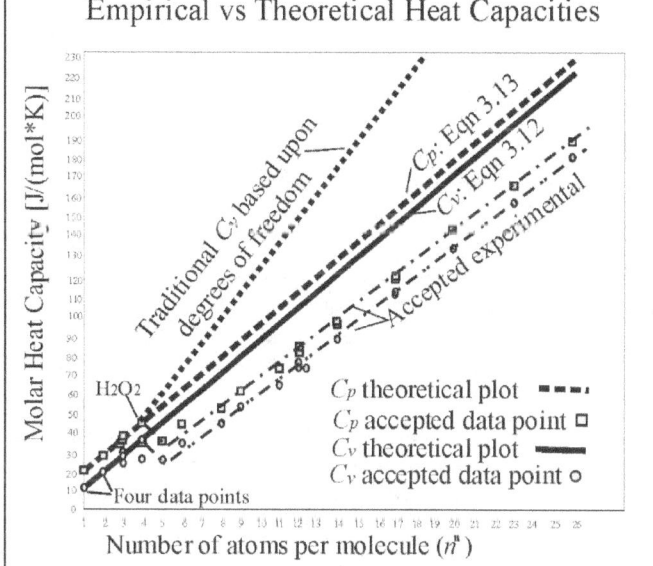

Graph 5.1: Theoretical isobaric (Cp) vs isometric (Cv) specific heats based upon Eqn 5.11 and Eqn 5.12 compared accepted experimental values as well as the traditional degrees of freedom based value for isometric specific heat.

the theory presented herein (Chapter 2 and herein) verges on inarguable. Remember the traditionally accepted theory is based purely upon mathematics (degrees of freedom) rather than being enshrined in inherent logic.

Possible Reason for Discrepancies

Why the discrepancy exists for $n'' > 4$ may be open for debate! Certainly preliminary questions arise, such as: Does the shape affect how a gas molecule absorbs surrounding thermal radiation? Or: Does the shape influence the exchange of various energies (translational & rotational plus vibrational) with wall molecules? Perhaps elongated linear molecules and/or large molecules tend to "flatline"[6] against the wall as is illustrated in Fig 5.3 at location A. The implication being that large and/or elongated gas molecules tend to hit several (few or more) wall molecules at once, with some vibrating wall molecules moving inwards, while others are moving outwards, thus affecting the cleanliness of kinematic energy exchanges.

To better understand consider the small monatomic gas molecule hitting location C. Here the wall molecule is moving outward from the wall thus instantly imparting momentum, hence pumping kinetic energy onto the gas molecule during collision. Next consider the gas molecule hitting at location B. Although the wall molecule and gas molecule are initially moving in the same direction, i.e. both into the wall, since the wall molecule is vibrating at such a high frequency then within a fraction of a nanosecond the wall molecule will start moving in the opposite direction. At which point the wall molecule imparts momentum hence kinetic energy (translational plus rotational) onto the now colliding gas molecule.

Accordingly, small gas molecules i.e. $n''<4$, should have the ability of interacting with a wall molecule cleanly, which is to say that the significantly larger vibrating wall molecule pumps its mean kinetic energy directly onto the small gas molecule. Seemingly, this is not the case for larger molecules, where kinetic (translational plus rotational) energies tends not be cleanly/clearly transferred.

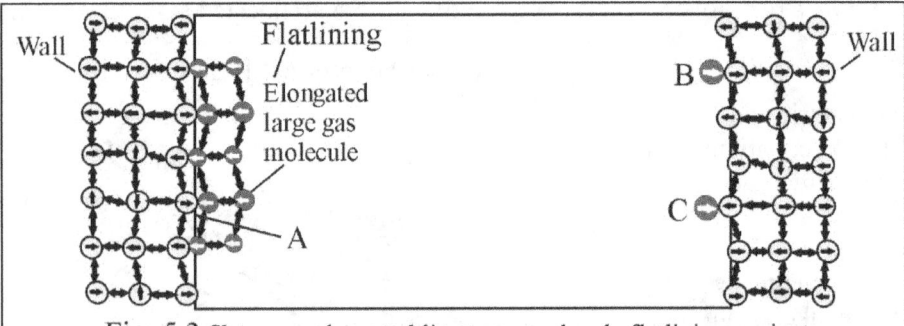

Fig. 5.3 Shows an elongated linear gas molecule flatlining against a wall at location A and the relative motions of the wall's molecules or atoms plus the relative motions of the gas's atoms. Also shown are smaller gas molecules hitting the wall at locations B and C

Furthermore, although our theoretical analysis does parallel the accepted experimental findings reasonably well for $n''>4$, as shown in Graph 5.1, the slope of accepted data is slightly less than our analysis predicts. This author further wonders if collisions between larger vibrating gas molecules with vibrating wall molecules can be considered as *completely inelastic*. By completely inelastic it is meant that a significant amount of heat/thermal energy is given off during such a collision but the net exchange of kinetic energies is indeterminable. This too would alter one's experimental expectations and may be best explained in terms of what is discussed.

Reconsider Joules weight experiment as was discussed in Chapter 1. Does the fact that the increased motions in the liquid mean that intermolecular friction has increased, hence even in the liquid state intermolecular collisions are inelastic? Seemingly, yes!

Another plausible explanation is that differing polyatomic gaseous configurations possess different vibrational energies at a given T? Of course this all may require extensive thought and/or modeling!

New Discussion

In the preface of this book it was stated that the first step to solving any problem is asking the right question. Seemingly one of the first questions that should have been asked concerning thermodynamics being; what are the fundamental differences between energy and work?

Work versus Energy

In Chapter 4 the essentials of work was discussed. Let us highlight some of what was discussed. Although work and energy can both be defined in terms of so many Joules of energy, one could say that energy has no sense of direction while work is performed with a unique direction in mind. Or that energy of a gas exists along all axes equally in all directions, while work is performed along a solitary axis and is powered by energy impacting a plane perpendicular to that axis. Of course there is more to it than this.

On Earth work moves machines and/or man in a particular direction whether it be up, down, right or left. When work is performed and directed upwards, it generally involves the movement of a mass against gravity, resulting in a potential energy increase onto that mass.

And when a system expands, it does not matter in which direction the expansion is, the net result is the upward displacement of our atmosphere ($W_{lost} = P_{atm}dV$), or at least the equivalent; whether that involves a regional pressure increase or the heating of the atmosphere (direct or indirect). Certainly the atmosphere cannot be displaced in any direction other than upward, because downward would require solid Earth's compression, while sideway would require the compression of the atmosphere, which is feasible but that too is a regional pressure increase.

Just consider the formation of our solar system or any other galactic entity. This involves energy with a sense of direction, which involved work! Again the sense of direction that gravity often differentiates work from pure stored energy.

Furthermore in our new perspective, the ability of a gaseous system to perform work, as a function of its temperature, is defined by the ideal gas law i.e. $PV = NkT$. While, the energy associated with a monatomic gas is defined by eqn 2.12: $E_{Tk(t,r)} = 3NkT/2$. Comparing the two; seemingly an ideal monatomic gas's ability to do work represents only 2/3 of its kinetic energy at a given temperature. Ditto for change, i.e. eqn 4.19: $d(E_{Tk}) = (3/2)NkdT$ vs eqn 4.20: $W = d(PV) = NkdT$. Seemingly the sense of direction that work has puts a 66.67% upper limit to the efficiency because all of the system's gaseous molecules cannot contribute all of their momentum to performing work.

Another way of viewing this; when a monatomic gas is heated, all the molecules within a closed system will experience an increase in kinetic (translational plus rotational) from all directions (+/- x,+/- y, +/-z). Only 2/3 of the gas's kinetic energy increase can impart all of the increased momentum can be used for work to be done.

What about Vibrational energy?

It makes for interesting conjecture to ask; to what extend can the vibrational energy of polyatomic molecule contribute to work? Assuming that the vibrational energy of diatomic, triatomic and other polyatomic molecules is strictly/primarily due to the absorption and radiation of surrounding blackbody/thermal radiation, the answer seemingly is that vibrational energy does not contribute to work.

Of course the above assumes that all parts of the system is always exactly in thermal equilibrium. It also assumes that polyatomic molecules receive as much vibrational energy from, as is gives onto, all walls during all energy exchanges/impacts. Such assumption may not be true at all times in all situations.

Proof

What is the proof for all of our reconceptualization of lost work? Part of the proof lay in the differences between isobaric and isometric specific heat of gases, which was briefly discussed in Chapter 4. Investigating further; consider Figs 5.3 and 5.4. Both show System 2 giving the same amount of heat (Q_2) into System 1, that being a piston-cylinder apparatus. For the isobaric case in Fig. 5.4, the heating of the gas causes a kinetic energy increase of that gas, which then drives the piston outwards by a distance, dx . Considering that the piston is both frictionless, and massless, then all the work that is actually done by the heating, is done onto the atmosphere's displacement; $W = P_{atm}dV$. Herein; $T_{system1} > T_{surroundings}$.

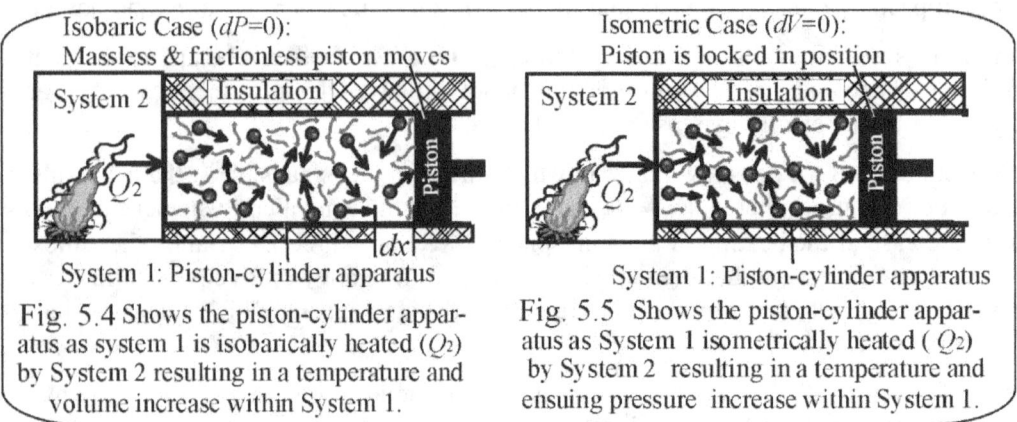

Isobaric Case ($dP=0$):
Massless & frictionless piston moves

System 2   Insulation

Q_2 Piston dx

System 1: Piston-cylinder apparatus

Fig. 5.4 Shows the piston-cylinder apparatus as system 1 is isobarically heated (Q_2) by System 2 resulting in a temperature and volume increase within System 1.

Isometric Case ($dV=0$):
Piston is locked in position

System 2   Insulation

Q_2 Piston

System 1: Piston-cylinder apparatus

Fig. 5.5 Shows the piston-cylinder apparatus as System 1 isometrically heated (Q_2) by System 2 resulting in a temperature and ensuing pressure increase within System 1.

Conversely, for the isometric case (Fig. 5.5), the piston is locked in place. Thus the heating of the gas results in a kinetic energy increase of the gas, which increases the pressure within the apparatus. Since no work is actually done, then the amount of heat required to raise the temperature of the gas inside the piston-cylinder must be less than it was for the isobaric case. Again; $T_{system1} > T_{surroundings}$.

If the input of energy is the same for both the isometric and isobaric cases, then the temperature increase in the isometric case will be greater than it is in the isobaric case. Again, this is because for the isobaric case work is done ($W = P_{atm}dV$) onto the surrounding atmosphere.

Of course, if the piston in the isometric case is unlocked, then work will be done onto the atmosphere, and either: 1) the gas will cool if the heat (dQ_2) is turned off. Or: 2) Extra heat will be required in order for the gas to remain isothermal. We now begin to understand the differences between isometric and isobaric heating of gases.

The heating process can be described in terms of the isometric heat capacity, i.e. eqn 1.31: $C_v = (1/n)(dQ/dT)_v$. Consider Q_1 to be the total thermal energy input for n moles of gas becomes:

$$Q_1 = nC_vdT \qquad 5.14$$

For n moles of monatomic ideal gas, in terms of gas constant (R), $C_v = 3nR/2$ hence

$$Q_1 = 3nRdT/2 \qquad 5.15$$

Equally, this process can be rewritten in terms of N molecules and Boltzmann's constant (k).

$$Q_1 = (3/2)Nk(T_{1f} - T_{1i}) \qquad 5.16$$

Next reconsider the isobaric heating as shown in Fig 5.4. The piston is not locked thus the system readily expands. If the piston is both massless and frictionless then in order for the piston to move P_1 only need be infinitesimally greater than 1 atm. In order for the piston to continually move right, then the temperature of the gas must be constantly increasing. The heating of the gas can be described in terms of the isobaric heat capacity, i.e. eqn 1.30: $C_p = (1/n)(dQ/dT)_p$. For n moles of gas:

$$Q_1 = nC_pdT \qquad 5.17$$

Herein, not only is the gas heated but work is also done onto the Earth's atmosphere, where the energy involved is quantified by eqn 3.28: $W_{done} = d(PV)_{atm} = P_{atm}dV$. Accordingly, the heat in (Q_1) in terms of isometric heat capacity plus work done onto the atmosphere is:

$$Q_1 = nC_vdT + P_{atm}dV \qquad 5.18$$

Rewriting in terms of N molecules and Boltzmann's constant (k).

$$Q_1 = (3/2)Nk(T_{1f} - T_{1i}) + P_{atm}(V_{1f} - V_{1i}) \qquad 5.19$$

Since the ideal gas law applies then eqn 5.19 can be rewritten as:

$$Q_1 = (3/2)(P_{1f}V_{1f} - P_{atm}V_{1i}) + P_{atm}(V_{1f} - V_{1i}) \qquad 5.20$$

In eqn 5.20: $P_1 > P_{atm}$. When the piston no longer moves, then: $P_1 \approx P_{atm}$ and eqn 5.20 for the system of monatomic gas becomes:

$$Q_1 \approx (5/2)P_{atm}(V_f - V_i) \qquad 5.21$$

Remember: In the above analysis the gas molecules were considered to be ideal and monatomic. For polyatomic molecules, one would need to also need to consider any changes to the gas's vibrational energy.

Adiabatic index

It was previously discussed for the adiabatic index; when comparing empirical values to our eqn 5.13 $\gamma = C_p/C_v = (n''+3/2)/(n''+1/2)$. The fit wis good for $n''<4$. Looking at Graph 5.2 we can see that the spread of values for the adiabatic index (γ) seems to increase for $4< n''<9$. Interestingly, for larger $n''>12$ our eqn 5.13 predicted values approximates the accepted values.

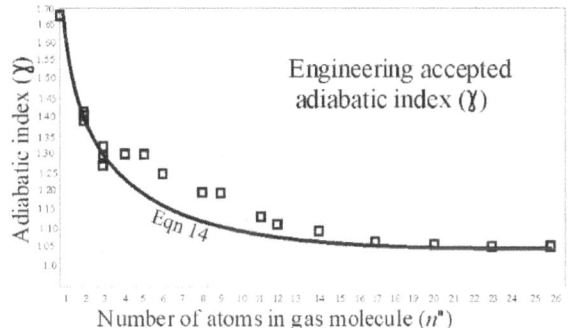

Graph 5.2: Plot of the theoretical adiabatic index vs number of atoms. Also shown is adiabatic index data points based upon engineering table for gases.

An interesting aspect of this is that this helps to affirm that the ideal gas constant (R) is independent of the type of gas. And this helps confirm that our explanation of work being done onto the surrounding atmosphere simply makes sense over any illogical association with the entropy increase within an expanding system of gas.

Other Temperature Regimes: Approaching Absolute Zero

It is also accepted that the heat capacity of gases is not constant through all temperature regimes. The traditional explanation for low temperature change to the heat capacity is that the rotational energy is frozen out as systems approach absolute zero. This author would be more comfortable with thinking that the vibrational energies are frozen out at low temperatures. It seemingly makes more sense to say that at cold temperatures there is little to no vibrational energy within a gaseous molecule, as well as condensed matter. Since both the translational & rotational energy are pumped into the gas from the wall's vibrational energy, then as this vibrational energy declines, then so to do the gas's kinematic energy. Where is the logic in thinking that any of this does not also occur at exceptionally low temperatures, as some seemingly like to profess?

There may be more to it than this. Assume that there are no unaccounted for seepages of energy into, or out, of the experimental apparatus. Now consider that as $T \to 0$ the thermal energy density within any substance is no longer directly proportional to its temperature. This is to say that the energy related to a molecule's translation, rotation, and vibration, as well as any surrounding thermal/blackbody radiation is no longer linearly proportional to temperature.

In order to improve our understanding, just consider low temperature blackbody radiation i.e. blackbody radiation at 3 K, wherein the energy density curve peaks at $\lambda \approx 1$ mm. Obviously, the thermal energy (i.e. infrared) would no longer be proportional to T, as it is for our Sun's radiation normally witnessed here on Earth. This is not to say that absolute zero does not represent the point that all matter is completely frozen. It is just to acknowledge that thermodynamic relations are no longer linear function of temperature, when near absolute zero. The exactness of this statement may vary depending upon the substance in question and in what manner it is viewed. This will be discussed in more detail in Chapter 8.

Even compare blackbody radiation of 3 K to 300 K, where the peak of the radiation curve occurs in the infrared, $\lambda \approx 10 \mu m$. Obviously, there must be a point between 3 K and 300 K wherein there is simply not enough thermal energy (i.e. infrared frequencies) to cause any real significant measurable molecular vibrations. Based upon published heat capacity curves, this seemingly happens when temperature drop below 100 K.

It must be emphasized that up to this point the thermal energy density within matter has unwittingly been considered to be linearly proportional to temperature, which is not the case for all T!

Thermal Conductivity & Gas

The thermal conductivity of a gas is a cause for an interesting discussion. Back in Chapter 1, the thermal conductivity was defined by eqn 1.36: $dQ/dt = -\kappa dT/dZ$. It was discussed that there resides thermal radiation within a vacuum. Accordingly, if in a vacuum one wall were at a higher temperature than another wall directly across from it, then net heat would be transferred from the hot to the cold wall, through the emission and absorption of thermal photons, within the blackbody radiation spectrum. This would continue until thermal equilibrium between the two walls, was attained.

Instead of a vacuum, consider that the vessel contains a gas where the gas molecules dramatically increase the rate at which thermal energy is exchanged between the two walls. This will decrease the time required for thermal equilibrium to be attained. Moreover, the rate of thermal energy exchange is not universal for all gases, i.e. it is gas dependent.

Theoretically, the kinetic energy of all monatomic gases at the same temperature should be equal. Thus, solely in terms of kinetic (rotational plus translational) energy it becomes awkward to understand why differing monatomic gases would possess different thermal conductivities. Unless it is based upon lighter molecules having higher mean velocities, hence requiring less time to travel a given distance, hence resulting in a quicker transfer of their kinetic energy. Conversely, polyatomic gases should have higher thermal conductivities due to their vibrational energy enhancing their ability to transfer energy.

Although measuring the thermal conductivity of gases is not an easy process, it is accepted that lighter gas molecules tend to have higher thermal conductivities. Furthermore, the more atoms a gas molecule has then the greater gas's the vibrational energy is at a given temperature, therefore the greater its thermal conductivity should be. This explains why argon possesses a low thermal conductivity hence is preferred for insulation purposes, e.g. used in thermal windows. I.e. argon being a relatively large monatomic gas will tend to be a relatively heavy slow moving molecule with no vibrational energy. Table 5.3 gives the thermal conductivity of some known gases[1].

Table 5.3

Gas	Ar	Air	CO	CO2	H2	He	N2	O2
$Kx10^3$	42.71	67.21	64.44	68.05	428.1	362.2	63.02	71.79

The apparatus used to measure the thermal conductivity of gases is similar to that used for heat capacity, as was illustrated in Fig 5.2. It is accepted that such thermal conductivity apparatus does not allow for the best reproducibility. This author can envision issues such as:

1) A hot wire produces blackbody radiation that is not 100% thermal radiation.
2) A wire does not represent a flat surface wherein gas molecules can readily bounce off, in a simple discernable linear fashion.

Accordingly, the apparatus for calculating a gas's conductivity may require a rethink. Perhaps, in the future one might try having two parallel walls at different temperatures and then measuring the rate of heat transfer. In order for this setup to work, one would want to shield the hotter of the two walls from the gases in question, until the moment of measurement, e.g. use a sliding insulating wall. Of course this will only address 2) i.e. it will not address the above issue 1).

Conclusions

Herein an understanding of heat capacities of gases and their application to heat transfer processes was investigated. Furthermore this can be considered as in part proof of the new perspective given concerning kinetic theory in Chapter 2.

References:

1) Rolle, Kurt C "Thermodynamics and Heat Power", Maxwell Macmillian Canada, 1993
2) Reif, F., "Fundamentals of Statistical and Thermal Physics", McGraw-Hill, New York, 1965
3) Reif, F., "Statistical Physics", McGraw-Hill, New York, 1967
4) Planck, Max "Treatise on Thermodynamics" Third edition, London, Logmans, Green and co., 1917
5) Mayhew, K.W. "A new perspective for kinetic theory and heat capacity", *Progress in Physics* Vol. 13 (**4**) 2017 pg 166-173
6) Mayhew, K.W. "Kinetic theory: Flatlining of Polyatomic Gases", *Progress in Physics* Vol. 14 (**2**) 2018
7) Giguere P.A. Heat capacities for water-hydrogen peroxide systems between 25° and 60° C. *J. Chem, eng Data* 1962 7(4) pp526-52

Table 5.4: Accepted isometric (C_v) and isobaric (C_p) heat capacities and theoretical; $C_v = R(n''+1/2)$ [eqn (5.11)] and $C_p = R(n''+1/2) + R = R(n''+3/2)$ [eqn(5.12)]. Note: Accepted science heat capacities were calculated from the engineer values given in Table 5.5, which was obtained from Rolle[1] exception being H_2O_2 which is marked with * and was taken from Giguere[7].

Substance		n''	Accepted Empirical C_p [J/(mol*K)]	Theoretical C_p [J/(mol*K)]	Accepted Empirical C_v [J/(mol*K)]	Theoretical C_v [J/(mol*K)]
Helium	He	1	20.80	20.78	12.48	12.47
Neon	Ne	1	20.79	20.78	12.47	12.47
Argon	Ar	1	20.81	20.78	12.46	12.47
Xenon	Xe	1	20.58	20.78	12.47	12.47
Hydrogen	H_2	2	28.83	29.09	20.52	20.78
Nitrogen	N_2	2	29.14	29.09	20.82	20.78
Oxygen	O_2	2	29.34	29.09	21.02	20.78
Nitric Oxide	NO	2	29.86	29.09	21.55	20.78
Water vapor	H_2O	3	33.58	37.40	25.26	29.09
Carbon Dioxide	CO_2	3	37.14	37.39	28.83	29.09
Sulfur Dioxide	SO_2	3	39.78	37.39	31.46	29.09
Hydrogen peroxide*	H_2O_2	4	46.05	45.71	37.73	37.40
Ammonia	NH_3	4	35.70	45.71	27.37	37.40
Methane	CH_4	5	35.72	54.0	27.4	45.71
Ethylene	C_2H_4	6	43.54	62.325	35.24	54.02
Ethane	C_2H_6	8	52.65	78.95	44.35	70.64
Propylene	C_3H_6	9	63.92	87.26	53.82	78.95
Propane	C_3H_8	11	73.51	103.88	65.18	95.57
Benzene	C_6H_6	12	81.63	112.19	73.50	103.88
I-Butene	C_4H_8	12	85.68	112.19	77.09	103.88
n-butane	C_4H_{10}	14	97.42	128.81	89.10	120.50
Isobutane	C_4H_{10}	14	96.84	128.81	88.52	120.50
n-Pentane	C_5H_{12}	17	120.20	153.74	111.91	145.43
Isopentane	C_5H_{12}	17	119.99	153.74	111.69	145.43
n-Hexane	C_6H_{14}	20	143.06	178.67	134.78	170.36
n-Heptane	C_7H_{16}	23	165.94	203.60	157.62	195.29
Octane	C_8H_{18}	26	188.83	228.53	180.60	220.22

Table 5.5: Engineering table for adiabatic index compared to eqn 5.13:
$\gamma = C_p / C_v = (n''+3/2)/(n''+1/2)$. Note: Data in first six columns was obtained from Rolle[1].

Substance		Mass per mole	R''	Engineer isobaric specific heat (25°C)	Engineer C_v isometric specific heat (25°C)	Accepted $\gamma = C_p/C_v$	n''	γ Eqn (5.13)
		[g/mole]	[J/(kg*K)]	[kJ/(kg*K)]	[kJ/(kg*K)]			
Helium	He	4.00	2079	5.196	3.117	1.667	1	1.667
Neon	Ne	20.18	412	1.030	0.618	1.667	1	1.667
Argon	Ar	39.94	208	0.521	0.312	1.668	1	1.667
Xenon	Xe	131.30	63	0.1568	0.095	1.667	1	1.667
Hydrogen	H_2	2.02	4124	14.302	10.178	1.405	2	1.4
Nitrogen	N_2	28.02	297	1.040	0.743	1.400	2	1.4
Oxygen	O_2	32.00	260	0.917	0.657	1.396	2	1.4
Nitric Oxide	NO	30.01	277	0.995	0.718	1.386	2	1.4
Water vapor	H_2O	18.02	462	1.864	1.402	1.329	3	1.29
Carbon Dioxide	CO_2	44.01	189	0.844	0.655	1.288	3	1.29
Sulfur Dioxide	SO_2	64.07	130	0.621	0.491	1.264	3	1.29
Ammonia	NH_3	17.03	488	2.096	1.607	1.304	4	1.22
Methane	CH_4	16.04	519	2.227	1.708	1.304	5	1.18
Ethylene	C_2H_4	28.05	297	1.552	1.256	1.236	6	1.15
Ethane	C_2H_6	30.07	277	1.751	1.475	1.188	8	1.12
Propylene	C_3H_6	42.08	198	1.519	1.279	1.187	9	1.11
Propane	C_3H_8	44.10	189	1.667	1.478	1.128	11	1.09
Benzene	C_6H_6	78.11	106	1.045	0.939	1.113	12	1.08
I-Butene	C_4H_8	56.11	148	1.527	1.374	1.111	12	1.08
n-butane	C_4H_{10}	58.12	143	1.676	1.533	1.093	14	1.07
Isobutane	C_4H_{10}	58.12	143	1.666	1.523	1.093	14	1.07
n-Pentane	C_5H_{12}	72.15	115	1.666	1.551	1.074	17	1.06
Isopentane	C_5H_{12}	72.15	115	1.663	1.548	1.074	17	1.06
n-Hexane	C_6H_{14}	86.18	96	1.660	1.564	1.062	20	1.05
n-Heptane	C_7H_{16}	100.20	83	1.656	1.573	1.053	23	1.04
Octane	C_8H_{18}	114.23	73	1.653	1.581	1.046	26	1.04

Rolle's[1] reference: Data Source J.F. Masi, Trans. ASME, 76:1067 (October, 1954): National Source of Standards (U.S.) Circ. 500, Feb 1952; "Selected Values of Properties of Hydrocarbons and Related Compounds," American Petroleum Institute Research Project 44, Thermodynamic Research Center, Texas, A&M University; College Station, Texas.

Chapter 6: **Gases and Walls**

Adiabatic

The word *adiabatic* comes from the Greek word *adiabatikos* meaning, "heat is not able to go through"[1]. When applied to a system, adiabatic implies that the system is 100% thermally insulated, thus it cannot exchange heat with its surroundings and/or other systems. Adiabatic systems are generally closed systems that contain a fixed amount of matter & energy. However, thermodynamics does allow for energy other than thermal energy to be exchanged between two systems, in what is termed an *adiabatic interaction*.

Heat Sink/Thermal Reservoir

Consider two systems are in thermal contact wherein System 2 exchanges energy with System 1. If changes to System 2's thermodynamic parameters remain immeasurable, then in terms of that process, System 2 is either a *heat sink*, or a *thermal reservoir*. Since a heat sink/thermal reservoir remains isothermal throughout a given process, then generally it is a significantly massive body when compared to the system with which it exchanges energy, i.e. a *heat bath* maintains the isothermal nature of the experimental system.

Another prime example would be our atmosphere, which generally acts as a heat sink/thermal reservoir to most systems that it surrounds. Too often energy exchanges between a system and our atmosphere goes unnoticed. Interestingly, the main source for thermal energy in our atmosphere is our Sun. Hence, the Sun controls much of the thermal energy density contained within our mother of all heat baths that being our planet, i.e. atmosphere, oceans and land.

Walls and Thermal Energy

Consider a non-insulated expanding system of ideal gas where the walls, gas and surroundings are in thermal equilibrium. Wherein:

1) The walls absorb as much blackbody/thermal radiation, as they radiate.
2) The gas molecules absorb as much blackbody/thermal radiation, as they radiate, hence maintain a constant vibrational energy (assuming they have it).
3) The gas molecules receive amount of given kinetic (translational plus rotational) from the wall's vibrational energy: Elastic or inelastic? At this point we do not know.
4) The gas molecules equally (give vs take) exchange vibrational energy with the walls.

Consider that the surroundings are either the atmosphere or some heat bath as illustrated in Fig 6.1. During expansion, the gas can remain isothermal $(T_{atm}=T_{sys}$ or $\Delta T = 0)$ when it expands quasi-statically, thus allowing infinirtesimal amounts of blackbody/thermal radiation to enter the system through its walls. In which case the blackbody/thermal radiation density within the system remains constant. The only energy change to the system is the addition of blackbody/thermal radiation into the expanding freespace. Therefore, in terms of radiation density (ρ_B), the change in energy within our expanding isothermal system becomes:

$$dE_{system} = dE_{blackbody} = \rho_B dV \qquad 6.1$$

Since the energy associated with blackbody/thermal radiation generally is extremely small, when compared to the kinematic energies associated with gas molecules, then the energy change as defined by eqn 6.1, generally can be approximated as zero during the quasi-static expansion. Remember, too often this energy is *freely given* into the expanding systems from the surroundings hence its transfer goes unnoticed. Note: This is similar to the issues concerning Joules experiment in Chapter 1.

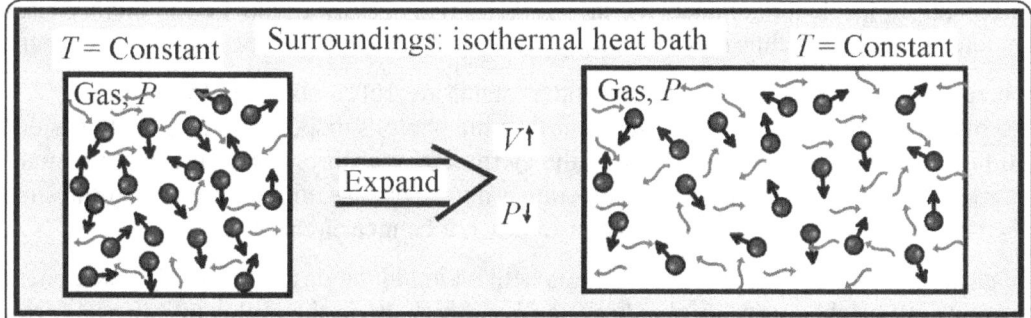

Fig 6.1 Shows the expansion of a volume of ideal gas. Quasi-staic expansion generally allows expanding systems to maintain a constant temperature due to the influence of the walls, whose temperature remains constant due to the influence of the surroundings.

Rather than quasi-static, consider what happens if the system experiences a rapid large volume increase? If the blackbody/ thermal radiation cannot be transferred from/through the walls quickly enough, then the blackbody energy density will not remain constant. If a thermometer is placed into the above rapidly expanding system, then the thermometer will still be giving as much kinematic energy onto its surrounding gas molecules, as it receives. But, the same can no longer be said of the blackbody radiation and its accompanying thermal radiation!

One could rightfully argue that given enough time then thermal equilibrium will again be attained. But this is missing the point. Furthermore, some might argue that the expanding gas must do work? Perhaps, but unless both the expanding system's surroundings, and expanding force, are clearly defined then any work cannot be determined. I.e., if an external force is expanding the system, then it is the external force, which is displacing the surroundings and hence is doing the work.

Expansion into a Vacuum Revisited: Version of Joule's Experiment

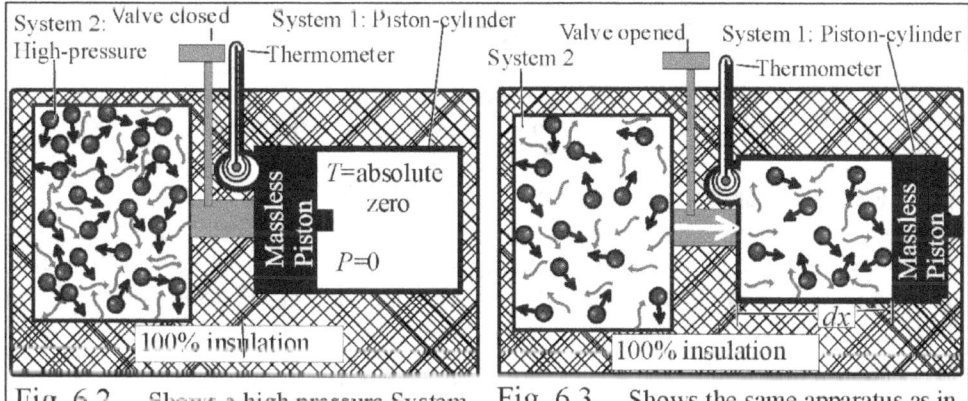

Fig. 6.2   Shows a high pressure System 2, separated by a valve from a zero volume zero energy piston-cylinder apparatus that being System 1.

Fig. 6.3   Shows the same apparatus as in Fig 6.2 but the valve has been opened allowing thermal and mechanical equi-librium to be attained.

Consider Fig. 6.2; System 2 is a high-pressure monatomic gas, which is 100% insulated and separated from System 1 that being a hermetically sealed piston-cylinder apparatus that contains neither mass nor thermal energy, i.e. an energy-less vacuum; $P=0$, $T=0$ K. Imagine that the valve is opened, then the gas molecules will push the massless, frictionless piston into System 1's vacuum and no work would be done.

Hence the gas molecules from System 2 maintain their kinetic energy[2]. If this were the only factor involved, then each gas molecule that impacts the thermometer will equally exchange its kinetic energy with the thermometer's molecules, thus they seemingly maintain their mean kinetic energy i.e. constant.

What about blackbody/thermal radiation? Since both systems are 100% insulated from the surroundings, no blackbody/thermal radiation can enter thus the photons disperse between both systems, hence the radiation density decreases. In which case, the thermometer absorbs less blackbody/thermal radiation than it radiates. Therefore, System 2's temperature must decrease, albeit by some infinitesimally small amount, hence the temperature change would generally not be measureable.

The astute reader may realize that the system's walls help maintain the gas molecule's kinetic energy, as well as the surrounding blackbody/thermal radiation, all of which are in thermal equilibrium. Since the density of blackbody radiation decreased, then the walls of System 2 must emit more blackbody radiation than they absorb during the expansion. Therefore, the wall's temperature decreases by an infinitesimally small amount, which in turn causes the walls to pass less kinetic energy onto the gas molecules than they receive. Even so, your thermometer generally cannot measure such infinitesimal temperature change.

Ultimately, the temperature associated with the gas molecules must eventually decrease by an immeasurable infinitesimally small amount. Is System 2 truly isothermal? No it is more a case that the thermometer is not accurate enough to measure the changes. Still, although the blackbody/thermal radiation is generally infinitely small, in order to fully comprehend what is going on, then that radiation cannot simply be ignored!

Pressurized Vessel

Consider a wall-less expanding system, e.g. the opening of a tank containing high-pressure gas, as shown in Fig. 6.4. High-pressure gas molecules from within the high-pressure system will disperse outwardly through the valve, colliding with atmospheric gases, and exchanging their molecular energies.

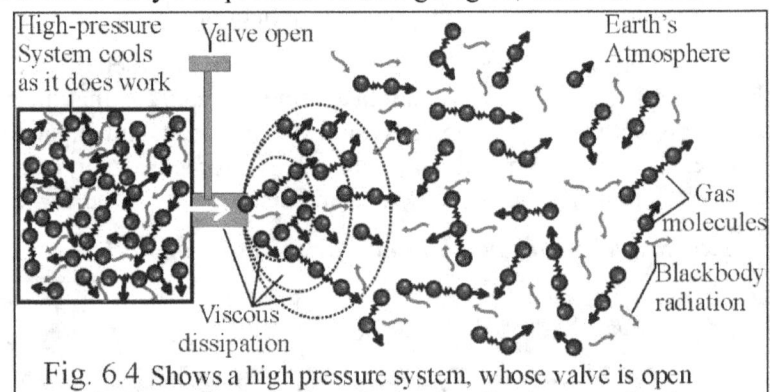

Fig. 6.4 Shows a high pressure system, whose valve is open does work onto the surrounding atmosphere's gas molecu

If their temperature hence mean molecular energies were the same before the valve was opened, as after the opening, then one might expect that the mean molecular energies will remain the same, after the valve was opened. From experience, it is known that this is not the case. The expanding gas does work onto the atmosphere in a similar manner that an expanding closed system would! Since the expanding gas is doing work, then its temperature should decrease. However reality is never that simple.

There will also be a zone of hotter molecules due to increased intermolecular friction (viscous dissipation) between the gases leaving through the tank's valve and the atmosphere. Accordingly, the expanding gas both upwardly displaces the surrounding atmosphere and gives heat into the surroundings, plus creates a regional high pressure zone in the surroundings. All of which can be viewed as part of the atmosphere's potential energy increase.

Consider the blackbody radiation. As the high-pressure gas molecules disperse out through the valve, then the blackbody radiation density associated with the volume occupied by each gas molecule must decrease, as the mean molecular volume of the gas molecules increases. The expectation is that the radiation from the surrounding atmosphere would rapidly fill this newly forming volume, hence any effect of the expanding gas molecules dragging thermal radiation with them, should generally be a minimal and possible immeasurable. Note: This may not apply to monatomic expanding gases in the same way that it would to polyatomic gases with vibrational energy.

Furthermore, the concept that all gases do such work equally really assumes that all gases interact with the atmospheric gases in an identical manner. The <u>net result will actually depend upon the extent that this process adheres to the ideal gas law, and Avogadro's hypothesis.</u> This will become more apparent in Chapter 7, i.e. all gases may not necessarily lift the atmosphere in the same manner, unless they are part of a closed sufficiently dilute system!

For example, EM potentials and/or scattering cross-sections may affect the interactions of the gases with the atmosphere, i.e. small gas molecules with low scattering cross-sections may not displace the atmosphere in the same manner as larger gas molecules would. Smaller scattering cross-sections may result in the enlargement of the cool zone, but with a smaller temperature drop. And, there may be other factors such as different gas molecules may alter the amount of heat created in *viscous dissipation*. Note: If heat is the result, then this heat ultimately raises the atmosphere's temperature, which may then cause expansion, which may give the same net result as simple displacement or visa versa. Of course this all depends upon how the interaction of the expanding gas with the atmosphere is modeled. Ultimately, since it depends upon the properties of the gas, the expansion of certain open gaseous systems may differ with the expansion of a closed system.

Commentary Concerning the Traditional vs Our New Perspective

Those well versed in thermodynamics would realize that both our, and the traditional, interpretation for the above-discussed cooling is based upon the expanding gas doing work. As previously stated, the traditional understanding of such work lacks the clarity concerning onto what the work is done. As previously discussed; this allows for various misinterpretations such as the work is done onto the expanding system's fictitious walls, and/or an increase in the system's molecule's entropy, wherein entropy is related to randomness.

With the clarity that work is done onto the surrounding atmosphere through those fictitious walls, this renders the approach given herein to be both simpler and more sensible. Note: There are those who now accept that the work is done onto the surrounding but have failed to rewrite the science with either clarity or purpose.

Fan affect

How does a fan cool you? Put your hand into the breeze coming from a fan, then a greater flux of gas molecules will impact your hand than impacts the rest of your body. So long as the atmosphere's temperature is lower than your body's temperature ($37^{\circ}C$), then each gas molecule that impacts your hand should absorb more energy from your body than they impart onto you. Since a greater flux of gas molecules are hitting your hand, and these molecules have a lower mean thermal energy than that associated with your body's temperature, then your hand feels relatively cool. In simpler terms, your hand is giving off heat at a faster rate than the rest of your body is.

Furthermore a fan increases the rate of evaporation (work onto atmosphere) of the moisture from your skin, which is nature primary method for cooling us down.

Walls & Gases

Begin with a quick review, thus enhancing our understanding of walls in thermodynamics. Reconsider an ideal gas in thermal equilibrium with its surroundings. Not only does this mean that the gas molecules obtain as much thermal kinematic energy from the walls as it gives, but it also means that both the gas molecules, and the walls absorb as much thermal/blackbody radiation as they radiate.

Since monatomic gases lack intramolecular vibrations they neither absorb, nor radiate radiation. Conversely, polyatomic gases have intramolecular vibrations therefore they absorb, and emit, thermal/blackbody radiation.

Consider a dilute monatomic gas wherein the blackbody radiation moves freely until it strikes a wall e.g. Fig. 6.5. In chapter 2, it was discussed that the mean speed (magnitude of the mean velocity) of each gaseous molecule will be a function of the system's temperature, and this has to do with the gaseous molecules hitting the container's walls, and bouncing back into the fray. Extrapolating this further; since the vibrational energy of the wall's molecules is a function of its temperature, then the mean energy impeded upon each gaseous molecule due to wall collisions, will also be a function of the wall's temperature!

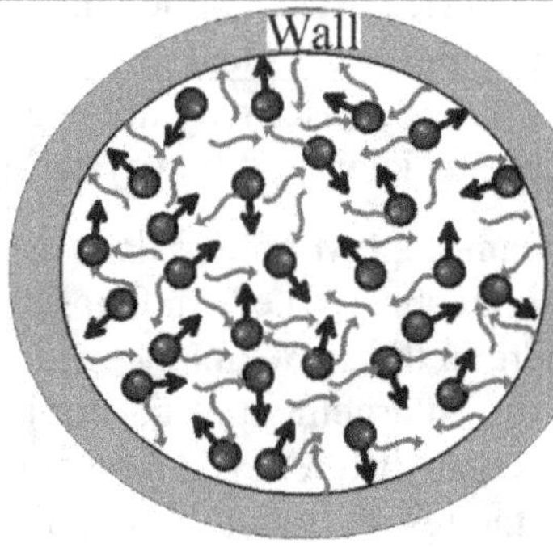

Fig 6.5 Shows an ideal gas with blackbody/thermal photons contained within a spherical isothermal wall.

Now consider a polyatomic gas. The above still applies in so far as kinetic (translational plus rotational) energy is concerned, except now the gas molecules also absorb and radiate thermal/blackbody radiation, such that their vibrational energies remain related to the system's temperature. The blackbody/thermal radiation density, vibrational energy and the kinetic energy within the gas's volume are interdependent. However, the fact remains that they are all dependent upon the wall's temperature hence correlations exist within systems that are bounded by condensed matter, e.g. walls. We now begin to better understand what Maxwell & others failed to understand, as was eluded to back in Chapter 2.

This all can be put another way; traditional teaching considers the momentum of the gas and then considers how they transfer their momentum/energy with the walls during impacts. However once you realize that the kinetics of the gas molecules are fundamentally controlled by the walls, i.e. walls act as kinetic energy pumps, then the traditional way of teaching puts the cart in front of the horse.

Note: What is stated here and in Chapter 2 means that: The distribution of the magnitude of a gas's molecular velocities about their mean value, as is traditionally taught is known as *Maxwell's velocity distribution*. This wrongly considers the kinematics purely in terms of translational energy, hence the currently accepted distribution will require much reconsideration.

True Adiabatic Expansion

Reconsider the gaseous molecules surrounded by blackbody/thermal radiation, except now imagine that both the walls and surroundings/atmosphere are magically removed, as is depicted on the L.H.S. of Fig 6.6. Without walls, the gas would disperse, expanding outwardly, as is depicted on the R.H.S. of Fig 6.6. Without any real surroundings, then the expanding gas can do no work.

Understandably, without the influx of blackbody/thermal radiation from/through the walls, the blackbody/thermal radiation density within the system must decrease, as the system expands. Over time, this radiation density decreases i.e. $\rho_b \downarrow$, hence the rate at which polyatomic gaseous molecules absorbs thermal radiation must decrease with time. Hence their vibrational energy must slowly decrease. Of course monatomic gases will not be affected in this way.

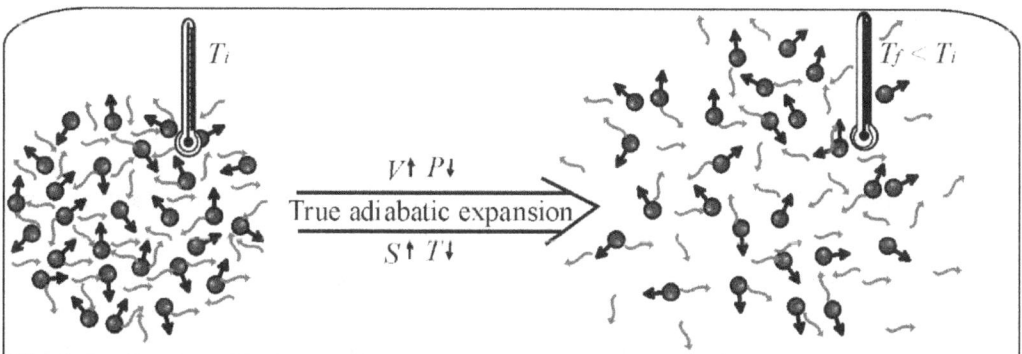

Fig 6.6   Shows an ideal gas undergoing "true adiabatic expansion" i.e. no walls, no surround-ings, maintaining a constant blackbody/thermal radiation density. Therefore both the radiation density and density of gas molecules decreases, resulting in a temperature decrease.

For both polyatomic and monatomic gases; if a thermometer is placed into that expanding wall-less gas, the thermometer should read a lower temperature due to the decrease in blackbody/thermal radiation density. Remember, the thermometer should start radiating more blackbody/thermal radiation than it absorbs, hence thermal equilibrium in terms of blackbody/thermal radiation no longer exists. One might ponder to what extent will the thermometer's reading drop? That would depend upon the kinetic energy of gas molecules, in comparison to the energy of the blackbody/thermal radiation at temperature (T_i).

Furthermore, as the system's temperature decreases, our expectation becomes: As polyatomic gas molecules become colder, the mean frequency of their emitted radiation will decrease, although their kinetic energy hence mean velocity remains constant. Again this is due to a thermal non-equilibrium condition between surrounding blackbody/thermal radiation and kinematics of matter. Ultimately, the gas molecules cool as the density of surrounding radiation decreases, although their velocities remain constant.

Kinetic Energy & Wall-less Expansion into Nothingness

How about the kinetic energy (translational plus rotational) of the gas molecules? Imagine that the gaseous molecules experience no external forces e.g. gravity, or EM, then conservation of momentum implies that the gaseous molecules will forever continue rotating and travelling along the trajectories that they attained after their last collision. Ultimately, the gaseous molecules will travel outwards towards infinity, as is illustrated in the R.H.S. of Fig 6.6. Since the gaseous molecule's momentum would be conserved, then so too is their kinetic energy.

If the total kinetic energy of the gas molecules within our system is constant, then for this truly adiabatically expanding ideal gas, Boyle's law holds, i.e.: $P_f V_f = P_i V_i$. Although $P_f V_f = P_i V_i$ remains valid for our adiabatically expanding gas, the system's temperature should be decreasing because the blackbody/thermal radiation density is decreasing. If the temperature is decreasing then the ideal gas law cannot apply, even though $P_f V_f = P_i V_i$ does! Remember the ideal gas law: $PV = NkT$.

To some, it may seem mind-boggling that $P_f V_f = P_i V_i$ applies to the expansion of a gas irrelevant of whether its expansion is isothermal in a system surrounded by walls, or, non-isothermal adiabatic expansion of a wall-less system into nothingness. Reconsider the ideal gas law. It must only apply to systems wherein the blackbody/thermal radiation is at the same temperature as the matter contained within that system. Therefore the validity of ideal gas law is truly limited to systems surrounded by walls.

An Expanding Universe has No Walls?

Without walls the mean kinetic energy, and consequentially the pressure of a gas, can decouple from the blackbody/thermal radiation's associated temperature, i.e. thermal equilibrium might not exist between matter and its contiguous blackbody/thermal radiation. Does this mean that the magnitude of gas molecule's velocity is no longer a function of the system's temperature? Perhaps! Therefore, temperature does not necessarily hold the same meaning for all systems! Of course such conceptualization is foreign to traditional theory wherein temperature is wrongly limited to molecular motions.

Can you think of a truly naturally occurring adiabatically expanding system? Perhaps, the only true adiabatically expanding system is our expanding universe, as was envisioned by Hubble. Now this statement assumes that our universe is actually expanding and that no energy is exchanged between our universe and its surroundings. Concepts that remain open for debate!

So far as an adiabatically expanding universe is concerned; as previously discussed the traditional consideration that work (PdV) being done into/onto the expanding system's walls[1,3,4] no longer applies. Interestingly, in commenting upon this traditional consideration: "Enrico Fermi once suggested"… "*the work done, which is equal to the energy expended by the expanding gas, goes "into the hands of god*", which is another way of saying that we do not know"[5]. Our new perspective that work involves the movement of a mass against a gravitational force, renders Fermi's commentary as illogical, e.g., if our universe is surrounded by nothingness then no work can be done onto such surroundings.

Traditional Misgiving

One may ask why is thermal/blackbody radiation so often omitted from analysis. First and foremost as stated the amounts of energy associated with such radiation is often infinitesimally small when compared to the quantities of energy associated within both condensed matter and polyatomic gases. Again this is part of the traditional assertion that temperature ois only associated with the kinematics of matter.

Secondly, this can be taken a step further with the realization that this is all part of traditional use of statistical analysis, which is really the strongest defence for maintaining the false postulate, i.e. that painful insistence of using the second law to explain all forms of irreversibility.

Forced Adiabatic Expansion

Consider the hermetically sealed, 100% insulated piston-cylinder apparatus filled with a monatomic gas, as illustrated in Fig. 6.7. A force is then applied to the piston, moving the piston outwards a distance, dx, e.g. expanding the piston-cylinder, as is illustrated in Fig. 6.8. Since it is fully insulated then no thermal energy (heat) can enter the apparatus. Therefore as its volume increases: $V \uparrow$, and its pressure decreases: $P \downarrow$, the blackbody/thermal radiation density must decrease: $\rho_b \downarrow$, resulting in a temperature decrease: $T \downarrow$, albeit this generally is an infinitesimally small temperature decrease, unless the expansion becomes comparatively massive.

Walls maintain equilibrium. Specifically, as the system massively expands, the blackbody/thermal radiation density decreases, then the system's walls will continually radiate more radiation than they absorb. The wall's temperature slowly decreases resulting in a decrease to the mean kinetic energy that the wall molecules pass onto the gas molecules. This results in a decrease to the mean kinetic energy of the gas molecules. And so the system's temperature slowly decreases, and as the temperature decreases the blackbody/thermal radiation mean frequency should decrease.

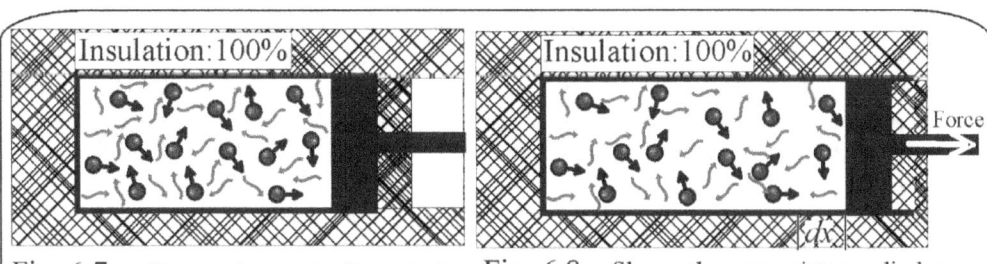

Fig. 6.7   Shows a hermetically sealed piston-cylinder apparatus which is completely insulated from its surroundings.

Fig. 6.8   Shows the same piston-cylinder as shown in Fig. 6.7 but an applied force results in the adiabatic expansion of the gas.

To what degree does the system's temperature actually decreases will depend upon the nature of walls and the gas (polyatomic vs monatomic), as well as the amount of expansion. Thick walls would tend to hold large quantities of thermal energy. Thus the trick would be to make the walls so thin that they hold little thermal energy yet can contain, and insulate, the gas within. Certainly designing an expandable system with walls that fully insulate yet holds little thermal energy and remains structurally strong requires ingenuity.

Our first intuition might be to think that the system's pressure will decrease as the volume increases i.e. Boyle's law: $P_f V_f = P_i V_i$. However, this would not be 100% correct. Specifically, unlike true adiabatic expansion, this adiabatic expansion occurs in a system surrounded by walls. Since Boyle's law is for isothermal processes, it cannot remain valid for a system whose wall's temperature is decreasing. Okay change will generally be infiniteseimally small but that is not to say that the change is not real.

One's initial thoughts maybe to use a natural logarithmic derivation as was used when calculating the work required for the syringe's expansion by an external force, i.e. our negative work conceptualization in Chapter 3. However, this too is generally based upon the isothermal Boyles law, which does not apply. To emphasize: In order to mathematically deal with our adiabatic expanding apparatus, the ideal gas law is differentiated, i.e.: $d(PV) \approx VdP + PdV \approx NkdT$. Therefore:

$$VdP \approx NkdT - PdV \qquad 6.2$$

$$PdV \approx NkdT - VdP \qquad 6.3$$

At temperatures and pressures that are experienced here on Earth, the amount of blackbody radiation within such a system would be infinitesimally small, therefore: $d(PV) \approx VdP + PdV \approx 0$. Therefore Boyle's law based logic will provide a good approximation. In cases wherein the approximation no longer holds then the system's pressure is not simply inversely proportional to the system's volume and now the polytropic equation is required.

Polytropic Equation

The *polytropic equation* was originally devised for dealing with fully convective gases, e.g. gases within stars. For such highly convective gases, wherein the density is so high that there is not enough time for radiation to be exchanged with the gas molecule's surroundings thus the process is approximated as being adiabatic. The polytropic equation is an offshoot of the ideal gas law (see Appendix B.10) that is written in terms of the polytropic exponent (n), and the proportionality constant (C):

$$PV^n = C \qquad 6.4$$

Although, derived for an adiabatic process, the polytropic equation can be applied to any process where the volume and pressure are not simply inversely proportional to each other, i.e. $PdV \neq VdP$. Note the polytropic equation has no deeply rooted theoretical basis; see Appendix B-7 for the traditional derivation. It was derived using the isometric heat capacity combined with the ideal gas law, both of which <u>does not</u> really apply to the adiabatic expansion. Although, poorly conceived, it correlates well with empirical data tool, e.g. is used by technologists, engineers and cosmologists.

Table 6.2 Process	Value of polytropic exponent (n)
Isobaric	0
Isometric	∞
Isothermal	1
Adiabatic	$\gamma = c_p/c_v$
Polytropic	$1 < n < 2$

Graph 6.1

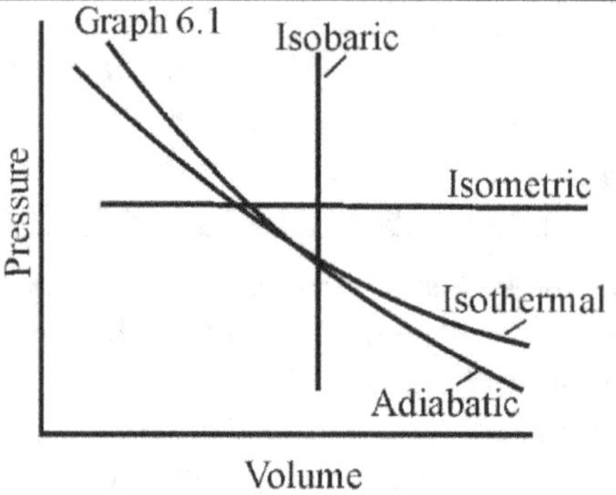

The polytropic exponent n is generally; $1 < n < 2$. Table 6.2 gives the often-used values of the polytropic exponent (n), while Graph 6.1 illustrates them. The two processes of interest are the isothermal ($n=1$), where the ideal gas law remains valid, and the adiabatic process where $n = \gamma$. In Chapter 1, the traditional adiabatic index was given as $\gamma = c_p/c_v$.

Closing Remarks

Herein the kinetic energy of a gas was separated from its surrounding thermal radiation arriving at interesting conclusions that were never previously beheld. Furthermore, the fact that walls affect one's thermodynamics perspective has also never been properly addressed. Similarly, the ramification of our atmosphere being the mother of all heat sinks hasn't been appreciated. Interestingly, traditional thermodynamics and its microcanonical ensemble become quirky when applied to non-isothermal processes, in part because the science was formulated upon the notion of heat bath (isothermal) based empirical data, while conveniently ignoring any freely given energy involved in the process.

References
1) Reif,F., "Fundamentals of Statistical and Thermal Physics", McGraw-Hill, New York, 1965
2) Mayhew, K.W., Phys. Essays ,24 vol 3, 338 (2011)
3) Reif F."Statistical Physics", , McGraw-Hill, New York, 1967
4) Carey, V."Statistical Thermodynamics and Microscale Thermophysics", Cambridge U 1999
5) Alpher, R.A. and Herman R.C. "Genesis of the Big Bang", Oxford University Press, UK 2001

Chapter 7: **Avogadro, Ideal gas law and Boltzmann**

Up to this point we have discussed how walls can have an influence upon what is witnessed. Let us now take a look at how this influences some of our traditional understanding.

Avogadro's Hypothesis/Conundrum

Traditional kinetic theory implies that at a given temperature all gases possess the same mean trnslational energy. This seemingly verified Amedeo Avogadro (1776-1856) hypothesis commonly known as *Avogadro's hypothesis*, which states: *"Equal volumes of different gases measured at the same temperature and pressure contain an equal number of molecules"* (circa 1811)[1]. An implication being; when comparing two gases in thermal equilibrium then those two gases will have both the same mean translational energy, per unit area.

To further exasperate the situation equal pressure implies that the gases also possess, the same mean momentum per unit volume i.e. mechanical equilibrium or if you prefer the same pressure. This brings us to a conveniently forgotten concern of Avogadro's hypothesis being that it seemingly confounds the conservation of momentum!

In order to understand; consider that two dissimilar objects, i.e. masses M_1 and M_2, with the same momentum collide. Conservation of momentum implies that their total momentum is conserved, therefore in terms of their velocities ($\vec{v}_1$ and $\vec{v}_2$):

$$\vec{v}_2 = (M_1/M_2)\vec{v}_1 \quad \text{and/or:} \quad \vec{v}_1 = (M_2/M_1)\vec{v}_2 \qquad 7.1$$

If one truly thinks about it, the conceptualization that different gases in thermal equilibrium will exert the same mean pressure, while possessing the same mean kinetic energy per unit volume is logically difficult at best. Specifically, realize that the pressure exerted by a gas molecule is directly related to its momentum. Now ask; how can all gas molecules (with various masses) at a given temperature exert the same pressure, and at the same time possess the same mean kinetic energy all while occupying the same mean volume? It seems improbable, and may even be impossible, at least mathematically speaking!

Investigating further; consider that the mean molecular kinetic energy of a gas equals constant A, while the mean molecular momentum equals constant B. For such a gas molecule with mass (M) and velocity (v):

$$2[M|\vec{v}|^2/2] = A \qquad 7.2$$

And

$$2[M|\vec{v}|] = B \qquad 7.3$$

In terms of velocity equations 7.2 and 7.3 can be respectively rewritten as:

$$|\vec{v}| = \sqrt{2A/M} \qquad 7.4$$

$$|\vec{v}| = B/M \qquad 7.5$$

Equating eqn 7.4 to eqn 7.5 gives:

$$\sqrt{2A/M} = B/M \qquad 7.6$$

Squaring both sides of eqn 7.6 gives:

$$2A/M = (B/M)^2 \qquad 7.7$$

Therefore in terms of constants A and B:

$$A = B^2/2M \qquad 7.8$$

To repeat: Eqn 7.8 is a solution for a pair of identical gas molecules both with mass (M) colliding with the same velocity resulting in all molecules possessing momentum and kinetic energy both of which remain constant. So ask: Now consider what happens when considering the collision of two gas molecules with different masses and/or velocities. The complexity becomes astronomical when compared to eqn 7.8 and solutions are in not easy. Other more complete and accepted solutions are given in Appendix B.8

You could rightfully argue that eqn 7.8 treated all the kinetic energy as being translational rather than both translational and rotational as was discussed in Chapter 2 and then proven in Chapter 5. Consider adding rotational momentum and energy to the above, then the complexity skyrockets. Remember traditional kinetic theory wrongly considers that all the kinetic energy of a gas is translational!

Back to reality; consider that two gas molecules collide with each other, then based upon eqn 7.1, the net result of a collision becomes the more massive object should attain a lower velocity in comparision to the less massive object. One can readily envision that over a period of time that this could allow the different gas molecules to all attain the same momentum hence exert the same pressure when they occupy the same mean molecular volume. Even so: Does this not lend itself to questioning how different gas molecules (with differing mass) maintain the same mean kinetic energy?

What a conundrum! Seemingly, Avogadro's hypothesis goes against the principles of conservation of momentum, all in order to maintain conservation of kinetic energy i.e. elastic collisions. Perhaps this helps explain why Avogadro's hypothesis was not appreciated when it was first formulated, i.e. this may explain why some 19th century researchers such as John Dalton, were against it. Okay, it is now accepted that our understanding of molecules and elements was naïve during that period. Even so, no matter how you view this; by circa 1860, poorly envisioned kinetic theory was accepted as confirmation of Avogadro's hypothesis. Why? Let us investigate beyond the wants of statistical physics!

Resolving Avogadro's Conundrum

Explaining the above; imagine a gas molecule traveling along the positive x-axis strikes a wall. As was previously discussed in Chapter 2, it was traditionally considered that the gas molecule exchanges momentum and a mean energy with the wall. As previously stated, this puts the cart in front of the horse.

Instead consider the wall as some massive immovable object then the gas molecule does not simply bounce off of the wall! Rather, the vibrating wall pumps/imposes a mean kinetic energy of $kT/2$ directed along the x-axis, onto that gas molecule with each and every collision. At the risk of sounding redundant; the walls act like an engine, continually exchanging/pumping/imposing a mean kinetic energy of $kT/2$, along each axis that is perpendicular to that wall. This occurs onto each and every gas molecule that is sufficiently small that it can interact cleanly with the walls. The net result being that each gas molecules will possess a total mean kinetic energy of $3kT/2$ that being rotational plus translational energy. And this explains why traditional kinetic theory of gases is has a certain validity, although it was far from correct!

Furthermore, when a gas molecule collides with another gas molecule, then it will adhere to the conservation of momentum that being eqn 7.1, which does not necessitate the conservation of kinetic energy. Of course eventually these very gas molecules might hit enough vibrating wall molecules to once more attain that specified mean kinetic energy along each axes, as predicted by kinetic theory (eqn 2.11).

In simple terms walls (and/or, other such surfaces of condensed matter) within a system are what enforces Avogadro's hypothesis validity[2], at least when talking about small-molecule gases near room temperature! A requirement becomes that such gas molecules tend to bounce of the walls much, more frequently than they collide with each other i.e. be sufficiently dilute. An implication being that laboratory findings where walls exist, cannot always be simply applied all systems e.g. those without walls[2,3].

Mean Free Path and Sufficiently Dilute

 The concept of sufficiently dilute gases was previously used without our providing clarity. Now, a *sufficiently dilute gas* can be understood as being a gas in a container whose dimensions are smaller than, (or at least not significantly greater) the gas's mean free path, which is also related to its concentration.

For a high velocity gas molecule relative to an ensemble of similar molecules at random locations, the mean free path (l) of a gas molecule is defined in terms of the gas molecule's cross-sectional diameter (d), and number of molecules per unit volume (n_v), i.e. by:

$$l = 1/\pi d^2 n_v \qquad 7.9$$

Interestingly, the mean free path does not directly depend upon the gas molecule's velocity, however in reality it does, because at a given pressure, gas molecules with the higher momentum may tend to occupy a greater mean molecular volume, hence a lower number of molecules per unit volume (n_v). Furthermore eqn 7.9 remains an approximation since molecules tend to have attractions at large distances and repulsion at short distances (the Lennard-Jones potential).

Elastic vs Inelastic Collisions

As previously discussed, traditional kinetic theory considers all collisions between gaseous molecules to be elastic, i.e. both energy and momentum are conserved. For any elastic collision, the relative velocity before of the two colliding masses equals the minus of the relative velocity after the collision. The following solution for an elastic collision is derived in Appendix B.8.

$$M_2/M_1 = 1 - [2\vec{v}_{1f}/(\vec{v}_{2f} - \vec{v}_{2i})] \qquad 7.10$$

Herein, a gas molecule is considered to behave like a superball when colliding with a wall, i.e. $\vec{v}_{1f} = -\vec{v}_{1i}$. Of course this fits with an elastic collision and relative velocities, but not necessarily with eqn 7.10 because the wall's mass is infinite compared to the gas molecule. The awkwardness goes beyond this, because no real mechanism for both the wall and gas molecules being related to $kT/2$ is given when implementing such logic. Accepting this author's assertion, that massive vibrating wall molecules pump a given energy onto the relatively minute gas molecules circumnavigates such issues. Thus the kinetic energy based relationships becomes understandable without the requirement of elastic collisions[2,4].

Now imagine; when gaseous molecules do collide that heat (or radiation/photons/phonons) are given off, hence although such collisions are inelastic, energy remains conserved. When such inelastic collisions occur within a closed system, then the other gas molecules and/or the surrounding wall molecules should absorb any collision derived heat that is given off. Accordingly, such collisional generated heat becomes part of the equilibrium state between molecular collisions and vibrations, plus the emission and absorption of radiation.

Importantly, this helps formulate an explanation for viscous dissipation, and/or natural *P-T* system relationships that being molecular collisions are generally inelastic therefore heat is readily given off. The implication being that intermolecular collisions even in the condensed matter may be inelastic. Again this would be in agreement with what is witnessed concerning both natural *P-T* system relationships and molecular viscous dissipation.

Furthermore it has been determined that collisions between photons and electrons are inelastic[5,6,7,8]. With that being the case then why would anyone still believe that collisions between molecules are anything but inelastic! This also bodes the question; to what degree are collisions between thermal radiation and condensed matter, elastic?

A Most Interesting Question

Does this all mean that the masses of colliding bodies needs to be equal in order for a collision to be elastic? If yes then this may open a whole new set of questions in physics to be contemplated.

Consider Newton's cradle where several steel balls are each hung on strings like several pendulums in series. If the balls are the same size and mass, then when you lift the first one and let it collide with the others, then the collisions are near elastic. Hence the last/seventh ball will rise to the first ball's intial height and the collision is considered almost elastic with the motion being near perpetual.

However: The collisions no longer seem to be elastic when the balls are not all equal mass i.e. Newton's cradle losses it's symmetric beautiful motion. Seemingly elastic collisions need to be the same mass. One may further ponder: Do the motions of center of mass have to be along the same plane for collisions to be elastic?

And: Does this help explain problematic relativity? Okay some see relativity as problematic. While most do not e.g. this includes those who control the reigns of science. Interesting questions for you and others to deliberate?

Radiation Surrounding a Gas

Throughout this text, the preferred terminology describing radiation has been either blackbody/thermal or thermal/blackbody radiation, rather than simply either blackbody or thermal radiation. The reason is actually two-fold. Firstly, not all frequencies associated with blackbody radiation inside of a void may actually be considered as thermal energy. Remember thermal energy consists of frequencies that readily absorbed by condensed matter resulting in both intramolecular and intermolecular vibrations.

Secondly researchers like Robitaille[9] have argued that voids do not necessarily contain what is considered as being true blackbody radiation. Reconsider the perspective given herein, that being that molecular collisions generally are not elastic hence often produce various frequencies of radiation, some of which can be considered as thermal. Understandably, gases are generally surrounded by radiation, some of it attributable to blackbody radiation, and other attributed to intermolecular collisions radiation (thermal?) and maybe even some attributable to photons inelastic collisions with electrons.

Obviously all of the above may require consideration by others. Again the energy associated with all this radiation will be infinitesimally small, when compared to the kinematics associated with most gases. Even so it can NO longer be ignored out of convenience! That is because it is part of a system being in, or out of, thermal equilibrium.

Mathematical Argument for Avogadro's Hypothesis

Start off with the ideal gas law: $PV = NkT$. Which can be rewritten:

$$V = NkT/P \qquad 7.11$$

The mean kinetic energy of N gaseous molecules is given by eqn 2.12: $E_{k(t,r)} = 3NkT/2$. Which can be rewritten as:

$$NkT = 2E_{(k,r)}/3 \qquad 7.12$$

Combining eqn 7.11 and eqn 7.12, gives:

$$V = 2E_{(k,r)}/3P \qquad 7.13$$

The point of eqn 7.13 is that the volume (V) occupied by N gas molecules does not depend upon the molecule's mass in closed systems. Rather it depends upon the gas molecule's kinetic energy, which was attained from its collisions with the walls. Hence, we could claim that Avogadro's hypothesis is confirmed but only for sufficiently dilute small-molecule gases enclosed by walls.

In reality the above mathematical analysis is a circular argument, because it started with the ideal gas law, whose validity must now be questioned when walls do not exist. Interestingly, a whole gambit of thermodynamic relations begins to falter when walls are removed and/or when gas molecules become relatively large. Remember as was discussed in Chapter 5, large gas molecules may not experience clean collisions with vibrating wall molecules i.e. flatlining

Avogadro's Hypothesis Limitations

Conditions must exist, wherein Avogadro's hypothesis becomes invalid! If the radius of the gaseous molecules/atoms were large enough, then the scattering cross-section of gaseous molecules within a system would become significant. In which case momentum conservation would dominate, and Avogadro's hypothesis would falter even if the gas was dilute. However, this remains unlikely since the freespace associated with most gases is so much greater than any gas molecule's radius.

More likely scenarios occur in high-density gases, wherein the mean free path becomes too small for Avogadros' hypothesis to hold! Or, when a gas has no walls i.e. our atmosphere. In such situations the gas molecules are most likely to collide with each other hence pressure equalization should dominate, as the velocities of gas molecules after collision will tend to be governed by eqn 7.1. Accordingly, kinetic theory the ideal gas law and Avogadros hypothesis all begin to falter! So too does Maxwell's velocity distribution, which also previously failed because it is limited to translational energies e.g. no rotation!

Taking this a step further and realizing that since both the ideal gas constant, and the ideal gas law are similarly based upon sufficiently dilute gases. This explains our acceptance that the ideal gas law is not valid for highly dense gases, as we now begin to understand why. This also helps explain the need for the polytropic equation when dealing with stars, as was previously discussed in Chapter 6. Certainly, an unrelenting traditionalist, may adhere to elastic collisions, but then they cannot explain why the polytropic is required, unless they argue that elastic collisions are limited to dilute gases, but then give how do they explain why dilute gases are elastic while dense gases are not!

Avogadro's Hypothesis at Low Temperatures

Interestingly, Avogadro's hypothesis also falters at low temperatures. Graph 7.1 is the plot for the "Number of Moles per Cubic Meter versus Temperature", for five elements: Hydrogen, helium, argon, nitrogen and oxygen. The graph for each element starts near their boiling point, and continues until 320 K. The data[5] for Graph 7.1 is located in Table 7.1, at the end of this chapter.

If the gases obeyed Avogadro's hypothesis for all temperatures, then the: "Number of moles per m³" would be a decreasing linear function of temperature. It appears to be approximately so from 100 K to 320 K, however, as the gas's temperature approaches absolute zero, the number of mols/m³ increases in an exponential-type fashion.

Low Temperatures Gases

The simple traditional argument for the behavior of gases at low temperatures is that the gas molecules have less kinetic energy, therefore other forces, e.g. the electro-magnetic forces (EM), become prevalent. However, this argument is troublesome because as just witnessed in Graph 7.1, Avogadro's hypothesis does not hold for the noble gas, helium! A gas wherein the EM interaction between molecules should be minimal. It could be a case of other forces such as gravity affecting what we see, because they become substantial in comparison to the translational and

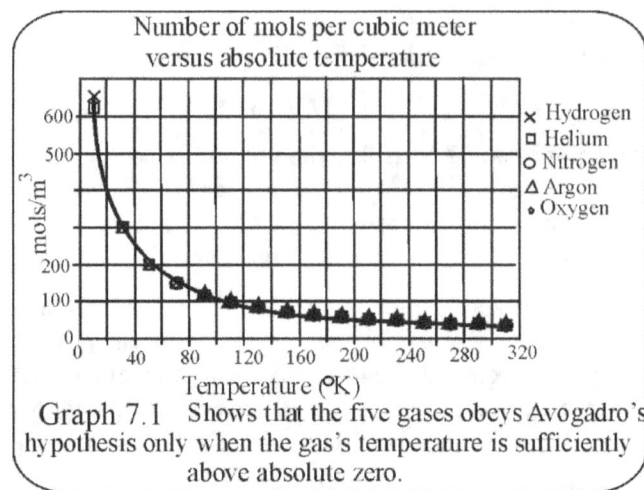

Graph 7.1 Shows that the five gases obeys Avogadro's hypothesis only when the gas's temperature is sufficiently above absolute zero.

rotational energies of the gases. So perhaps, although probably unnecessary!

There are also quantum mechanical arguments used to help explain the behavior of gases at low temperature, e.g. the Bose-Einstein condensation (BEC). Basically equipartition is considered to falter at low temperatures. The reasoning is based upon quantum effects becoming significant; "*when thermal energy kT is smaller than the quantum energy spacing in a particular degree of freedom, the average energy and heat capacity of this degree of freedom are less than the values predicted by equipartition. Such a degree of freedom is said to be frozen out when the thermal energy is smaller than the spacing. For example: The heat capacity of a solid decreases at low temperatures as various types of motion become frozen out, rather than remain constant as predicted by equipartition*"[18]. Such quantum arguments tend to be rather complex and out of the scope of this text.

Interestingly, you should realize that part of the reason quantum was embraced throughout the 20th century is the fact that traditional kinetic theory could not properly explain empirical findings. Certainly the fact that a gas's Hamiltonian in quantum physics is often wrongly solely written in terms of the translational energy, which is then equated to a function of kT, implies the need for new contemplations.

It is the translational plus rotational energy of the gas that should be equated to the function of kT, and this should have ramification to how quantum physics is conceived and rewritten. Interestingly the total energy of the gas remains the same. Does the fact that our new kinetic theory better explains empirical findings mean that quantum needs a complete overhaul? Or just several minor insignificant alterations. That is left for others to decide.

Is there a simpler argument for Avogadros hypothesis faltering at low temperatures? This author cannot help but ponder if the simplest explanation resides in the fact that the thermal radiation density would no longer simply be directly proportional to temperature at low T. This will be discussed in the next Chapter. Of course the fact that a gas's energetics is no longer proportional to T means that kinetic theory would no longer be valid at low T.

Reality vs Kinetic Theory

Seemingly kinetic theory is an idealistic simplification of reality; a gas' energy is pumped in from vibrating wall molecules (condensed matter) on all sides. this certainly allows us to understand the fundamentals, as well as perform experiments within our new context i.e. heat capacity of gases.

Contemplating real life scenarios; consider a gas that is not surrounded by walls hence the dominant interactions are intermolecular gas collisions, i.e. kinetic theory, ideal gas law and/or Avogadro's hypothesis all lose some of their integrity. As discussed in the previous chapter; now place a thermometer into that gas and consider what you are actually measuring.

In order to answer, consider that you take a box of sufficiently dilute polyatomic gas into remote outer space and punch a hole through one of the box's walls. As the polyatomic gas molecules stream out through the hole their energies (i.e. translational, rotational and vibrational energy) are all defined by our new kinetic theory, i.e. remain a function of the box's previous uniform temperature. However, in the cold outer space these streaming polyatomic gas molecules continually emit more blackbody/thermal radiation than they adsorb, so their vibrational energy decreases but their kinetic (translational plus rotational) energy remains constant. An interesting question arises: What is the real temperature of these polyatomic gas molecules? Clearly the whole accepted science crumbles as the radiation energy and vibrational energies are no longer isothermal with the kinematic energies!

Sauna

Consider a sauna where kinetic theory does not strictly apply because:

1) It has a heat source, hence different temperature regimes within.
2) The air is not sufficiently dilute because the sauna's dimensions exceed the mean free path of the gas/air molecules within.

The air molecules coming off the sauna's heater generally have velocities that are greater than the air molecule's mean velocity within the sauna. Why is it significantly hotter at the top of the sauna than it is along the floor? Certainly air molecules with higher mean velocity will have a higher mean kinetic (translational plus rotational) plus vibrational energy. One might think that this implies that the hotter gas will possess a higher associated pressure, but this wrongly assumes that all the gas molecules in the sauna occupy the same mean molecular volume!

A better analogy is that gas molecules with a higher mean kinetic energy a given pressure, tend to occupy a greater mean molecular volume i.e. the mechanical principle of pressure equalization is applied. Since the sauna is isobaric (1 atmosphere), then within this volume, molecules that occupy a greater mean molecular volume tend to end on top of molecules that want to occupy a smaller mean molecular volume. In other words, hotter gases tend to float on top off colder molecules. However, such an explanation retains a certain awkwardness, when one thinks in terms of each gas molecule acting independently.

The awkwardness is avoided by understanding that gas molecules with greater kinetic energy are not as readily influenced by gravity, as slower less energetic ones would be. In order to fully understand consider that you magically release a stationary gas molecule into the atmosphere, then barring this gas molecule colliding with another gaseous molecule, the stationary molecule will plummet downwards due to the influence of gravity. Accordingly, it is understood that the greater a gas molecule's mean velocity is, the less the influence that gravity will have on its motions.

The sauna remains a dynamic system, with hot air rising, some of which impact the ceiling, then plummet downwards, eventually colliding with cooler air molecules below. Remember such collisions are inelastic hence often result in the emission of energy/photons, which all remains part of the heat balance.

Schlieren Phtography

On the internet great videos concerning the following technique can be witnessed. Schlieren photography is a great way to view the flow of gases into and out of containers amongst other things i.e.

cold air can be seen to flow downwards, while hot air can be seen to flow upwards, using this technique.

It must be understood that the high school level answer is that the heavier/denser gas tends to flow downwards, while lighter/(less dense) gases rises. As previously discussed; gases that have higher molecular mass, and/or slower mean molecular velocities are more strongly influenced by gravity when compared to those with lower molecule mass, and/or slower mean molecular velocities.

So although the cold air in Fig 7.1 tends to sinks and appears to flow like a cohesive liquid, it has more to do with the fact that the cold air molecules all share similar kinematics, hence similar net gravitational influences. So although there are no actual cohesive forces between the various air molecules, the appearance of cohesion may be a false one. Ditto for hot air rising!

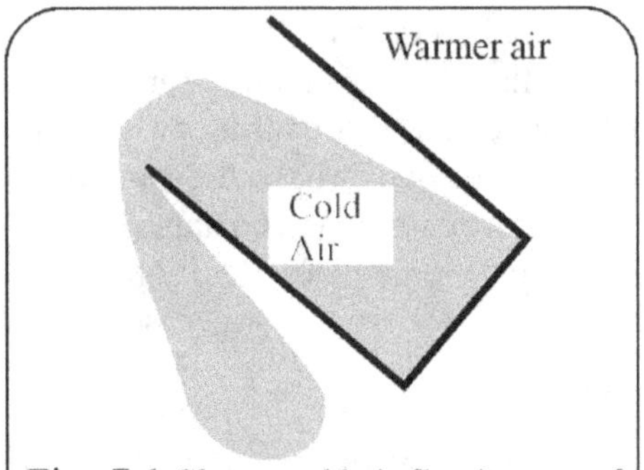

Of course the greater a gas molecule's mean velocity is, then the greater its mean molecular volume will be, at a given temperature and pressure. Accordingly, hotter gases with a given mean molecular mass generally tend to behave as a less dense fluid would!

Fig. 7.1 Shows cold air flowing out of a tilted container. It appears to be the flow of a cohesive liquid, when in reality it is no such thing.

Atmosphere Physics

Based upon our new insights, atmospheric physics will also require an overhaul. No longer does entropy, the second law, kinetic theory, ideal gas law, all simply apply with impunity. Since high pressure atmospheric systems tend to be colder, mean molecular volumes seemingly prevail over any association with the mean kinetic energy of gas molecules, i.e. mechanical equilibrium dominates!

Do simple associations between measured air's temperature and the air's kinetic energy remain? Dynamic systems like our atmosphere are never overly simplistic. Yet the difference between isobaric and isometric specific heats remains the ideal gas constant (R) and as will be shown, this can be viewed as a relationship based upon how much work is required to displace our atmosphere. Does this not imply that our atmosphere adheres to kinetic theory functionality to some degree? Yes but not necessarily completely as the ideal gas constant is more associated with potential than kinetic energy.

Perhaps most importantly, significant energy changes when putting atmosphere gases into a box, have not been readily witnessed! Accepting that walls influence what is measured in closed gaseous systems one might expect this to be the case! So how do we explain that atmospheric gases can readily put into, or removed from a closed system without any real witnessed significant changes?

Our atmosphere (air) is 99% diatomic gases [nitrogen (78% N_2) & oxygen (21% O_2)], with the larger molecules namely water (H_2O) and methane (CH_4) making up most of the final percent while monatomic Argon constitutes most of the remainder. Accordingly our atmosphere's energetics is dominated by the absorption of thermal radiation by diatomic gases along with H_2O and CH_4. Since these polyatomic gases adsorb our Sun's thermal energy (heat), then the atmosphere's n''-molecule polyatomic gas's vibrational energy will be a function of the temperature, as defined by eqn 2.16 i.e. $\overline{E_v} \cong (n''-1)kT$, e.g. for diatomic gases $n''=2$ hence: $\overline{E_v} \cong kT$.

Due to inelastic intermolecular collisions, the gas's vibrational energies influence the translational energies of any surrounding gases. Remember; kinetic theory only precisely applies to sufficiently dilute gases, which are also sufficiently small gas molecules ($n''<4$) and are in enclosed systems! This implies that larger polyatomic molecules ($n''>4$) do not obey kinetic theory even when in a closed system as has been confirmed by empirical findings[4]. Certainly the vast majority of our atmosphere's gases would fall into the sufficiently small category, implying that when inter-gas collisions dominate, then the vibrational energies strongly influence all the kinematics of such gases.

Accordingly, for gaseous systems without walls; if the various molecules actually do occupy the same mean molecular volume, as described by Avogadros hypothesis, then they all should possess the same mean momentum, at a given pressure. The reality is more likely that; a given pressure means mechanical equilibrium exists between gas molecules with conservation of momentum applying to the gas's center of mass translational and rotational motions during inter-gas molecule collisions exist dominate. But the polyatomic gases vibrational energies must have a strong influence during collisions especially upon any similar or smaller sized gas molecule's motions. The exact final mean kinetic energy may take some thought and perhaps computer modeling to determine.

Furthermore, air molecules close to Earth's surface will receive kinetic (translational plus rotational) energy from collisions with Earth's angular surfaces. If the Earth's surface was perfectly flat rather than angular then the mean kinetic energy pumped into small gas molecules ($n''<4$) could be defined in terms of along one axis i.e. y-axis by:

$$\overline{E}_{ky} = kT_{earth}/2 \qquad 7.14$$

Of course the rough nature of Earth's surface means that the kinetic energy pumped onto small air molecules colliding with Earth should be significantly greater than that defined by eqn 7.14. Even so it must be that any of the larger polyatomic air molecules absorb thermal radiation which then dominates (influences) the kinematics of smaller air molecules.

Ideal Gas Constant (R) Reconsidered

All of this renders the fundamental question as to what is the ideal gas constant? Substituting eqn 1.22: $C_p = (1/n)(dQ/dT)_p$, and eqn 1.23: $C_v = (1/n)(dQ/dT)_v$, into eqn 1.24: $R = C_p - C_v$, gives for the ideal gas constant:

$$R = (1/n)[(dQ/dT)_p - (dQ/dT)_v] \qquad 7.15$$

If the temperature change in the denominator is the same for both the isobaric and isometric case, then eqn 7.15 becomes:

$$R = [(dQ)_p - (dQ)_v]/(ndT) \qquad 7.16$$

Since the difference between the energy required for isobaric heat capacity, and isometric heat capacity, is the work done onto the atmosphere, therefore:

$$W_{done} = W_{lost} = [(dQ)_p - (dQ)_v] \qquad 7.17$$

Substituting eqn 7.17 into eqn 7.16 gives:

$$R = W_{done}/(ndT) \qquad 7.18$$

Based upon our new perspective that work can be: $W = d(PV)$. Substituting back in gives:

$$R = d(PV)/(ndT) \qquad 7.19$$

Eqn 7.19 could have been simply calculated by differentiating the ideal gas law ($PV = nRT$), and then dividing both sides by: ndT. Eqn 7.18 is a reiteration that the ideal gas constant is nothing more than a relation for a mole of gas's ability to perform work $[d(PV)]$ per degree temperature change (dT).

Again it must be emphasized that work, and the thermal energy of a gas, are two different things. Expressing the total energy contained within a volume of gas by:

$$E_T = \varepsilon + 3PV/2 \qquad\qquad 7.20$$

The total thermal energy change then becomes:

$$dE_T = d\varepsilon + 3d(PV)/2 \qquad\qquad 7.21$$

For an ideal monatomic gas: $d\varepsilon = 0$, and eqn 7.21 becomes:

$$dE_T = 3d(PV)/2 \qquad\qquad 7.22$$

Ideal Gas Constant, Boltzmann's Constant, and Work

Eqn 7.19 implies: The "ideal gas constant (R)" is the change in the ability of an ideal to perform work, per mole, per degree Kelvin temperature change. Rather than change one could equally perceive that the ideal gas constant represents the ability of an ideal gas to perform work per mole, per degree kelvin.[7]

Either way the ideal gas constant (R) relates to the total energy contained in a volume of ideal gas within a closed system, because the ability of an ideal gas to perform work is 2/3 of that gas's total kinetic energy[7,8]. The ideal gas constant "R", is equivalent to Boltzmann's constant "k", for a mole of gas molecules, i.e. $R = Nk$, where: N=1 mole. Therefore: "Boltzmann's constant (k)" is the mean ability of a solitary ideal gas molecule, to perform work, per degree Kelvin. Again this relates to an ideal molecule's mean kinetic energy in a closed system by the same factor 2/3.

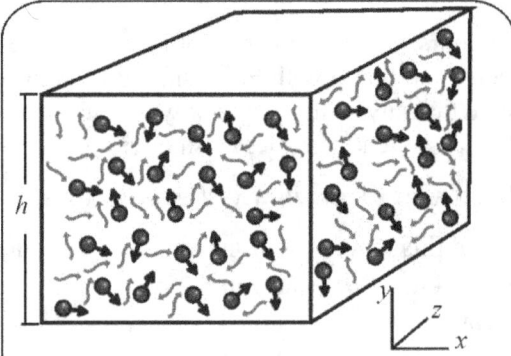

Fig. 7.2 Shows a unit cube, containing N molecules of gas wherein: $x=y=z=h$.

Emphasizing: Boltzmann's Constant & Work

Reconsider the constant volume thermometer as was discussed in Chapter 1. The temperature change was measured by the pressure change of the ideal gas in the thermometer's glass bulb, which was measured by the height of mercury. The height of mercury is the displacement of its mass against gravity, which is really what work is all about. Although we are now starting to entertain circular arguments, the point remains, if we know how many gaseous molecules (N) are in the thermometer's glass bulb, and the pressure (P), volume (V), and temperature (T), then by rearranging the ideal gas law, k can be calculated:

$$k = PV/NT \qquad\qquad 7.23$$

Consider a unit cube whose volume is "V" with surface area:" A", as is shown in Fig 7.2. If "M" represents the mass of overlying atmosphere, and "$\vec{g}$" is gravitational constant. Then the pressure exerted by the Earth's atmosphere on the top surface of the unit cube is:

$$\vec{P} = M\vec{g}/A \qquad\qquad 7.24$$

Accordingly, eqn 7.23 can be rewritten:

$$k = (Mg / A)(V / NT) \qquad 7.25$$

Remember that for a unit cube $(V / A) = h$, wherein: "h" is the height of the unit cube. Therefore eqn 7.25 can be re rewritten as:

$$k = (Mg / T)(h / N) \qquad 7.26$$

Thus:

$$kTN = Mgh \qquad 7.27$$

Limit the volume change to only vertical expansion, i.e. along the y-axis and then differentiating both sides, gives the change in temperature with height as:

$$NkdT = Mgdh \qquad 7.28$$

Thus

$$k = (1 / N)Mgdh / dT \qquad 7.29$$

Eqn 7.29 also implies that Boltzmann's constant (k) is the proportionality for the work required to displace the overlying Earth's atmosphere overlying mass (M) by a height (dh), per degree of temperature change (dT), on a per molecule basis. This remains independent of the type of gas doing the work so long as that gas is sufficiently dilute that they obey Avogadros hypothesis, ideal gas law etc.

Ideal Gas Constant and Mass

Interestingly, engineering texts on thermodynamics prefer to use an ideal gas constant based upon the mass, whilst physics and chemistry texts on this subject do not. Let R'' be the universal ideal gas constant (R) divided by molecular weight (MW):

$$R'' = R / MW \qquad 7.30$$

where MW is molecular weight, and R is the ideal gas constant: R= 8.31 J/mol·K

Eqn 7.30 considers the ideal gas constant (R) to signify a gas whose atomic mass is unity (1 g/mole). Thus, the ideal gas law can be rewritten in terms of the total mass (m) of the gas:

$$PV = mR''T \qquad 7.31$$

Obviously, engineer's preference is to perform their calculations based upon the total mass of the gas.

Closing Remarks

In this chapter numerous ideas came together, and especially how they relate to work and total energy contained within a volume of gas. We were enlightened as to the relevance of both Boltzmann's constant (k), and the ideal gas constant (R) and how they relate to work and gravity. New understandings and the limitations for Avogadro's hypothesis, the ideal gas law and kinetic theory were discussed. Once more traditional conceptualizations were challenged, often allowing a simpler understanding to prevail in its place.

References

1) Reif,F. "Fundamentals of Statistical and Thermal Physics", McGraw-Hill, New York, 1965
2) Mayhew, K.W. "A new perspective for kinetic theory and heat capacity", *Progress in Physics* Vol. 13 (**4**) 2017 pg 166-173
3) Mayhew,K.,W., Phys. Essays **19**, vol 4, 604(2013)

4) Mayhew, K.W. "Kinetic theory: Flatlining of Polyatomic Gases", *Progress in Physics* Vol. 14 (**2**) 2018
5) P. Marmet IEEE Transactions on Plasma Science vol 18 issue 1 (1990)
6) P. Marmet, Phys. Essays, vol 1, pp 24-32 (1998)
7) J. M. Jauch and F. Rohlich, The theory of Photons and Electrons. Cambridge, MA:Addison-Wesley,195
8) H. A. Bethe and E. E. Salpeter, Quantum Mechanics of One and Two electron Atoms. New York: Springer-Verlag, 1957,
9) Robitaille Pierre-Marie, A critical analysis of the universality and Kirchoff's Law: A return to Stewarts law of Thermal Emission, Progress in Physics Vol 3 pp 30-35 July 2008
10) "Handbook of Chemistry and Physics", 91st Edition, Editor W.M. Haynes CRC Press, New York, 201

Density vs Temperature: Data for Graph 7.1

Table 7.1

T (°K)	Density at 1 Bar pressure (kg/m³)					Number of Moles per m³ at 1 Bar pressure (mols/m³)				
	H_2	He	Ar	N_2	O_2	H_2 2.016	He 4.002	Ar 39.95	N_2 28.01	O_2 32
20	1.32	2.50				656.25	624.19			
40	0.62	1.20				305.41	299.85			
60	0.41	0.80				201.09	200.05			
80	0.30	0.60		4.38		150.55	150.10		156.3	
100	0.24	0.48	4.92	3.44	3.94	120.78	120.11	123.03	122.6	123.16
120	0.20	0.40	4.06	2.84	3.25	100.20	100.12	101.58	101.3	101.63
140	0.17	0.34	3.46	2.42	2.77	85.86	85.83	86.64	86.53	86.69
160	0.15	0.30	3.02	2.12	2.42	75.10	75.11	75.60	75.53	75.63
180	0.13	0.27	2.68	1.88	2.15	66.77	66.77	67.09	67.04	67.09
200	0.12	0.24	2.41	1.69	1.93	60.07	60.09	60.30	60.26	60.31
220	0.11	0.22	2.19	1.52	1.75	54.61	54.65	54.80	54.40	54.78
240	0.10	0.20	2.01	1.41	1.61	50.10	50.10	50.19	50.15	50.19
260	0.09	0.19	1.85	1.30	1.48	46.58	46.25	46.31	46.30	46.31
280	0.09	0.17	1.72	1.20	1.38	42.93	42.93	42.98	42.98	43.00
300	0.08	0.16	1.60	1.12	1.28	40.06	40.08	40.13	40.09	40.13
320	0.08	0.15	1.50	1.05	1.20	37.56	37.58	37.60	37.59	37.59

Handbook of Chemistry and Physics

Editor: W.M. Haynes (CRC Press, New York, 2010)[10]

AW signifies Atomic Weight

Note: mols/m³ calculated by: Density*1000/AW

Chapter 8: Thermal Radiation

Again:*Thermal radiation* shall be considered as any spectrum of *thermal photons* in freespace, those being photons, which are readily absorbed by condensed matter resulting in molecular vibrations. Accordingly, thermal radiation consists of a spectrum of photons at various specified frequencies. Since thermodynamics concerns the interaction of matter and energy, it only makes sense that our primary concern is with thermal radiation.

EM radiation can be considered as being a wave at a specified wavelength. However, our preference will be to consider it in terms of photons, whose energy is defined by Planck's constant (h=6.6262x10^{-34} joule-sec), and the photon's frequency (v), i.e.:

$$E = hv \qquad\qquad 8.1$$

Understandably, energy is associated with any thermal photons moving about the freespace between gas molecules. Our interest becomes, how much and what relevance do they have. Consider some generalities when dealing with the interaction of radiation and matter. When a spectrum of photons strikes matter, some photons are:

1) Reflected by the matter's surface.
2) Transmitted through the matter.
3) Absorbed into the matter.

Reflected photons from the visible part of the spectrum are what determine an object's color. For example, if a substance reflects most violet photons (λ = 413 nm) while absorbing the other incident frequencies, then those reflected violet photons are what give the substance its violet color. Transmitted photons do not exchange energy with matter, rather they simply refract as they pass through matter. Hence, neither reflected, nor refracted photons result in temperature changes within condensed matter.

Accordingly, our main concern must be photons that are absorbed by the matter and interact as thermal energy (heat) within that matter. Understandably, not all absorbed photons will interact thermally with matter, e.g. high-energy photons such as gamma rays and x-rays tend not to absorbed as thermal energy hence generally do not contribute to the temperature associated with condensed matter. Note gamma rays are associated with the energy of the atom's nucleus, while x-rays are associated with the states of energy of electrons.

Blackbody Radiation

For most readers it is suffice to understand the basics, as is discussed herein. However, readers who want a more comprehensive understanding can read Appendix B.1. A *blackbody* is considered to be the perfect absorber, which is to say that all incident radiation is absorbed into that matter thus explaining its lack of color i.e. black, e.g. a perfect blackbody material is carbon black, whose absorptivity approaches unity. Since it is a perfect absorber then by definition it is also a perfect emitter i.e. its emissivity also approaches unity. Substances, whose emissivity is less than unity, are called *graybodies*.

Fig 8.1 shows a blackbody curve for T=6,000 K. The curve gives the actual power emitted [$\rho_B(v)$] at each frequency (v), which is the *spectral radiancy*[1]. The area enclosed between the curve and frequency axis is the *radiancy*. It can get confusing; The *radiance* (W/sr·m^2)(A.K.A. *intensity*) of a substance is the radiant flux emitted or received by a surface in a given direction per unit solid angle per unit area. *Irradiance* is the power received per unit surface area (W/m^2) (confusingly A.K.A. *intensity*). The *spectral irradiance* is the irradiance per unit wavelength or frequency. Note the peak resides in the visible spectrum, between 0.4 and 0.8 micrometers.

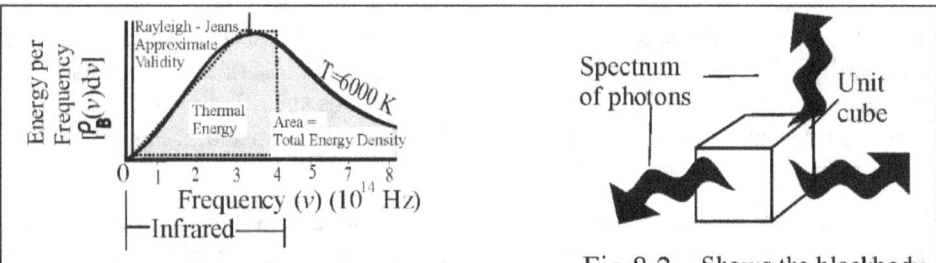

Fig. 8.1 Sketch of the blackbody radiation spectrum for 6000 K The radiancy is the area enclosed
by the blackbody radiation curve and the frequency
axis. Energy density (E/V) is proportional to radiancy.

Fig 8.2 Shows the blackbody
radiation spectrum being emitted
outwards in all directions through
a unit cube, with unit a surface
area, each and every second.

A simple way to envision a blackbody spectrum is to consider it in terms of a flux through a unit surface area that one would measure over a second of time. The measurement in a vacuum's cavity should be identical no matter what direction is chosen to measure the spectrum. Fig 8.2 shows a unit cube emitting a spectrum of photons, equally in all directions. Radiancy becomes the power of that spectrum coming out of one of the unit cubes faces.

The *Stefan-Boltzmann law* states that the total emitted power is proportional to the matter's temperature to the fourth power[1]:

$$Power \propto T^4 \qquad 8.2$$

Eqn 8.2 can be rewritten, as:

$$Power = A\varepsilon\sigma'T^4 \qquad 8.3$$

where ε = emissivity. Note: $\varepsilon = 1$ for blackbody, while $\varepsilon < 1$ for graybody.

$\sigma' = 2\pi^5 k^4/15c^2h^3$ = 5.670400x10^{-8} J/sm^{2}k^4, and A = surface area
At 300 K: One obtains a power per unit area of 459.30 J/sm^2

Realizing that radiancy must be proportional to the energy density, we can now write[1]:

$$\rho_B = aT^4 \qquad 8.4$$

where ρ_B is the blackbody radiation energy density (energy per volume) and a is a blackbody radiation constant; $a = 8\pi^5 k^4/15c^3h^3$. At 300 K: Energy density becomes; $\rho_B = 6.124 \times 10^{-6}$ j/m^3

Thermometer in a Vacuum

Let us reinforce why an object's *emissivity*, must correlate to its corollary that being its *absorptivity*. The *absorptivity* is the total radiation energy absorbed by the substance, per unit area, per unit time. An object, with no internal heat source is in thermal equilibrium with its surroundings, when that object emits as much radiation energy as it absorbs, i.e. a good absorber is equally a good emitter of radiation. This was first recognized by Balfour Stewart in 1858, who stated when in equilibrium absorption equals emission (known as Stewart's Law)[2].

Consider a volume of freespace containing blackbody radiation. Common sense dictates that the total energy of the blackbody radiation residing within that freespace must be its volume (V) multiplied by the *blackbody radiation density (ρ_B):*

$$E_{blackbody} = V\rho_B \qquad\qquad 8.5$$

Consider Fig. 8.3, where a thermometer at temperature (T_2) with surface area (A_2) is illustrated. The thermometer is in a cavity that is surrounded by walls at temperature (T_1), whose surface area is (A_1). Assume that both the condensed matter and thermometer act as perfect blackbodies, equally absorbing and emitting all frequencies. Accordingly, the total energy emitted per unit time by the surrounding walls (condensed matter) is:

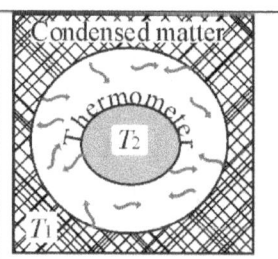

Fig 8.3 Shows a thermometer at temperature T_2 within a cavity surrounded by condensed matter at T_1.

$$\dot{Q}_{emit1} = A_1 a T_1^{\,4} \qquad\qquad 8.6$$

where $\dot{Q}_{emit1}$ is the rate of energy emission.

Consider that the surrounding condensed matter is in equilibrium with the cavity and then consider what the thermometer absorbs, i.e. the total energy absorbed per unit time is:

$$\dot{Q}_{absorb2} = A_2 a T_2^{\,4} \qquad\qquad 8.7$$

When in thermal equilibrium $T_2 - T_1$. Obviously, there is a temperature associated with the blackbody radiation residing within the cavity, as is measured by the thermometer.

Rayleigh-Jeans Approximation

When; $hc/\lambda = h\nu \ll kT$, the energy density as defined by eqn 8.4: $\rho_B = aT^4$, can be obtained by using the *Rayleigh-Jeans approximation*:

$$\rho_R = a'T \qquad\qquad 8.8$$

where ρ_R is the energy density and a' is the Rayleigh-Jeans constant, which differs from the constant (a) in eqn 8.4. The units for a' are "energy per volume per degree", which in SI units would be written as: J/K·m³. Note: The Rayleigh-Jeans approximation for 6000 K is sketched in Fig 8.1. It is also discussed in more detail in Appendix B.1.

Remember thermal radiation is the part of the radiation spectrum that interacts with matter as thermal energy, i.e. the spectrum of relatively long wavelengths e.g. infrared and other such frequencies. In Table 8.1, hc/λ is compared to the value of kT, for two infrared frequencies and two microwave frequencies, all at room temperature (20°C = 293 K), followed by at 1000°C (1,273 K) and finally 5,700 K i.e. the approximate temperature of our Sun.

The majority of what is considered as being thermal energy are spectrums consisting of mid to long infrared frequencies. When considering a thermal energy spectrum; the Rayleigh-Jeans approximation would not be valid for blackbody radiation from a body unless those bodies are extremely hot. Thus importantly, the Rayleigh-Jeans approximation is valid for the infrared part of the spectrum from blackbodies whose temperature is above several thousand degrees Kelvin. In which case the emitted thermal energy density becomes prorportional to temperature as witnessed by eqn 8.8

Table 8.1

	Wavelength (λ) nm	hc/λ	kT, when, T=293 K	kT, when, T=1,293 K	kT, when, T=5,700 K
Infrared	10^3	2.0 x 10^{-19}	4.0 x 10^{-21}	1.8 x 10^{-20}	8.0 x 10^{-20}
Infrared	10^4	2.0 x 10^{-20}	4.0 x 10^{-21}	1.8 x 10^{-20}	8.0 x 10^{-20}
Infrared	10^5	2.0 x 10^{-21}	4.0 x 10^{-21}	1.8 x 10^{-20}	8.0 x 10^{-20}
Microwave	10^6	2.0 x 10^{-22}	4.0 x 10^{-21}	1.8 x 10^{-20}	8.0 x 10^{-20}
Microwave	10^8	2.0 x 10^{-24}	4.0 x 10^{-21}	1.8 x 10^{-20}	8.0 x 10^{-20}

Our Sun's temperature (T=5,700 K) is slightly too low for the Rayleigh- Jeans approximation to be a great approximation but even so it remains a rather good approximation. Specifically at 5,700 K: $hc/\lambda \approx kT$, when λ=2500 nm. Thus for all wavelengths greater than 2500nm, $hc/\lambda << kT$, and the Rayleigh-Jeans approximation is valid!! Accepting that wavelength greater than 2500nm constitutes the majority of wavelengths that behave as thermal energy the conclusion becomes that our Sun's thermal radiation density incident upon Earth can be roughly approximated as being directly proportional to the temperatures, as measured here on Earth. This can be seen in Fig 8.1, where the apex for T=6,000 K is located in the short wavelength infrared: By Wein's Laws it is at 3.5×10^{14} Hz. See appendix B.1 for more on Wein's displacement law.

The above helps explain why most thermodynamic relations are directly proportional to temperature here on Earth. Specifically, realize that our atmosphere not only acts as a thermal blanket holding in the Sun's heat and acts as a massive heat bath surrounding both us, and our systems including our experiments.

In plain English; our biggest experimental heat bath has its thermal energy density directly proportional to temperature, because our Sun's radiation makes it so. Remember in thermodynamics the thermal energy within any system is often readily compared to thermal energy contained in the surrounding heat bath/sink/reservoir, which ultimately is our atmosphere/planet as a system.

To make matters more understandable, take: $TS = \varepsilon + PV$, and divide by V, i.e.:

$$T(S/V) = \varepsilon/V + P \qquad 8.9$$

Comparing eqn 8.9 to eqn 8.8: $\rho_R = a'T$, there is a correlation between a' and traditional (S/V) i.e.:

$$S/V = a' \qquad 8.10$$

Since eqn 8.9 concerns a system of molecules, while eqn 8.8 concerns radiation eqn 8.10 is questionable at best. Furthermore, $TS = \varepsilon + PV$ is questioned in this book. The fact remains that in equilibrium, the thermal energy density within condensed matter must be in equilibrium with its surroundings. Importantly, any generality relates to the thermal energy density within our atmosphere that is directly due to the thermal radiation density from our Sun, all of which is directly proportional to the measured temperatures. This is undeniable.

Asserting that most thermal energy consists of wavelengths from the microwave through infrared portion of our Sun's spectrum surely fits with our understanding of thermal energy here on Earth. Importantly, having the thermal energy density directly proportional to temperature bodes well with most of our empirically verified relations. The implication being that the thermal energy density, which is readily absorbed by condensed matter, and/or exchanged through molecular vibrations/collisions should generally be directly proportional to the measured temperature.

Interestingly, Planck's work showed that the Rayleigh-Jeans approximation is valid, when: $hc/\lambda = h\nu \ll kT$. Note: Planck further realized that the energy of photons must be quantized i.e. by realizing that the number of photons decreases with increasing frequency, Planck prevents the *ultra-violet catastrophe*. Note: Blackbody, radiancy, Rayleigh-Jeans approximation, UV catastrophe and thermal energy is all discussed in more detail in Appendix B.1

Limitations of Temperature in Thermal Radiation

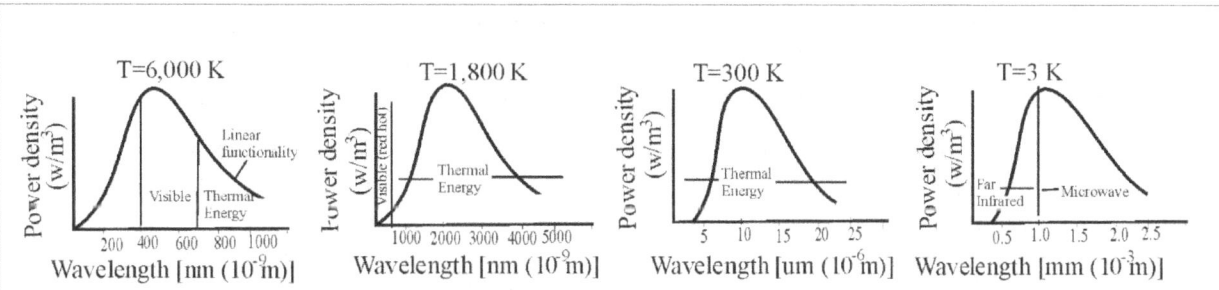

Fig 8.4   Shows sketches of the blackbody radiation power density versus wavelengths for temperatures of 6,000 K (Sun's temperature), 1,800 K (blast furnace temperature), 300 K (room temperature) and 3 K (CBR temperature). Obviously, the power density associated with thermal energy is not always linear function of temperature.

Ultimately, the thermal energy density found here on Earth tends to be proportional to temperature all because the thermal energy density from our Sun's blackbody radiation makes it so, which is defined by the Rayleigh-Jeans Approx. Limitations of this statement are illustrated in Fig 8.4, e.g. sketches for the power density of blackbody radiation per wavelength. Obviously, at our Sun's temperature (T=5,700) the radiated thermal energy density over time i.e. radiated thermal power density, has an approximate linear functionality that does not exist at significantly lower temperatures.

Accept: The primary reason that Earth bound systems adhere to this linear functionality is our atmosphere, lands and oceans, all behaving as a massive heat bath/sink, from which other systems must exchange thermal energy i.e. matter absorbs and condenses incoming thermal energy from our Sun, whose energy density is proportional to temperature.

At room temperature (300 K) the thermal energy dominates the blackbody spectrum hence matter reemits that energy as blackbody radiation with a peak around 9 micrometers e.g. an infrared spectrum; see T=300 K in Fig. 8.5 (Note: Earth's blackbody radiation peaks at 9.7 micrometers).

It should be stated that herein it is considered that in equilibrium the total absorbed energy equals that emitted, but this is not saying the power at each

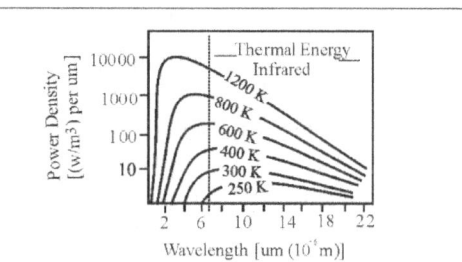

Fig 8.5   Sketches of the blackbody versus wavelengths for various T. The radiated thermal energy starts to lose its linear functionality in relation to T as the temperature decreases i.e. T<250 K [Sketched from "Radiation Principle" (Feb 2016)] (http://www.r-s-c-r.org/node/114)

wavelength necessarily is the same for emission as it is adsorption.

Fig 8.5 shows the power density per wavelength for temperatures from 1800 K to 250 K. For $T > 300$ K the infrared part of the spectrum roughly approximates a simple linear decreasing function in relation to the power density per increasing unit of wavelength. For temperatures below 250 K this apparently is not the necessarily case, with the any linear association between temperature and thermal energy density being lost. This becomes increasingly apparent since temperatures decrease when microwaves frequencies start to dominate over the infrared.

To further complicate things the power density per unit wavelength increases logarithmically with temperature. This helps explain why when dealing with a blast furnace i.e. 1200 K plus, the radiated thermal energy is no longer infinitesimally small in comparison to the thermal energy associated with matter. Accordingly, thermodynamic relations based upon the temperature of matter (or kinematics of matter) can no longer be used to explain what is witnessed. Note: Radiated thermal energy is accepted as being proportional to temperature to the fourth power, i.e. eqn 8.4: $\dot{Q}_{emit1} = A_1 a T_1^4$.

Therefore our conceptualizations of the thermal energy density being proportional to temperature loses it validity at both very high temperatures, and temperatures approaching absolute zero. This fits with discussions in previous chapters i.e.thermodynamics seemingly falters as temperatures approaching absolute zero. We now begin to understand why this is so, irrelevant of either entropy or quantum based arguments and all their inherent complexities. All that can be said at this point is that the linear functionality between temperature and the thermal energy density must be limited to temperature regimes that we are generally exposed to here on Earth's surface. Moreover this functionality is not universal!

Radiation within a Cavity

It is generally accepted that any cavity within condensed matter or even any volume surrounded by walls will be filled with blackbody radiation. This is primarily based upon 19th century experiments with blackened (lamp black) cavities, as well as fully reflective cavities. Accordingly, if a hole was put into such an enclosure then any emitted light out of that enclosure will be blackbody in nature. In other words the complete spectrum of radiation within a vacuum cavity is believed to be blackbody radiation, irrelevant of the type of walls.

Interestingly, Robitaille has challenged the above assertion claiming that the 19th century studies were flawed and the reason that reflective cavities were thought to emit blackbody radiation is that a piece of lamp black (soot from lamp) was inside of the enclosures hence it was the lampblack that caused the enclosures to appear blackbody in nature. Moreover Robitaille[2] asserts that the sciences wrongly choose to adhere to Kirchoff's claims over that of Balfour Stewart. Stewart's interpretation fits better with what is known of colors and most surfaces namely that differing surfaces tend to absorb and reflect differing frequencies. If right, this demonstrates how a science can be misled for over a century.

Furthermore, if Planck had realized that total radiation followed Stewart's guidance i.e. the total radiation witnessed was the sum of reflected plus absorbed/radiated then the universality to blackbody radiation within cavities may not have been accepted. Specifically, is Stewart correct then the radiation spectrum within an enclosed vacuum should include the reflected as well as any adsorbed and radiated frequencies.

Throughout this book, it has been professed that the understanding of energy within enclosures containing gases should also include any blackbody plus any thermal radiation that resides within the enclosure. At this point it is not fully understood what all constitutes this radiation. No matter, it should include:

1) Thermal radiation emitted from vibrations
2) Radiation associated inelastic molecular collisions as discussed in Chapter 7. Is this thermal?
3) The radiation that resides within cavities i.e. is this blackbody as Kirchoff and Planck professed?

Whatever the true nature of the radiation within a given gas filled enclosure actually is, at this point we simply refer to it as blackbody/thermal because it is assumed to be some combination of the above and possibly other. When clarity is finally determined then and only then can proper terminology be given. It should be emphasized that thermal equilibrium does not mean that matter emits the exact same radiation spectrum that it absorbs rather it just means that the flux/rate of energy emitted equals the rate of absorption.

A Discussion: Blackbody vs. Kinematics

As was previously stated; the energy density of blackbody/thermal radiation tends to be significantly less than the energy density associated with the kinematics of matter in most systems. So much so, that the energy associated with the radiation often can be approximated as being zero. Therefore, the total energy contained within a volume of gas, generally can be calculated solely in terms of the gas's molecular kinematic energies. Accordingly the traditional approach is understandable, especially since radiation does not fit in with beloved statistical arguments in the same manner as matter does.

Our reality is that due to the speed of the light, over time a significant energy exchange can occur even though the radiation's energy density is comparatively small. It must be emphasized that experiments can only disprove a theory rather than prove a given theory i.e. more than one theory may claim the similar results.

Closing Remarks

Thermal radiation exists in freespace and remains relevant to understanding thermodynamics. It was determined that the thermal energy density (predominately infrared) from our Sun's blackbody radiation can be approximated by some linear function of temperature. Hence our position in our solar system along with our sun radiance helps us understand our temperature's structure as witness here on Earth. This doesn't provide us with a full understanding rather it is simply part of our new improved perspective.

Furthermore inelastic molecular collisions means that heat/radiation is created by molecular inelastic collisions! Therefore for a gaseous system, thermal equilibrium exists when the temperature associated with radiation (thermal/blackbody or otherwise) equals the temperature associated with the gas' kinematics, which also equals the temperature associated with the wall's molecular vibrations. In other words the heat created by molecular collisions is also continually adsorbed and re-radiated by matter.

When talking about the radiation spectrum due to inelastic collisions, is it blackbody, or thermal, or both? At this point, who knows. All we are saying is that the radiation (whatever you want to call it) is present and it constantly tries to attain thermal equilibrium with its surroundings. Certainly some sort of spectrum would be expected in systems/cavities dependent upon all the possible variations associated with such collisions, plus any radiation that naturally resides within whether it is blackbody or otherwise.

Lots of new insights by others will be required. This becomes an opportunity to be taken.

References:

1) R. Eisberg, R. Resnick "Quantum Physics", John Wiley & Sons Toronto 1974
2) Robitaille Pierre-Marie, A critical analysis of the universality and Kirchoff's Law: A return to Stewarts law of Thermal Emission, Progress in Physics Vol 3 pp 30-35 July 2008

Chapter 9: **Engines, Efficiency, Cycles and Refrigeration**

Cyclic Devises

Cyclic devices start out in one state, proceed through a series of steps to finally ending back at its original state. An *engine* is a cyclic device that extracts energy from a source, transforms that energy into mechanical work enabling the movement of man, and/or machine. The source of this energy can be anything from a fire, to a thermal reservoir, to some chemical reaction and/or to an explosion.

A "cycle" is the combination of steps that takes the engine/system from its initial to final state, e.g. the 4-stroke combustion engine cycle consists of:

1) The intake stroke: Fuel-air mixture is drawn into the cylinder, as the piston drops.
2) The closing of cylinder's intake valves.
3) The compression stroke: A rising piston compresses the fuel-air mixture.
4) Explosion of the fuel-air mixture.
5) The power stroke: The piston is driven downward by the explosion.
6) The opening of the cylinder's exhaust valves.
7) The exhaust stroke: The rising piston pushes the spent exhaust out of the cylinder
8) The opening of the cylinder's intake valves, and the closing of its exhaust valves.

At the end of the 8th step the engine returns back to the 1st step, therefore completing the cycle. The reason it is called a 4-stroke engine is that the cycle has four steps that involve the movement of the piston, namely steps 1), 3), 5) & 7).

Understanding Lost Work

Lost work goes by many names, such as *lost heat, lost energy, dissipated energy* and *non-sensible energy*. In its broadest context, it is energy that is lost by a system/machine thus preventing reversibility in processes hence efficiency in machines. Lost work by expanding systems is pertinent too most engines and other *useful processes*, because at some point in their cycle, there is system expansion that upwardly displaces the Earth's atmosphere's mass against gravity, thus adding to our atmosphere's potential energy and/or its contained heat.

Atmosphere displacement is different than the reversible work of lifting a rock upwards wherein human energy is transformed into a potential energy increase of the rock. Importantly, a rope can be tied to the raised rock, enabling us to harness its increased potential energy as the rock falls back to the ground. Unlike the reversible work onto the rock, an increase to Earth's atmosphere's potential energy is not readily harnessed. Specifically, if the expanded system collapses back to its original state, then the gained potential energy is transformed into atmospheric molecular kinetic energy, which eventually generally results in an atmospheric temperature increase, which may or may not radiate away. Rather than only resulting in an isobaric volume increase, expanding systems may equally cause a regional pressure increase, resulting in an isometric temperature increase, which then results in an isothermal volume increase. Irrelevant of the actual path, the end result remains that it is lost work, hence useful processes tend to be irreversible!

Reversibility

A process can be reversible if no energy is lost, i.e. lost work = 0 and there is no friction or other loss. There is another aspect to this that should be discussed. Reversible processes must be capable of having a net energy that naturally can equally flow either forward or reverse, in a process. Accordingly reversibility is best suited to idealistic processes where any changes are deemed infinitesimally small.

Moreover, the closer the change of any parameter associated with energy (temperature, volume, pressure) is to zero, then the closer to reversible that process can be. This renders reversibility into the realm of illusion, when contemplating real processes. Remember infinitesimal change is a cornerstone for traditional thermodynamics!

Certainly, a process that involved lost energy could conceivably be reversible if that lost energy could be controlled and redirected back into the process itself. Again, this becomes an idealistic rather than realistic conceptualization because the control and redirection of energy more often than not requires the input of resources, which is to say the process is no longer reversible.

Consider a mechanical process wherein friction is generated. Friction generates heats that naturally radiates outwards in all directions therefore the idea of completely controlling and then directing such heat verges upon impossible without some input of resources/energy. And if you require an input of resources/energy to manipulate the lost energy, then the energy of manipulation must be deemed energy that is lost in the process, hence the process remains irreversible!

Traditional Explanation

The fact that useful processes generally are irreversible has traditionally been wrongly explained in terms of entropy production and/or the second law, i.e. only isentropic ($dS=0$) processes are reversible. Isentropic processes are considered to be adiabatic and internally reversible.

Certainly expanding systems tend to do work onto their surroundings, hence cannot be truly adiabatic. And because of this thermodynamics have fundamentally become a complication of the simple, all so that the second law remains the postulate supreme. We shall show that other explanations are more exacting hence the second law is not required to explain inefficiencies i.e. second law is the false-postulate!

Efficiency

No manmade mechanical device or process is 100% efficient. Fig. 9.1 illustrates an engine that requires an energy input (Q_{in} ,or dQ_{in}), to power a bus that being W_{out} . Since the engine is not 100% efficient, the energy input is always greater than the work out, i.e.:

$$dQ_{in} \geq W_{out} \qquad 9.1$$

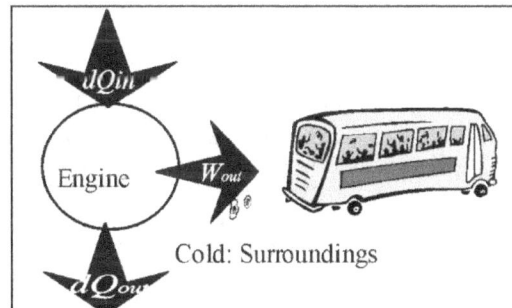

Let dQ_{out} represent the energy lost due to our engine's inefficiency, therefore:

$$W_{out} = dQ_{in} - dQ_{out} \qquad 9.2$$

The engine's efficiency (η) is simply the work out divided by the energy input:

Fig 9.1   Shows an input of energy dQ_{in} into the engine, which is then used to power the bus that being work (W_{out}). As with all real engines heat (Q_{out}) is also lost into the cold surroundings

$$\eta = W_{out} / dQ_{in} = (dQ_{in} - dQ_{out}) / dQ_{in} \qquad 9.3$$

Remember: $W_{required} = W_{atm} + W_{friction} + W_{out}$, can be used to define the work required, which equates to the energy input (dQ_{in}). If all the energy changes involved could be readily defined in terms of work done then eqn 9.3 could be rewritten as:

$$\eta = W_{out} / (W_{atm} + W_{fiction} + W_{out}) \qquad 9.4$$

In writing 9.4 $W_{friction}$ is assumed to include all the sources of heat that is generated within and dissipates away from any mechanical device e.g. the engine. The greater the work out (W_{out}) is, in comparison to the work required to overcome dissipative energy loses e.g. $W_{friction}$, plus work to required displace the atmosphere (W_{atm}), then the more efficient the mechanical device should be.

Understandably, no cyclic engine that experiences friction and/or displaces our atmosphere, can ever power the elusive perpetual motion machine. For those of you not familiar with the concept: A perpetual motion machine is a machine that keeps on going and going, without the need of the addition of energy from an external source, into one (or more) of its cycles. As we are about to find out there is more to understanding efficiency than this i.e. the above described efficiencies are idealistic, let us now investigate the reality.

Isometric versus isobaric energy input

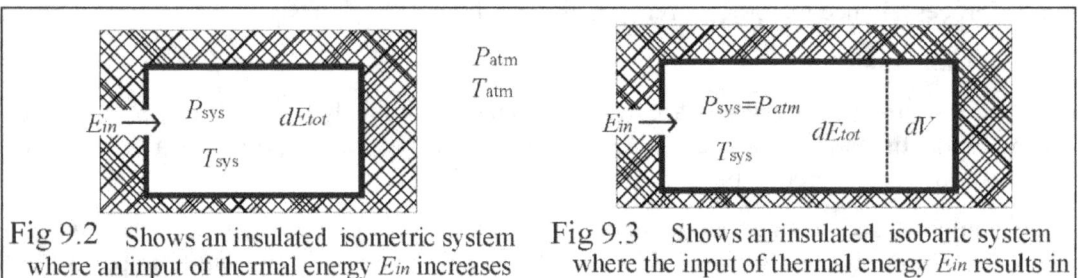

Fig 9.2   Shows an insulated isometric system where an input of thermal energy E_{in} increases the system's total thermal energy (dE_{tot})

Fig 9.3   Shows an insulated isobaric system where the input of thermal energy E_{in} results in an increase to the system's volume (dV)

Consider that thermal energy (E_{in}) is input into an insulated isometric system, as is shown in Fig. 9.2. The increase in the system's total energy (dE_{tot}) can be defined in terms of the system's molar isometric heat capacity (C_v) multiplied by the temperature change (dT). For n moles:

$$E_{in} = dE_{tot} = nC_vdT \qquad 9.5$$

Instead of being isometric, imagine that the above system experiences isobaric expansion, hence does work onto the Earth's atmosphere ($W = P_{atm}dV$), as is shown in Fig. 9.3. Now the energy input (E_{in}) must equal the increase to the system's thermal energy plus the work done by that system. Therefore:

$$E_{in} = dE_{tot} + P_{atm}dV \qquad 9.6$$

Rewriting in terms of heat capacity and internal temperature change as:

$$E_{in} = nC_vdT + P_{atm}dV = nC_pdT \qquad 9.7$$

Where C_p molar isobaric heat capacity. If the energy input (E_{in}) is the same, then internal temperature increase in the isometric case (eqn 9.5) must be greater than it is in the isobaric case (eqn 9.7).

Free Energy: Isobaric vs Isometric

Free energy is a bit of a misnomer. It might be considered as related to the energy that can be extracted from a system, hence is available to do work. Consider, that there is an energy input (E_{in}), and we are extracting work out (W_{out}) of an isometric system, as shown in Fig.9.4. Assuming 100% efficiency, then the maximum amount of work that can be extracted from such an isometric system is:

$$W_{out} = E_{in} - nC_vdT \qquad 9.8$$

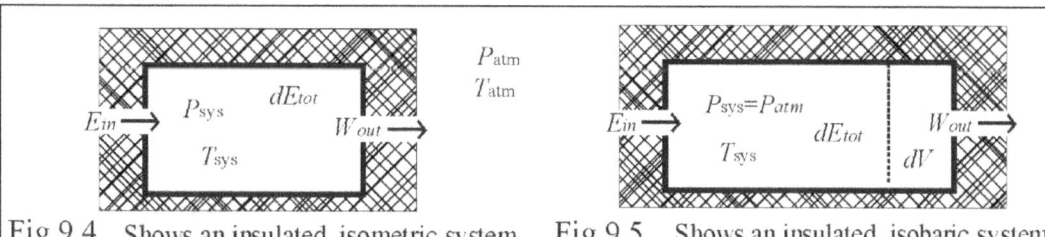

Fig 9.4   Shows an insulated isometric system with a thermal energy input (E_{in}) and work output (W_{out})

Fig 9.5   Shows an insulated isobaric system with a thermal energy input (E_{in}) and a work output (W_{out})

Next consider, an energy input (E_{in}), along with the extraction of work out (W_{out}) of an isobaric system, as shown in Fig.9.5. Assuming 100% efficiency, then the maximum amount of work that can be extracted from such an isobaric system is:

$$W_{out} = E_{in} - nC_v dT - P_{atm} dV \qquad 9.9$$

Obviously there are differences between isometric and isobaric free energy. Since isobaric free energy involves the displacement of our atmosphere, it requires the previously discussed lost work. While isometric free energy would not. Note: It becomes awkward at best to envision how a gaseous isometric system performs mechanical work.

Certainly 100% efficiency would be limited to all the energy of the gas purely being kinetic, hence may be limited to ideal monatomic gases. Understandably, polyatomic gases will have significantly lower efficiencies because of the fact that thermal energy must also go into any vibrational energy increases.

Comment about Traditional

Those familiar with traditional thermodynamics may recognize the isobaric eqn 9.9 as a version of the Helmholtz free energy [Just substitute dF for W_{out}, E_{in} for dU, and S for C_v]. This will be discussed again in Chapter 14 (physical chemistry).

One may ponder why the values for entropy and specific heat are not exactly the same. Perhaps the reason is that the specific heat (C_y) measure at 100% efficiency. One may think that this implies entropy is the heat capacity for inefficiency, or the heat capacity of a system that work is extracted from! It is not that simple, nor does this properly/fully define all the various meanings that entropy may possess.

We are not claiming to embrace traditional understanding of free energy, but it will be shown that it may give the right answers but for the wrong reasons, i.e. constructive logic based upon simpler reasoning will be presented.

Maximum Efficiency 66.67%

As good as eqn 9.9 is in showing that perpetual motion is unrealistic, it also generally underestimates the amount of energy required. As previously discussed, the increase in ability of an ideal monatomic gas to perform work, is seemingly limited to 2/3 of the gas's kinetic energy increase. This was shown mathematically by dividing the increase in ability to work: $W = d(PV) = NkdT$, by eqn 2.12: $E_{Tk(t,r)} = 3NkT/2$. The implication being as previously discussed: The ratio of "work out" to "energy in" for a system of gas possesses a maximum efficiency (η_{max}) of:

$$\eta_{max} = W_{out}/dQ_{in} = NkdT/(3/2)NkdT = 2/3 = 66.67\% \qquad 9.10$$

The 66.67% upper limit to the efficiency exists[1] because not all of the system's molecules will be able to contribute all of their momentum to the system's expansion. And in terms of energy this also assumes the gases are ideal monatomic gases.

Efficiency of a piston-cylinder

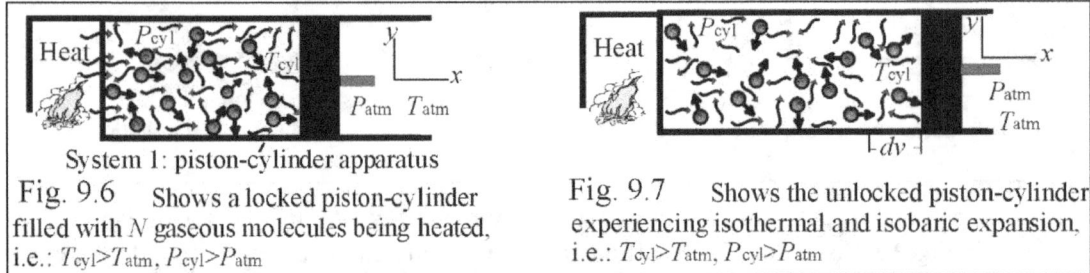

System 1: piston-cylinder apparatus

Fig. 9.6   Shows a locked piston-cylinder filled with N gaseous molecules being heated, i.e.: $T_{cyl} > T_{atm}$, $P_{cyl} > P_{atm}$

Fig. 9.7   Shows the unlocked piston-cylinder experiencing isothermal and isobaric expansion, i.e.: $T_{cyl} > T_{atm}$, $P_{cyl} > P_{atm}$

Investigating this in more detail; consider heating of a closed piston-cylinder apparatus filled with N monatomic gaseous molecules, as illustrated in Fig. 9.6. If the piston is locked in position and thermal energy (dQ) is added, then based upon kinetic theory the energy required to raise the cylinder's internal temperature (T_{cyl}) is:

$$dQ = 3NkdT_{cyl}/2 \qquad 9.11$$

As the gas' temperature increases, its kinetic energy increases, hence its pressure increases i.e. $P_{cly} > P_{atm}$ and this increases its ability to do work. When written in terms of infinitesimal change to the ability to do work can be written as:

$$dW_{ability} = NkdT_{cyl} \qquad 9.12$$

Eqn 9.12 is 2/3 of eqn 9.11, i.e. $\eta < 66.67\%$. Assume, $W_{machine} = W_{friction} = 0$. If the piston is unlocked allowing the piston-cylinder's volume to increase (dV_{cyl}), then the work done becomes:

$$W_{done} = P_{atm}dV_{cyl} \qquad 9.13$$

Consider that the heat is continuous during expansion, as illustrated in Fig 9.7. The piston-cylinder can expand such that the gas's pressure in the piston-cylinder remains greater than the surrounding atmosphere's pressure, thus driving the piston outwards. Assuming frictionless and massless, hence the pressure inside the piston-cylinder only needs to be infinitesimally greater than the surrounding atmosphere. Then the minimal thermal energy (dQ_{min}) that is required would be:

$$dQ_{min} = 3NkdT_{cyl}/2 + P_{atm}dV_{cyl} \qquad 9.14$$

For our ideal monatomic gas, in terms of ability to do work, as defined by eqn 9.12, then eqn 9.14 becomes:

$$dQ_{min} = 3NkdT_{cyl}/2 + NkdT_{cyl} = 5NkdT_{cyl}/2 \qquad 9.15$$

If the piston cylinder's ideal monatomic gas also does work for man, such as power a mechanical device, then:

$$dQ_{min} = 5NkdT_{cyl}/2 + W_{device} \qquad 9.16$$

In terms of efficiency:

$$\eta_{max} = W_{device}/dQ_{min} \qquad 9.17$$

Hence:

$$\eta_{max} = W_{device} / (5NkdT_{cyl} / 2 + W_{device}) \qquad 9.18$$

Is eqn 9.18 correct? If some of the heat inside of the expanded piston cylinder could be reused, then the efficiency as defined by eqn 9.18 could increase.

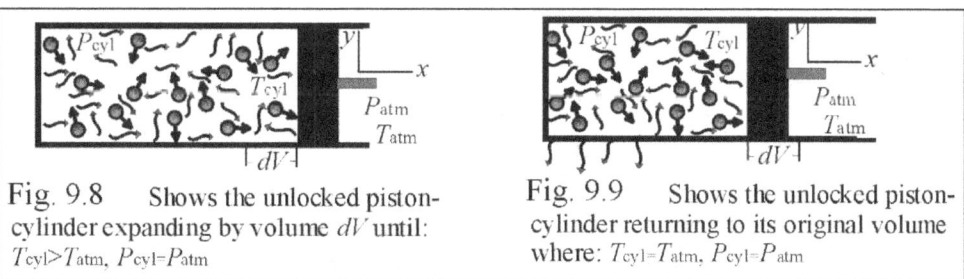

Fig. 9.8   Shows the unlocked piston-cylinder expanding by volume dV until: $T_{cyl} > T_{atm}$, $P_{cyl} = P_{atm}$

Fig. 9.9   Shows the unlocked piston-cylinder returning to its original volume where: $T_{cyl} = T_{atm}$, $P_{cyl} = P_{atm}$

Now consider that the locked piston-cylinder is heated as was shown in Fig 9.6, hence its pressure becomes significantly greater than that of the surrounding atmosphere at which point the heat source is removed and the piston is unlocked, as shown in Fig 9.8.

As the piston-cylinder expands (dV), the ideal monatomic gas within the piston-cylinder does work, thus cools down until the gas's pressure within the piston cylinder approximates the surrounding atmospheric pressure i.e. $P_{cly} \approx P_{atm}$. Although it is at the same pressure as its surroundings, and has cooled somewhat its temperature is still higher than the surrounding atmosphere's temperature, i.e. $T_{cly} > T_{atm}$. The added heat (thermal energy) that remains in the expanded piston-cylinder should be:

$$dE_{remains} = 3NkdT_{cyl} / 2 \qquad 9.19$$

If the piston's walls are not insulated, then heat will escape. As this happens, the pressure inside the piston-cylinder will decrease e.g. $P_{cly} < P_{atm}$. This will cause the piston to return to its initial position where $P_{cly} = P_{atm}$, and $T_{cly} = T_{atm}$ as illustrated in Fig. 9.9. At which point the thermal energy increase as was defined by eqn 9.19 will have radiated out through the piston-cylinder's walls into the surrounding atmosphere.

Many mechanical devices will suffer similar thermal energy loss as the above piston-cylinder system. However often the heat associated with eqn 9.19 is actually exhausted out through of a valve i.e. the exhaust stroke. The point becomes that the efficiency as defined by 9.18 is realistic.

Idealistically the efficiency could be increased by recovery of the hot gas's thermal energy that remains inside of the piston-cylinder (eqn 9.19) but this would often requires an input of resources, beyond the opening of a valve.

Compression Stroke

During the compression stroke of a cyclic engine, energy is requires to compress the gas, resulting in a pressure increase. Associate with the pressure increase is a temperature increase, which combined with heat generated by friction often results in thermal radiation (heat) dissipating into the surroundings, which ultimately heats of our atmosphere. This is often added to any heat created during the explosion of the fuel-air mixture, e.g. power stroke.

What is not so obvious is what else happens to the atmosphere while the fuel-air mixture is being compressed. The compression stroke is likened to magically removing a volume of gas located here upon the Earth's surface. Hence, during the compression stroke, the atmosphere's gaseous molecules will plummet downwards hence some of their potential energy converts into kinetic energy. Again this can be envisioned as ultimately resulting in an increase to our atmosphere's temperature, even though energy is not necessarily directly lost by the engine.

Inefficiency of Steam Engines

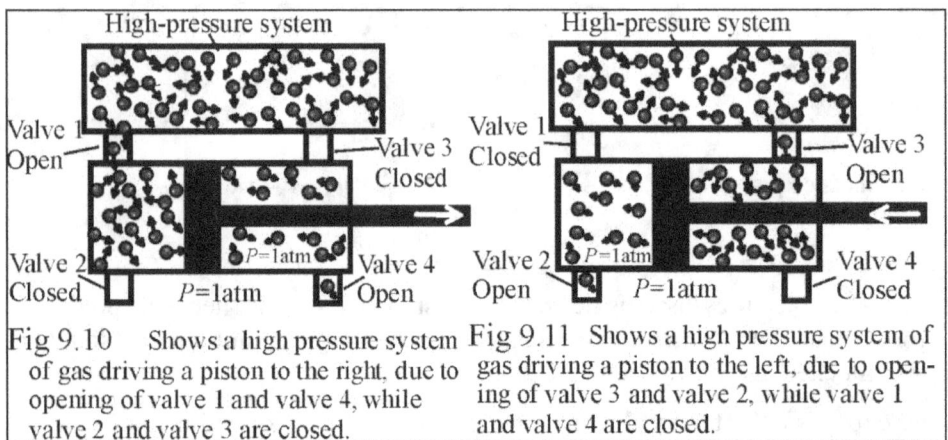

Fig 9.10   Shows a high pressure system of gas driving a piston to the right, due to opening of valve 1 and valve 4, while valve 2 and valve 3 are closed.

Fig 9.11   Shows a high pressure system of gas driving a piston to the left, due to opening of valve 3 and valve 2, while valve 1 and valve 4 are closed.

Often an engine functions as a closed system during the power stroke and an open system during other parts of the cycle. Consider, the cyclic engine illustrated Fig. 9.10, wherein; a high-pressure gas first drives the piston in one direction, and then drives the piston back in the opposite direction. The system of high-pressure gas can be due to: 1) A lower mean molecular volume occupied by the gas molecules, and/or, 2) a higher mean kinetic energy of the gas molecules than that associated with the surrounding atmospheric gas.

An engine, whose cycle is illustrated in Figures 9.10 and 9.11, where water is boiled to create the high-pressure gas, is commonly known as a "steam engine". Such engines function by starting with valves 1 and 4 being open, causing the high-pressure gas to be on the L.H.S. of the piston thus driving the piston to the right, as shown in Fig. 9.10. When the piston makes it to the end of the cylinder, valves 1 and 4 close, while valves 2 and 3 open. With the high-pressure gas now on the R.H.S. of the piston, the piston is driven back in the other direction, i.e. towards the left.

Consider that the piston is massless and frictionless, and that the engine performs no actual mechanical work. In this case, the higher pressure simply pushes the hot gas' kinetic energy through the piston onto the surrounding atmosphere, resulting in a continuous addition of potential and/or thermal energy into our atmosphere, i.e. W_{atm}. This is to be added to the continuous exhausting of warm gases into the atmosphere, which happens irrelevant of which direction the piston is moving.

In real life, such cyclic engines also experience extensive friction while running. Combine this with the fact that only a portion of the input energy can be involved in the contribution of momentum onto the piston. Plus the fact the gas is water vapors, which are not monatomic hence the gas's thermal energy is not simply kinetic. This all helps explain why steam engines tend to have such low efficiencies, e.g. $\eta \approx 10\%$.

Cyclic Inefficiency & Combustion

Of course in a multi-cylinder combustion engine, one piston-cylinder is compressing while another is expanding. How that affects the overall scheme of things will require some modeling and due

New Thermodynamics: Say No to Entropy

consideration. Even so some conclusions can be made.

Does the internal combustion engine (ICE) suffer the same demise as the steam engine? Reconsider the previously discussed cycle of the four-stroke engine. Obviously, unlike the steam engine, work is not continually lost in each step, due to the continuous displacement of the atmosphere. But the energy loss is still substantial!

Just consider the various steps, realizing that temperature increases ($T\uparrow$) will result in energy loss due to the consequential outflow of thermal radiation. During both the intake stroke (Step 1) the only energy loss should be due: $T\uparrow$ due to any viscous dissipation of a flowing gas. Similarly for the exhaust stroke (Step 7), except herein energy associated with hot gases is also lost as was defined by eqn 9.19. During the closing and opening of valves (Steps 2, 6 & 8), expect no loss.

During the compression stroke [Step 3)]: $P\uparrow$, $T\uparrow$, a volume of the atmosphere is compressed. Again, the surrounding atmospheric gases should experience a change to both their potential energy and kinetic energy, as the pistons move in their cylinders. This can be readily viewed as continuous increase in molecular dissipation that being continual increases to intermolecular friction, resulting in the direct heating of the atmosphere! Remember intermolecular collisions are now realized as being inelastic.

During Step 4, the explosion of fuel-air mixture creates heat: $T\uparrow$. During the power stroke [Step 5)], the rapid expansion of gas occurs in a closed system causing an increase to the atmosphere's potential and/or thermal energy i.e. lost work.

Global Warming

The implication of all of the above discussed inefficiencies. All piston-cylinder cyclic engines must contribute to global warming, with the motions of every stroke. Although, relatively small, over a sufficient amount of time, with a sufficient number of cyclic engines running, this likely is a significant contributor to global warming.

Although more analysis is needed, it may actually be the case that cyclic engines are more significant contributors to global warming than previously thought, irrespective of the greenhouse gases associated with the exhaust.

It must be emphasized that any motion that disturbs our atmosphere's gases will result in an increase to molecular dissipation, which ultimately becomes heat onto to our atmosphere. Accordingly based upon the insights provided throughout this text, global warming models may require a rethink.

Heat Engine

One of the simplest types of engine is a *heat engine* wherein power is obtained by the extraction of thermal energy from a heat reservoir. Consider an engine wherein heat-in (dQ_{in}) is required for an isobaric expansion process, e.g. the isobaric expansion of a piston-cylinder apparatus filled with an ideal gas. The issue becomes how can this be done with 100% efficiency? It would need to expand due to a heat source and more importantly any realistic system would have to displace our atmosphere, resulting in lost work! Thus the 19th century contemplation of such an engine being fully reversible was delusional, let us investigate.

Carnot Engine/Cycle

In 1824, Sadi Carnot envisioned the classic idealistic theoretical heat engine, which is traditionally used to educate young aspiring scientists, known as *Carnot Engine/Cycle*. Conceptually, his cycle was envisioned as a methodology for extracting energy/work from a constant temperature heat reservoir to be

Kent W. Mayhew

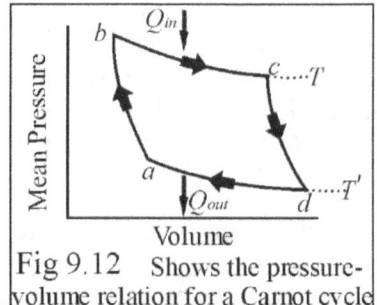

used in some cooler temperature surrounding, or, system. The cycle illustrated in Fig 9.12, is based upon traditional interpretations for the compression/expansion cycle of an ideal gas[2] operating between two constant temperature regions, wherein all processes are deemed reversible. Poor traditional conceptualization is as follows:

Fig 9.12 Shows the pressure-volume relation for a Carnot cycle

Step 1). "Reversible isothermal expansion"; $1 \to 2$: Put the engine in contact with the hot heat reservoir at; $T_1 = T_2 = T_{hot}$, and then allow it to isothermally expand from; V_1 to V_2. The gas's temperature remains constant by being in contact with the hot reservoir e.g. requires an input of energy; Q_{in} .

Step 2). "Isentropic expansion" as an adiabatic process; $2 \to 3$: The hot engine is thermally re-isolated and its volume is allowed to expand from; V_2 to V_3 until its temperature decreases back to; $T_3 = T_4 = T_{cold}$.

Step 3). "Reversible isothermal compression"; $3 \to 4$: The engine is now put into thermal contact with a cooler heat reservoir at $T_3 = T_4 = T_{cold}$ and the gas is now compressed; V_3 to V_4 .

Step 4). "Isentropic compression" as an adiabatic process; $4 \to 1$: The engine/cycle starts at a colder temperature; $T_3 = T_4 = T_{cold}$ and is thermally isolated (100% insulated). The gas is slowly compressed hence its volume decreases from; V_4 to V_1 and its pressure increase causes to the gas's temperature to increase back to $T_1 = T_2 = T_{hot}$. And thus completes the cycle.

Each of the steps in Carnot's cycle traditionally is considered as being reversible, hence the cycle as a whole is deemed reversible. Rather than adhere to traditional dogma let us now re-investigate the Carnot cycle to see what we can conclude.

Our new analysis of the steps

Step 1). "Isothermal expansion". It verges on lunacy to think that this can be properly analyzed without more information. First question should be what is the expanding force? If it is an external force pulling, as was the case of the piston-syringe shown in Fig 3.4, then the process is reversible however now $Q_{in} = 0$. This allows for the process to be deemed reversible, e.g. as traditionally claimed.

It is doubtful that the original intent of reversibility is for $Q_{in} = 0$. Consider the case of the energy input being real and greater than zero ($Q_{in} > 0$). In order to determine reversibility then the surroundings need to be clearly defined. Defining the surroundings as our atmosphere allows us to properly quantify this step. As the system expands, our atmosphere's energy (potential and/or thermal) increases, thus in terms of the pressure (P_{atm}), the energy input (Q_{in}) and lost work (W_{lost}):

$$Q_{in} = W_{lost} = P_{atm}dV \qquad 9.20$$

However eqn 9.20 is idealistic rather than realistic because it assumes that all the energy input can be used for work. If the system is defined as a gas then only 2/3 of the energy input can be used for work.

What if the surroundings were an ideal monatomic gas-filled closed isometric box? Now there would be no displacement of our atmosphere and the pressure in the surrounding isometric box would increase. If it is in thermal contact with the same high temperature isothermal heat reservoir ($T_1 = T_2 = T_{hot}$), then the temperature would remain constant. Is this reversible? If expansion means that the pressure in the

116

expanding system becomes less than that of the isometric surrounding, then yes it is reversible, at least from the perspective of mechanical equilibrium. Again this is probably not the intent.

Obviously, the work required depends upon the system's surroundings. Accordingly without clarity, one cannot expect to know how much work is required. Obviously, there are inherent problems with Step 1) of this Carnot cycle being deemed reversible, especially if the goal is to extract anything useful from it.

Step 2). "Isentropic expansion" as an adiabatic process. This step is deemed adiabatic because it is insulated. Herein the gas continues to expand hence cools; therefore the expanding gas must be doing work onto something. Again there is no clear clarification as to what the surroundings are. Assuming that the surroundings are our atmosphere and that in step 2) the system is cooling down because it is expanding, then it is doing work onto our atmosphere, i.e. eqn 9.20.

Now ask is this process adiabatic? Adiabatic means no exchange of energy whatsoever with its surrounding. Since the system/engine supposedly cools because it is doing work onto its surroundings then by strict definition, Step 2) is not adiabatic!

Step 3) "Isothermal compression". Since the compression is isothermal at the colder temperature ($T_3 = T_4 = T_{cold}$), then the ability of the gas to do work does not change I.e. [$d(PV) = 0$]. Is step 3) reversible? Again clarity is required. Assuming that the force of compression is external to the system, then as the system is compressed its temperature increases and heat can be extracted. If this extracted heat can then readily be returned into the compressing system, then yes it can be a reversible step in this Carnot cycle. If the return of heat requires some type of manipulation, then it is not reversible!

Step 4). "Isentropic compression" as an adiabatic process. Since the compression is not isothermal, then the gas's temperature increases, as does the ability of the gas to do work does change i.e. $d(PV) > 0$. Again this is probably not the original intent.

Is this step 4) reversible? Assuming that the force of compression is external to the system then yes this is a reversible step in this Carnot cycle.

Traditional Demise

Without going into further detail we should now realize why nobody has actually managed to build the idealistic Carnot engine, which basically assumes that the cycle is magically reversible. Obviously, even if the engine was perfectly designed, lost work into our atmosphere occurs in Step 1) & 2) hence, this cycle cannot be reversible, at least here on Earth. Due to not properly understanding lost work in the above Carnot cycle, both the second law of thermodynamics and entropy's attributes were misconceived.

It further remains interesting that Planck[3] understood that work is often done onto our atmosphere, but failed to realize that such work is irreversible for reasons discussed throughout this text. Certainly any realization that any work done onto our atmosphere is lost work would have changed the traditional analysis of the Carnot cycle. Note: Some contemplations of the cycle do discuss the exchanges of energy with the surroundings but they fail to properly define the surroundings, which again enables the critical errors of logic.

Rankine Cycle

The traditional interpretations of the Rankine cycle can be found in texts or on-line, and for the most part are also complications of reality, all due to the traditional entropy based considerations. Herein a simpler interpretation based upon our new perspectives, will be sought.

The Rankine cycle is associated with power plants, wherein electricity-generating steam turbines are powered by anything from a nuclear reaction to the combustion of oil, natural gas, or coal, with water commonly being the working fluid. This cycle generally consists of four steps:

Step 1): Water is pressurized via a pump. Since water is incompressible, the energy required is minimal in comparison to power generated. The work required for the water's isometric pressure increase is:

$$W_{Step1)} = VdP \qquad 9.21$$

Step 2): High-pressure water enters a constant pressure & volume boiler at which point it is heated into a dry saturated steam. Energy is required to break the liquid water's intermolecular bonds that being; $U = dE_{bonding}$.

Step 3): The dry saturated steam is then allowed to expand thus spinning a turbine, which generates power. As the steam expands it not only turns the turbine but it also does work onto Earth's atmosphere, hence both the steam's pressure and temperature decreases. This can be treated in terms of changes to the steam's ability to do work. Assuming 100% turbine efficiency thus the negative change in steam's ability to do work equals the work done:

$$W_{step3)} = -d(PV)_{steam} = P_{atm}dV_{atm} + W_{Turbine} \qquad 9.22$$

It may be more useful to rewrite eqn 9.22 in terms of mean molecular volume ($\bar{v}$), number of molecules passing through the turbine (N), as well as final (f) and initial (i) states, i.e.:

$$W_{step3)} = -N[(P\bar{v})_{f(steam)} - (P\bar{v})_{i(steam)}] = P_{atm}dV_{atm} + W_{Turbine} \qquad 9.23$$

Generally the steam's final pressure will be atmospheric pressure thus:

$$W_{step3)} = N[(Pv)_{i(steam)} - P_{atm}v_{f(steam)}] = P_{atm}dV_{atm} + W_{Turbine} \qquad 9.24$$

Step 4): The vapors then enter a condensate chamber, wherein they return to the liquid state. Here energy from changes to the water's intermolecular bonding energy could be extracted that being; - $dE_{bonding} = -U$.

If the Rankine process were as efficient as possible, the efficiency would still be limited because of step 3). Again this is due to the lost work ($P_{atm}dV_{steam}$) into our atmosphere i.e. potential/temperature↑, as the steam that drives the turbine creates energy ($W_{Turbine} = E_{Turbine}$). The optimal efficiency ($\eta_{optimal}$) for the Rankine cycle becomes:

$$\eta_{optimal} = E_{turbine}/(P_{atm}dV_{steam} + W_{turbine}) \qquad 9.25$$

In a real power plants there would be additional energy loses such as frictional loses in flowing fluids (molecular dissipation), other heat loses into the surroundings, as well as frictional loses in machinery. Grouping the other energy losses together, E_{other}, more realistic efficiencies are obtained by:

$$\eta_{optimal} = E_{turbine}/(P_{atm}dV_{steam} + W_{turbine} + E_{other}) \qquad 9.26$$

Such processes are most efficient when the steam is supercritical because the work required for atmosphere's potential/temperature increase remains constant. So the greater the overall energy within the steam is, the higher the efficiency should be. Accordingly, steam turbines typically operate around the supercritical regime [838 K (565 °C)], and condensation occurs at slightly above room temperature i.e. above 300 K.

Real Engines

Combustion engines can be approximated as the expansion of the fuel air mixture. In the gasoline engine the fuel-air mixture is compressed, which is then ignited by a spark. For the higher compression diesel engine, the fuel is injected into the engine at the end of the compression stroke, resulting in combustion. Omitting the opening & closing of valves, and the injection of fuel, as steps, then one can consider that an idealistic diesel engine has four steps, as follows:

1) A to B: Isothermal compression stroke: Compression of the cool fuel-air mixture.
2) B to C: Isobaric combustion: Explosion of the fuel-air mixture.
3) C to D: Isothermal power stroke: Expansion of the hot gases from combustion.
4) D to A: Isometric exhaust stroke.

Fig 9.13 illustrates the idealistic curves for both the compression stroke (A to B) and the power stroke (C to D), which are considered as being isothermal, e.g. the following drawing of isothermal lines on a P-V diagram. Note: $T_{cd} >> T_{ab}$.

Reality: During the compression stroke, the cylinder's gaseous molecule's temperature increases with pressure. Conversely, during the power stroke, the exploded fuel-air mixture performs work hence its temperature decreases. Also, combustion is not isobaric. Accordingly, Fig.9.14 shows more realistic curves for a diesel engine. Note: In Fig.9.14 point E was added, which represents a fifth step, that being the intake stroke (E to A).

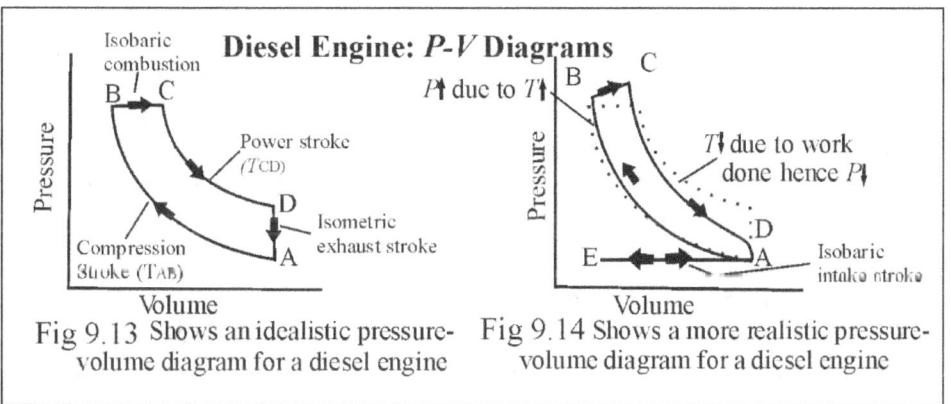

Fig 9.13 Shows an idealistic pressure-volume diagram for a diesel engine

Fig 9.14 Shows a more realistic pressure-volume diagram for a diesel engine

It should be stated that often the idealistic curves for both the compression stroke and power stroke are discussed as being adiabatic curves. This is simply not the case, isothermal maybe conceptually better, but even then that is simplification over a complete cycle at the operating temperature e.g. heat is created then carried and radiated away.

Refrigeration

How does a refrigeration system work? The refrigeration cycle for a standard household fridge, basically is:

Step 1): A compressor compresses a gas, e.g. ammonia. Hence; $P\uparrow$, $T\uparrow$.

Step 2): The hot gas then passes through coils located behind the fridge, cooling down below its boiling point (T_b at P_b), thus becoming a liquefied gas.

Step 3): The liquefied gas is then forced through an expansion valve, turning it into a low-pressure mist. Note the valve is located within another coil. As this low-pressure mist evaporates, it absorbs thermal energy through the coils inside of the freezer. This thermal energy/heat is extracted from the freezer as well as its contents, therefore keeping the freezer compartment frozen. As the evaporating gas draws heat from the freezer, it warms up becoming moderately cool. This moderately cool gas then travels to coils inside the fridge. Since this gas is still cooler than the surrounding fridge, it now extracts heat from the fridge, and its contents.

Step 4): The gas then gets drawn back into the compressor, where its cycle begins i.e. returns to Step 1).

Note: Although both interpretations agree that an expanding gas generally absorbs heat, again traditional explanations often are based upon work done onto walls. Our realization is that this is a closed system, so no real work is done onto the surrounding atmosphere. Note: This means herein the enthalpy of evaporation equals enthalpy of condensation, which is **not** the case as will be discussed in Chapter 11.

Refrigerants

In order to improve our understanding, consider what makes a gas a good refrigerant. The best gases to use as a refrigerant are the ones that are:

a)   Chemically non-reactive
b)   Safe i.e. low toxicity and low flammability
c)   Possess a boiling point somewhat below target temperature
d)   High latent heats (i.e. enthalpy of evaporation)
e)   High critical temperature
f)Non-corrosive in its application

Numerous gases have been engineered as refrigerants for various applications. Chloro-fluorocarbons are well suited, however they deplete the ozone layer, therefore most are being phased out. There are also more natural refrigerants in use such as ammonia, carbon dioxide and non-halogenated hydrocarbons. A few examples of refrigerant gases are shown in Table 9.1, taken from: "List of refrigerants" in Wikipedia (Apr., 2010)[4].

Although it is only a partial list, Table 9.1 shows that good refrigerants tend to be polyatomic polar molecules. As was discussed in Chapters 2 and 5, polyatomic molecules can contain significant amounts of thermal energy as vibrational energy e.g. as is approximated by eqn 2.16: $\bar{E}v \cong (n''-1)kT$, i.e. high heat capacity. Also, carbon bonds tend to have a 120-degree bond angle along a flat plane that may influence the way in which the molecules exchange kinematic energies with the wall as well as absorb radiation. However, these concepts do not fully explain why they function so well.

Table 9.1: List of Refrigerants

ASHRAE	Name	Formula
R-10	Tetrachloromethane	CCl_4
R-11	Trichlorofluoromethane	CCl_3F
R-12	Dichlorodifluoromethane	CCl_2F2
R-12B1	Bromochlorodifluoromethane	$CBrClF_2$
R-13	Chlorotrifluoromethane	$CClF_3$

Possibly the most important factor is their strong polarity allowing for substantial quantities of energy to be stored in the form of electro-magnetic potential energy (U), thus explaining the preference for extremely short bond (only 1.69 angstroms) with a strong polarization, e.g. the carbon-chlorine bond. Understandably, the electromagnetic potential associated with polar molecules may warrant more thought and study (see Appendix A.1).

Closing Remarks

For the most part, we dealt with engines/systems and their inefficiencies without any consideration to entropy change and/or the second law, as is traditionally accepted. Highest efficiencies come from monatomic gases rather than polyatomic ones. Based upon simple, constructive reasoning conceptualizations that resembled traditional Helmholtz's free energy were discussed, showing that our new perspective arrives at similar traditional results. And finally a new understanding of refrigerants were discussed.

References:

1) Mayhew, K.W., "Improving our thermodynamic perspective" Phys. Essays ,24 vol 3, 338 (2011)
2) "Fundamentals of Statistical and Thermal Physics", F. Reif, McGraw-Hill, New York, 1965
3) Planck, Max "Treatise on Thermodynamics" Third edition, London, Logmans, Green and co., 1917
4) http://www.Wikipedia.org/wiki/List of refrigerants (2010)

Chapter 10: **Latent Heat**

Latent heat is the energy, associated with isothermal phase transformations of matter, e.g. boiling/condensation, or melting/freezing. Herein we will be concerned with the differences between vaporization and condensation.

Latent Heat of Vaporization/Condensation

The energy required to go from the liquid to the gaseous state is the *latent heat of vaporization* (A.K.A. *enthalpy of vaporization*). Conversely, energy is released when going from the gaseous to liquid state that being the *latent heat of condensation* (A.K.A. the *enthalpy of condensation*).

The *boiling point* defines the maximum temperature a bulk liquid can be, while remaining in the liquid state at a given pressure (T_b at P_b). It is well defined because the thermal energy associated with the vibrating liquid molecules is now sufficient to break the intermolecular bonds holding the liquid molecules together. The boiling point applies to the bulk of the liquid, where the liquid as a whole begins to boil. Prior to reaching its boiling point, some liquid molecules still escape through the liquid's surface thus becoming vaporous in the process of *evaporation*. A commonly used definition: The boiling point is the temperature wherein the liquid's vapor pressure equals the system's pressure.

Calorimetry is the science of measuring the heats of reactions such as the latent heat, in a calorimeter. Latent heat of vaporization measurements are performed upon liquids whose tensile layer is relatively a flat, and constant, hence experiences no energy change associated with the tensile layer. Therefore, the only energy changes within the calorimeter are those due to the changes in intermolecular bonding, and/or, *PV* space. Furthermore, calorimeters are generally designed to measure latent heats, as being isobaric processes.

As was previously discussed, when heat/thermal energy is added to a liquid, then the liquid's temperature increases at a rate dependent upon the quantity of liquid, the liquid's heat capacity (C_y), and the rate at which the heat is applied. Accordingly, such energy is often referred to as *sensible energy* as is shown in Fig. 10.1 where the temperature (dT) increases as heat (dQ) is applied, with the slope being: eqn 1.2.9: $C_y = (dQ/dT)_y$.

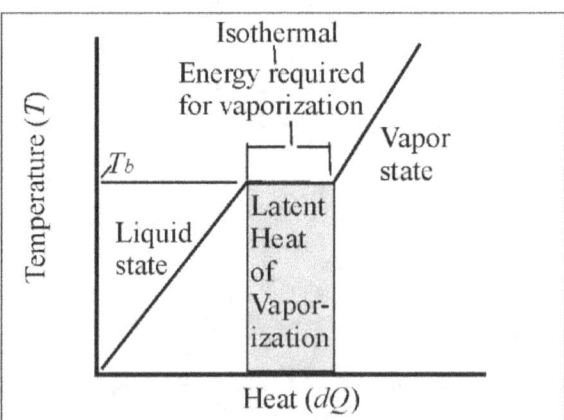

Fig. 10.1   Shows: Apply heat, the liquid's temperature increases until its boiling temperature (T_b). At T_b the latent heat of vaporization (W_{atm}) is required thus the liquid's temperature remains constant. After boiling, in the vapor state the molecule's temperature continue to increases.

Once the liquid reaches it boiling point, the liquid's temperature remains isothermal even though heat is continually applied, e.g. *non-sensible energy*. The reason being: The thermal energy goes into the phase change, i.e. liquid molecules that become vaporous experience a significant molecular volume (v) increase, resulting in the displacement of the surrounding Earth's atmosphere, hence requires work, which to an onlooker may appear as a non-sensible required energy. Furthermore, the work required as the latent heat is actually extracted from the thermal energy contained within the liquid, thus explaining why although the liquid is being heated, the bulk liquid remains isothermal.

Only after the molecules have gone through the phase transition from the liquid into the vapor state, can those, now vaporous molecules experience a temperature increase, as heat is added. This is illustrated by the positive sloping line in the vapor state, as shown in Fig 10.1: Slope being the heat capacity.

In Chap 4, it was previously discussed that if a sufficient quantity of water (e.g. an ocean) boils then there would be a significant mass transfer into the atmosphere, in which case both the volume, and pressure, of the Earth's atmosphere would increase, i.e. Eqn 4.42: $W_{atm} = d(PV)_{atm}$.

In most boiling processes, the mass of the created water vapor is infinitesimally small when compared to total mass of the Earth's atmosphere. Thus the weight, hence pressure exerted by the atmosphere, is approximated as being constant, i.e. an isobaric process ($dP = 0$). In which case the lost work required is given by eqn 1.6.15: $W_{atm} = P_{atm}dV_{atm}$.

Understandably, the latent heat of vaporization can be considered as an isothermal and isobaric process, i.e. $\Delta P = \Delta T = 0$. In this text "L" shall be used to signify the latent heat. For non-ideal substances, the energy required for boiling, must also consider any changes to bonding potentials [$dU = d(N\mu)$]. Therefore:

$$L = (dU + P_{atm}dV_{atm})_{P,T} \qquad 10.1$$

The subscripts "P, T" show that the latent heat is considered, as being an isobaric isothermal process. In order to keep the nomenclature simple, these subscripts will be omitted in the following equations. Furthermore, the subscript "atm" is removed with the understanding that the work is done onto the surrounding atmosphere. In order to enhance the clarity, the subscripts "g" and "l" will be used to respectively represent the gaseous state, and liquid state. Furthermore, we shall use the sign "$\rightarrow$" in the subscripts, to signify the direction of transformation. Employing this nomenclature, the latent heat of vaporization becomes:

$$L_{(l \rightarrow g)} = dU + PdV \qquad 10.2$$

where $L_{(l \rightarrow g)}$ is the latent heat of vaporization for a mole of molecules, which is often traditionally written as ΔH that being the molar enthalpy for vaporization.

The Problem with Latent Heat of Condensation

Traditionally, the latent heat of condensation is simply the negative sign of eqn 10.2, i.e.: $L_{(l \rightarrow g)} = -L_{(g \rightarrow l)}$. This is problematic, e.g. during condensation, the potential energy of the atmosphere's gaseous molecules is transformed into kinetic energy, but this is more of an energy transformation than losing work, i.e. lost work has little to do with condensation. Therefore the latent heat of condensation should be written in terms of bonding potential change:

$$L_{(g \rightarrow l)} = -dU \qquad 10.3$$

Eqn 10.3 contravenes traditional thermodynamics, as it simply states that the latent heat of condensation is the energy released as the liquid's intermolecular bonds form.

Molecular Volume

Since the molecular volume of a gas is of the order of 3 magnitudes greater than the molecular volume of a liquid, we can approximate the liquid's volume as zero, i.e. $V_g - V_l \approx V_g$. Thus the latent heat for vaporization and condensation respectively becomes:

$$L_{(l \rightarrow g)} = U_g - U_l + P_g V_g \qquad 10.4$$

$$L_{(g \rightarrow l)} = -(U_g - U_l) = U_l - U_g \qquad 10.5$$

Eqn 10.4 and eqn 10.5 are applied to macrostate that being the changes associated with an ensemble of N molecules, which are either, vaporizing, or condensing. Equations 10.4 and 10.5 can be rewritten in terms of N molecules, i.e.:

$$L_{(l \to g)} = N(u_g - u_l + P_g v_g) \qquad 10.6$$

$$L_{(g \to l)} = -N(u_g - u_l) = N(u_l - u_g) \qquad 10.7$$

where u_l is the bonding energy per molecule in a liquid state, u_g is the bonding energy per molecule in the gaseous state, N is the number of molecules vaporizing, and v_g is the mean molecular volume in the gaseous state.

In terms of the microstate, the mean molecular energy change is written:

$$l_{(l \to g)} = u_g - u_l + P_g v_g \qquad 10.8$$

$$l_{(g \to l)} = u_l - u_g \qquad 10.9$$

Equilibrium

Equilibrium is a zero net energy exchange state. This occurs when the rate of condensation, equals the rate of evaporation, over a given period of time. One may wonder why this results in a zero energy exchange, if the latent heat of vaporization involves lost work whilst the latent heat of condensation does not. The reason is simply that no work is actually being done when in equilibrium, i.e. there is no displacement of the surrounding atmosphere.

Types of Liquid Bonds

There are three general classes of bonds in the liquid state:

1) London dispersion bonds: Weak bonds
2) Hydrogen bonds: Moderate bond
3) Ionic bonds: Strongest bonds

London dispersion bonds are the weakest, hence require the lowest amounts of energy to break, therefore liquids with dispersion bonds have the lowest latent heats. Hydrogen bonds are moderate strength bonds due to the attraction of the dipole moment of polar molecules, i.e. water hence its latent heat is moderate. Ionic bonds are the strongest bonds, therefore, liquids whose molecular bonds are ionic, have exceptionally high latent heats. In Fig.10.2, we model a vapor above a hydrogen-bonded liquid. The liquid has a strong alignment of dipole moments, while a much weaker alignment exists for the polar vaporous molecules.

Fig. 10.2 Shows polar molecules have a strong alignment of their dipole moments in the liquid state and a weaker alignment of dipole moments in the vaporous state.

Latent Heat of Fusion

The energy required for the transformation between the liquid and solid state is commonly referred to as the latent heat of fusion (A.K.A. enthalpy of fusion). The energy required to go from the solid to liquid state is the latent heat of melting. Conversely, the energy given off when matter transforms from the liquid to solid state is the latent heat of freezing. Dealing with the transformations of matter between the liquid and solid state (freezing/melting) can be somewhat complicated because one often has to consider the energy associated with a crystalline structure. Accordingly, we shall leave the consideration of the energy associated with crystalline structures to other texts, which specialize in such matters.

Again the calorimeter measures the latent heat of fusion, as some isobaric process. Accordingly, eqn 10.1: $L = (dU + PdV)_p$ is the preferred equation. Utilizing the same previous nomenclature, the latent heat of melting becomes:

$$L_{(s \to l)} = U_l - U_s + P_l(V_l - V_s) \qquad 10.10$$

Traditionally, the latent heat of freezing would be:

$$L_{(l \to s)} = U_s - U_l + P_s(V_s - V_l) \qquad 10.11$$

For the case of the latent heat of freezing, the work term $[P_s(V_s - V_l)]$ can be ignored for most substances. I.e. generally the mean molecular volume in the liquid state is slightly greater than it is in the solid state exception being water. Again our reasoning is that any volume decrease does not require work. Therefore, eqn 10.11 becomes:

$$L_{(l \to s)} = U_s - U_l \qquad 10.12$$

The above is valid for the general case wherein the mean molecular volume in the liquid state is greater than it is in the solid state.

Obviously some substances, such as water expand during freezing. For such substances, the latent heat of freezing should be given by eqn 10.11, and then the latent heat of melting (eqn 10.10), becomes in terms of bonding potential change:

$$L_{(s \to l)} = U_l - U_s \qquad 10.13$$

If the molecular volume change is relatively small when contemplating either melting or freezing, then either, eqn 10.12, or 10.13, could be used. E.g. the volume change is approximated as being zero, then the latent heat of freezing can be considered as being equal and opposite to the latent heat of melting.

Isometric Latent Heat of Vaporization

The fact that calorimeters are designed to measure latent heats as an isobaric process has wrongly reinforced both, the traditional interpretation of work, and the concept that phase transformations are strictly isobaric processes. Consequentially in literature, latent heats are only listed as energy required/released, at a given pressure.

Instead of isobaric vaporization into an open/expandable system, now contemplate what will happen in isometric vaporization with the change in bonding potential between the two states. However, rather than an isobaric volume increase, there is an isometric pressure increase. Accordingly, for isometric latent heat of vaporization:

$$L_{(l \to g)} = (dU + VdP)_V = dU + VdP \qquad 10.14$$

We have to be careful here. Unlike the isobaric latent heat of vaporization, any temperature increase associated with the pressure increase may also need consideration. Accordingly, what we measure may depend upon how the experimental apparatus is setup, i.e. what happens to the temperature increase as a result of the pressure increase? Furthermore, the change in bonding potential should have volume dependence and therefore it may not exactly be the same for the isometric, as it would be for isobaric latent heats. This may affect how we write the isometric latent heat of condensation, e.g. a system's pressure decrease, may result in a temperature decrease, if the apparatus is insulated. Interesting conjecture!

What would happen if the vaporizing molecules caused both a pressure and volume, increase? In this case, the latent heat of vaporization becomes:

$$L_{(l \to g)} = dU + d(PV) \qquad 10.15$$

Again consideration to our apparatus's set up may be required, before precise analyze of empirical data can be determined.

Latent Heat of Vaporization vs Pressure

Consider an isometric pressure cooker. As the water boils, the pressure increases resulting in the water's boiling point (T_b, P_b) increasing, thus increasing the thermal energy density provided by the boiling water.

As the pressure increases, the latent heat of vaporization decreases as illustrated in Graph 10.1. The most likely reason being that the gaseous intermolecular bonding potential (dU) changes with pressure. Specifically, the energy of intermolecular bonding depends upon the mean molecular volume (v), i.e. intermolecular distances.

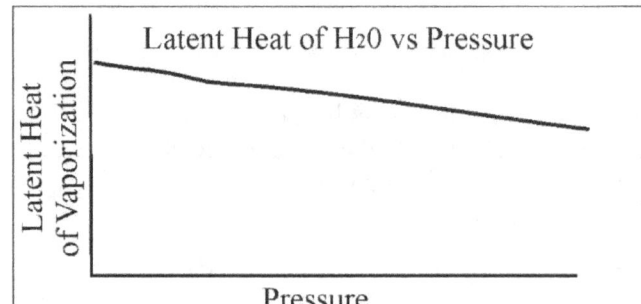

Graph 10.1 Is a sketch for the latent heat of vaporization for water vs an increasing pressure. A more precise graph is found in Appendix A.1.

Liquids are fundamentally incompressible, thus their intermolecular distance hence a liquid's bonding energy is pressure independent. Accordingly, changes in bonding potential (ΔU) must be due to changes in the vaporous state, wherein intermolecular distances are pressure dependent. Specifically, as its pressure increases, the vapor's mean molecular volume (v) must decrease. Therefore, for water vapors the density of polar molecule's dipole moments must increase. The electromagnetic attraction between such vaporous polar molecule's negative poles and their neighboring polar molecule's positive poles must increase with increasing pressure. Therefore, our expectation would be that the change in water's intermolecular bonding potential (ΔU) between the liquid and the vaporous state should decrease, as the pressure increases. This author's empirical proof for changes to water's latent heat of vaporization is based upon an analysis for a cloud of dipoles, and is detailed in Appendix A.1.

Higher vapor pressures means that molecules are physically packed closer together thus are more strongly bound to one another. Therefore, it should take less energy for a liquid molecule to vaporize in a high-pressure system than a low-pressure one. An analogy to what is being said can be seen in Fig 10.3. In A there are 12 polar molecules, whose positioning is relatively random based upon the following imposed constraint; *no two similarly charged ends of any two polar vapor molecule's dipole moments will be close to one another, without an oppositely charged end of a third polar vapor molecule, being equally close, or closer, to it.*

In B, of Fig 10.3, the same 12 molecules (1 thru 12) are all located in exactly the same positions. Another 12 polar molecules (13 thru 24) are then added to our system. Although, 12 molecules were added, much consideration concerning how the additional 12 molecules are located, had to be made. I.e. in order for the above constraint to be maintained the molecules numbered 13 through 24 were not as randomly drawn, as the first 12 molecules were in A.

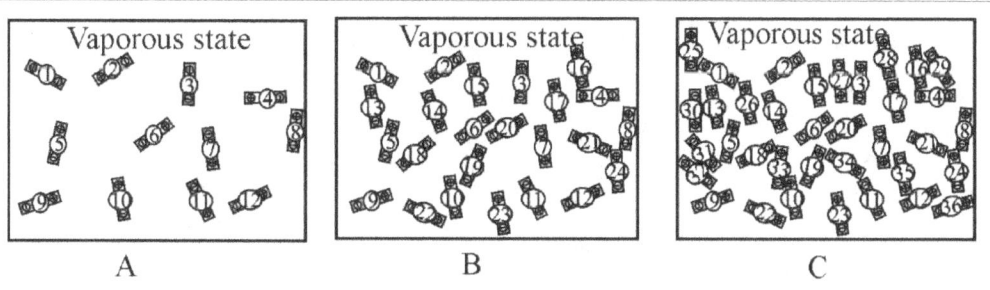

Fig. 10.3   Shows 12 polar molecules loosely bound in the vaporous state in A. In B, 12 more polar molecules are added and in C another 12 more are added again making 36.

In C, the same 24 molecules (1 thru 24) are all located in exactly the same positions, as they were in B. Another 12 polar molecules (25 thru 36) are again added to the system. Bearing in mind the same constraint, the amount of consideration concerning the allocation of molecules 25 thru 36 was significantly increased when drawing C, than it was when drawing B.

If another dozen polar molecules are added into C of Fig 10.3, the original positions of molecules 1 thru 36 could not be maintained. Specifically, we would have to start rotating, all the various molecules dipole moments, so that many of them are more parallel to each other, as happened to molecules 3, 15 and 27, in C. Such an alignment of dipole moments is the basis of how the polar molecules align themselves as states condense, e.g. hydrogen bonding in the liquid state as was illustrated in Fig 10.2. Is this a traditional entropy based argument? No, it simply common sense based on electromagnetism.

Trouton's rule

Trouton's rule states that for most (not all) liquids, the latent heat of vaporization can be approximated by a simple relationship, that being a constant (C") times the boiling temperature (T_b) i.e.

$$L_{(l \to g)} = C'' T_b \qquad\qquad 10.16$$

Trouton's rule is traditionally explained in terms of entropy change associated with latent heat, i.e. $\Delta S = C'' = 10.5R$. This applies to liquids with relatively weak bonds. The constant in eqn 10.16 is accepted to be in the range $85 < C'' < 87$ J/mol·K.

For liquids with hydrogen or stronger bonds, Trouton's rule generally does not readily apply. This is because the change to bonding potential is significant, accordingly liquids like water have higher latent heats of vaporization than that defined by equation 10.16.

Interestingly some ordered gases such as formic acid, have a negative deviance from Trouton's rule, as the dimerization in formic acid actually reduces the volume change. In the liquid state the monomers are held in place by the hydrogen bonds, while in the vaporous state dimers form, thus reducing the volume change. It is interesting that in traditional thermodynamics this is explained in terms of dimmers lowering entropy change. In the next chapter a new understanding of Trouton's rule will become obvious.

Closing Remarks

The biggest difference from tradition was our realization that while the latent heat of vaporization requires work (upward displacement of the atmosphere: PdV), whilst condensation does not. Hence the absolute magnitudes of the two are not necessarily equal unless considering some equilibrium state wherein no actual work is actually done.

Chapter 11: **Probability & Latent Heat: A Rethink**

Thermodynamics cannot be fully appreciated without discussing probabilities that being a basis of statistical thermodynamics. Herein, probabilities and distributions are treated in simple terms. Appendix B.2 and other texts[1,2], provides extensive traditional treatments.

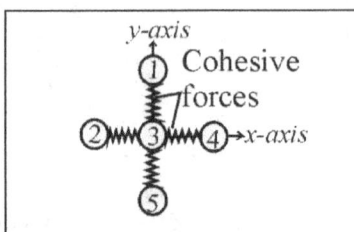

Fig. 11.1 Shows molecule #3 surrounded by four other molecules. All the molecules are cohesively bound, i.e. spring.

Consider a system of molecules where all the molecules are cohesively bound to each other through electromagnetic attraction (binding energy). A simple analogy is a system of springs connecting the neighboring molecules, as illustrated in Fig 11.1.

There exists 3 orthogonal directions, as defined by Cartesian space. In any one direction, i.e. along the x-axis in Fig 11.1, the mean quantity of vibrational energy that can be passed onto a neighboring molecule is kT. As was discussed in Chapter 2, the total mean energy of a one-dimensional harmonic oscillator is the summation of its kinetic, and potential energies, that being defined by eqn 2.4:

$$\overline{E_v} = \overline{E_k} + \overline{E_p} = kT .$$

If a neighboring molecule passes its kinetic energy onto molecule #3, then the kinetic and potential energy of molecule #3 will increase. The mean value of the energy exchanged between any two cohesively bound molecules is taken to be kT, as defined by eqn 2.4. Define kT as the *mean accessible energy* that being the mean amount of energy that a molecule can access from one of its bound neighboring molecules, along a given direction.

Cohesively Bound (Bonded) Molecules

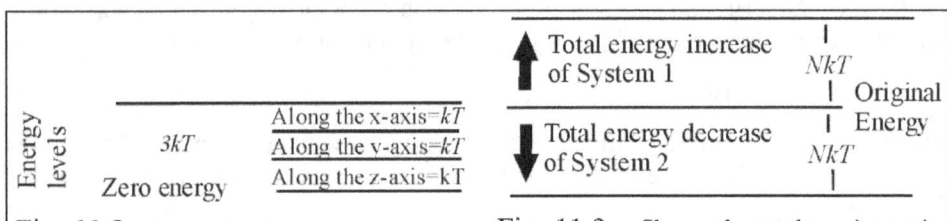

Fig. 11.2 Shows total mean energy level of each molecule in thermal contact and cohesively bound as: $3kT$. The mean accessible energy each molecule can pass along any one particular axis is: kT.

Fig. 11.3 Shows the total maximum instantaneous change in energy of System 1 from N molecules of System 2, is: NkT. This assumes that all N molecules are in thermal contact with System 1.

When contemplating cohesively bound molecules within condensed matter, the total mean thermal energy of each molecule is the summation of kT along the three orthogonal directions, as illustrated on the L.H.S. of Fig. 11.2. Obviously, the total thermal energy of a system containing N identical molecules, all in thermal contact with one another would be $3NkT$.

Fig 11.4 shows System 1, consisting of 48 identical molecules in thermal contact with System 2, which also consists of 48 molecules. The molecules in System 1 may be similar or different than those in System 2. We shall keep track of the molecules by numbering each one. As illustrated, only 12 molecules can physically pass thermal energy between the two systems. The mean amount of energy that molecules #37 through #48 of System 1 can instantaneously pass on to molecules # 1 through #12 of System 2, then the answer would be $12kT$.

Instead of a simple exchange of thermal vibrational energy, some of the molecules of System 2 try to enter a higher energy state (level), such as:

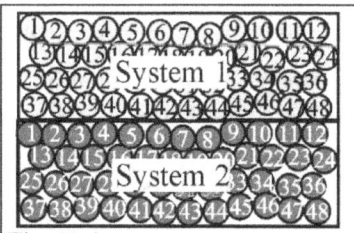

1) Another phase.
2) A chemical reaction.

In order to attain that higher energy state, the molecules must not only access sufficient energy, but they must also be physically capable of entering that higher state. Conditions that prevent molecules from readily entering a higher energy state are referred

Fig. 11.4 Shows System 1 in thermal contact with System 2, wherein each system contains 48 identical molecules

to as *constraints*, i.e. pressure is a force of constraint against molecules attaining a higher molecular volume. If such constraints are insurmountable, then no matter how much energy a system of molecules absorbs, they may never enter the higher energy state.

For illustration purposes consider molecule #3 of System 2, in Fig 11.4. There exists a probability of it being able to extract the thermal energy required to enter a higher energy state from any combination of its neighboring molecules, namely molecules #38 and #39 of System 1 and molecules #2, #14, #15 and #4 of System 2. Although at any instant in time, the mean thermal energy of any molecule in condensed matter is $3kT$, the actual thermal energy of a molecule, e.g. #39, may be greater than, or less than, that mean value. Therefore, the actual energy that a molecule, e.g. #39, can instantaneously pass along a given direction onto one of its neighboring molecules has a range about its mean value of kT.

Specifically, a simple Gaussian distribution that is symmetric about their mean value $3kT$, approximates the energies associated with each individual molecule in System 1, as is illustrated in Graph 11.1. The mean accessible energy that a molecule can pass on to, or access from, a neighboring molecule is kT, as is shown in Graph 11.2. Note: The slightly more complex Boltzmann distribution (resembles blackbody curves. See: Appendix B2) gives a more exacting analysis.

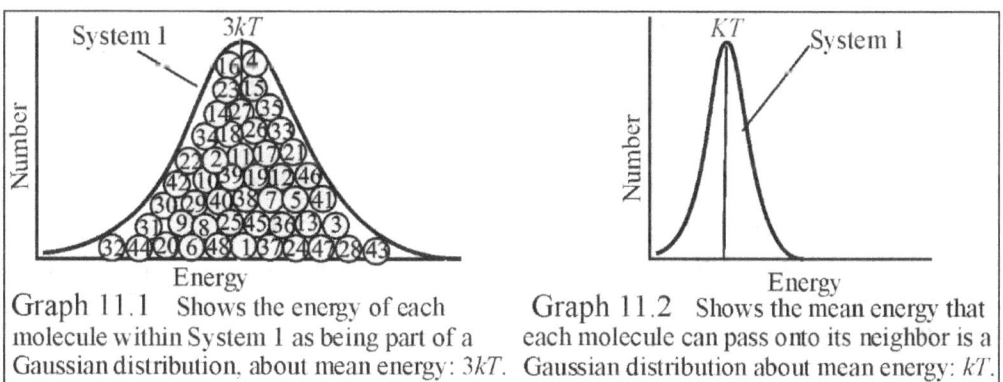

Graph 11.1 Shows the energy of each molecule within System 1 as being part of a Gaussian distribution, about mean energy: $3kT$.

Graph 11.2 Shows the mean energy that each molecule can pass onto its neighbor is a Gaussian distribution about mean energy: kT.

Reconsider Fig 11.4; the probability of molecule #3 accessing the energy required to go into a higher energy level, from any one of its neighboring molecules, namely molecules #38 or #39 of System 1, as well as molecules #2, #14, #15 and #4 of System 2, would similarly be based upon a Gaussian/Boltzmann distribution of its neighboring molecule's energies.

Probability

The probability [$P'_{(E)}$] of a neighboring molecule possessing energy E, is:

$$P'_{(E)} \propto \exp(-\beta E) \qquad 11.1$$

where E is the energy required to go into a higher energy level and $\beta = 1/(kT)$, k is Boltzmann's constant, and T is the absolute temperature of the surrounding molecules.

The probability $[P'(E)]$ as given by eqn 11.1, is often referred to as a canonical ensemble, as was first devised by Josiah William Gibbs, which is based upon the random walk analogy; which is briefly discussed in Appendix B2, and is very well detailed in many texts on statistical thermodynamics[1,2]. Such probabilities form the basis of statistical thermodynamics, as was first envisioned by James Clerk Maxwell, and Ludwig Boltzmann, in the late 19th century. The exponential factor $\exp(-\beta E)$, or $e^{-\beta E}$, is commonly referred to as the *Boltzmann factor*. The Boltzmann factor is proportional to the probability that a given molecule has a single state of energy, as defined by E. In a broader context, quantum mechanics/theory was founded in the 20th century by intense analysis from numerous greats, such as A. Einstein, N. Bohr, L. de Broglie, E. Shroedinger and W. Heisenberg. Based upon probabilities, quantum theory has enabled us to question the universe by comparing certain probabilities to observed realities.

The implication being that the probability of a molecule being able to extract energy (E), from one of its neighboring molecules, which is traditionally written[7]as:

$$P'_{(E)} = B\exp(-\beta E) \qquad 11.2$$

where B is the *proportionality constant*. The proportionality constant (B) can be determined by the normalization condition, which entails the consideration that the system has a 100% probability (unity) of being in the elevated state, i.e. $P'_{(E)} \to 100\%$, as; $kT \to E$. It often turns out that $B = 1$.

We are more interested in how the probability behaves. If the mean accessible energy kT, is greater than or equal to the energy (E) required for molecule #3 to go into a higher energy state, then given sufficient time molecule #3 will generally access the required energy from <u>one</u> of its neighbors, allowing it to enter that higher energy state. Accordingly, if there are no insurmountable constraints upon molecule #3, then the probability of molecule #3 going into a higher energy level, becomes unity: i.e. $P'_{(E)} = 100\%$ when $kT \geq E$.

For the case of $kT < E$. Then molecule #3 could only access the energy required to go into a higher energy level, from <u>one</u> neighboring molecule, if a neighboring molecule happens to have enough energy to give. Graph 11.2, shows that at a given instant, there may exist neighboring molecules whose accessible energy is sufficiently higher than the mean accessible energy (kT), thus allowing molecule #3 to access enough energy to enter into the higher energy state.

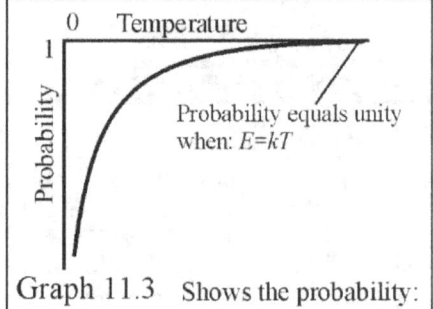

Graph 11.3   Shows the probability: Eqn 11.2, as a function of temperature When: $E=kT$: the probability is 100%.

The probability of occurrence is graphically illustrated in Graph 11.3. As the molecule's temperature increases, the probability of molecule #3 being able to access the required thermal energy from one of its neighboring molecules increases exponentially, as defined by eqn 11.2. Although each molecule within a system has a unique energy located about their mean, it is realized that this is only an instantaneous event, and that the mean energy of that molecule over time will still be: $3kT$.

An analogy would be a room full of people. Everyone has their own hot potato and starts randomly passing their hot potato to one of their neighbors. Over time people will average one hot potato in their hands. However, at some instant, anyone may have two or more, hot potatoes in their hands, while some of their neighbors have none. Rather than hot potatoes, molecules are exchanging packets of thermal energy (kT) amongst their neighbors.

In so far as heat (thermal energy) transfer is concerned, the real issue is one of thermal contact, which is to say that unless two molecules are close enough to exchange vibrational energy then there is no way for those two molecules to readily exchange packets of thermal energy. As previously stated, when molecules are bonded to each other, then they are certainly in thermal contact with each other. Of course one molecule could radiate a spectrum of energy into freespace, e.g. blackbody/thermal radiation, and a second molecule could absorb that spectrum, but that is exactly not what these probabilities are concerned with.

Traditional Probability and Latent Heat

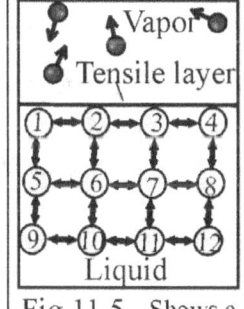

Fig.11.5   Shows a dozen liquid molecules near an interface

Traditional approach; consider a liquid molecule located near the surface of a gas/liquid interface, such as molecule #6 in Fig 11.5, and that the liquid is at its boiling point (T_b, P_b). The vaporizing molecule can extract a mean accessible energy of kT from any of its neighboring molecules, whilst still in the liquid state.

Letting the subscript "(v)" signify vaporization, then based on eqn 11.2, the traditional probability ($P'_{(v)Trad}$) of molecule #6 extracting enough energy from one of its neighboring molecules e.g. molecules #2,5,7 or 10, can be written:

$$P'_{(v)Trad} = B\exp(-\beta l_{(l \rightarrow g)}) \qquad 11.3$$

where $l_{(l \rightarrow g)} = L_{(l \rightarrow g)}/N$ is the molecular latent heat of vaporization.

What applies to a solitary molecule, equally applies to a mole of molecules. Therefore, there should be no transgression if the top and bottom of the exponential in eqn 11.3, was multiplied by Avogadro's number (6.02×10^{23} molecules/mol). In terms of the ideal gas constant (R), the probability can equally be expressed on a molar basis is:

$$P'_{(v)Trad} = B\exp[-L_{(l \rightarrow g)}/RT] \qquad 11.4$$

where $L_{(l \rightarrow g)}$ is the molar latent heat of vaporization. Note: Eqn 11.4 is traditionally often written in terms of the molar enthalpy for vaporization (ΔH).

Problematic Traditional Approach: Eqns 11.3 and 11.4

Examining the latent heats of vaporization for any substance, it is found to be significantly greater than the mean thermal energy of any of the molecules, at the liquid's boiling temperature (T_b). For example, water at 1 atm pressure; its boiling point is 373°K. Therefore; RT_b= 8.31 (J/mol·K) x 373 (K) = 3,099.6 Joules. The molar latent heat of vaporization [$L_{(l \rightarrow g)}$] for water is 40.65 KJ/mol at 298 K. Thus, the approximate ratio of latent heat of vaporization of water to mean molecular energy becomes:

$$L_{(l \rightarrow g)}/RT_b = (40,650\text{ J})/(3,040\text{ J}) = 13.1$$

On a per molecule basis, the same ratio: $l_{(l \rightarrow g)}/kT_b$ =13.1 is determined[3]. Obviously, latent heat of vaporization for water is approximately thirteen times the mean accepted accessible thermal energy of the water molecules, when water is at its boiling point. This is in no way unique to water, as similar ratios are found for all elements. Moreover in order for $\beta l_{(l \rightarrow g)}) \rightarrow 1$, then the water's temperature would need to be close to its latent heat divided by Boltzmann's constant (k), i.e.:

$(6.75 \times 10^{-20}$ J$)/(1.38 \times 10^{-23}$ J/K$)$ = 4,891 K

Based upon their fundamental conceptualizations, neither eqn 11.3, nor eqn 11.4 apply to vaporization, unless water's boiling temperatures approached temperatures found within stars[3]. Certainly, at such temperatures there are no expectations of the water being in the liquid state, i.e. water cannot be in its liquid state at temperatures above its critical temperature (T_c= 647.1 K, at 217.7 atm) yet alone at temperatures several times that.

Obviously, there is something conceptually wrong[3] with equations 11.3 and 11.4. Although exceedingly well studied the traditional explanation for the vaporization is wrong. Such thought has led many to consider latent heat as being strongly dependent upon surface effects, i.e. explanations are often based upon surface energy, e.g. the paper by Jozsef Garai[4].

Pondering Latent Heat

This author believes that there is another plausible way of envisioning the problem. Reconsider a liquid molecule, located near the surface of a gas/liquid interface, e.g. liquid molecule #7 in Fig 11.5. And that the liquid is at its boiling point. Are there other paths for molecule #7 accessing the thermal energy ($k\overline{T}_b$) required for vaporization, other than from traditionally conceived <u>one-single</u> neighbor?[3]

There must exist a probability of molecule #7 accessing thermal energy from two of its neighbors at some instant, e.g. molecule #3, and molecule #11, both simultaneously passing their thermal energy onto molecule #7. Although, this would be analogous to cutting the required temperature in half, the required temperatures would still be more than six times higher than the combined molecular energies at its boiling point[3].

Consider that the above argument applies to molecule #7 attaining a mean energy of $k\overline{T}_b$ from all six of its neighbors at once, e.g. molecule #7 acquires a packet of thermal energy from each of molecules #3, #6, #8, and #11, as well as the two molecules along the z-axis, which are not illustrated. The probability of molecule #7 vaporizing when all six neighboring molecules provide it with thermal energy $k\overline{T}_b$ at some instant would be[3]:

$$P'_{(v)} = B\exp(-l_{(l \rightarrow g)}/6kT) \qquad 11.5$$

This makes more sense especially if oone realizes that the vaporizing liquid molecule is bound to its sizx neighbors. Although closer, the latent heat of vaporization for water still cannot be readily explained, because now $l_{(l \rightarrow g)})/6kT_b = 2.2$. Perhaps the issue is with the concept of six neighbors.

Actual Number of Neighbors

The logic of six neighbors is probably valid due to intermolecular bonding in most liquids. However, the molecule's shapes may affect the number of neighboring molecules with which a vaporizing molecule exchanges its energy, e.g. if liquid molecules were perfectly round and closely packed as shown in

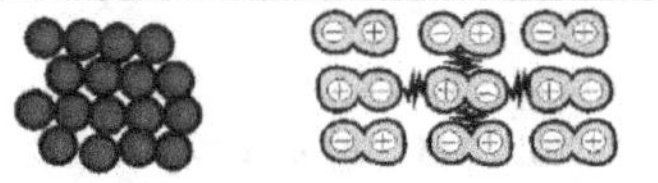

Fig. 11.6   Shows closely packed round molecules on the left with more than 6 neighbours versus dipole molecules on the right which only have 6 neighbours in thermal contact.

Fig 11.6, then it probably only has six neighbors.

However, for the various other shapes and molecular sizes the number of neighbor may vary. Conceivably, every molecule may require a separate analysis. Also we may have to question whether or not, we include the vibrational energies within molecules.

Some interesting research by Gerald Pollack's lab[5] has postulated a fourth phase of water, wherein there exists an exclusion zone (EZ layer). Herein, water molecules near interfaces form a helix type structure, which may also alter the number of neighboring molecules a vaporizing water molecule has. And this may alter one's analysis for such dipolar liquids.

Ideal Liquids

Perhaps the issue is with water itself. Consider ideal substances and their ratios for:

(latent heat of vaporization)/ (the boiling temperature times Boltzmann's constant).

For liquids becoming ideal gases the ratios are listed in Table 11.1. An example calculation of the ratio for argon (Ar):

$$l_{(l \to g)}/kT_b = [1.07 \times 10^{-20} \text{ (J)}]/[\{1.38 \times 10^{-23} \text{ (J/mol·K)}\}\{(87.3 \text{ (K)})\}] = 8.8$$

With the exception being neon (Ne) and helium (He), for liquids becoming ideal gases the mean accessible energy $k\overline{T}_b$ is approximately 1/9 of the energy required for the latent heat.

The poor correlation for the two smallest ideal gases (Ne & He) is actually expected. Specifically, as was discussed in Chapter 7 (graph 7.1), low boiling temperatures gases such as helium (He) and/or neon (Ne) do not obey Avogadro's hypothesis, when at/near their boiling points. Furthermore, as the temperature approaches absolute zero, thermal energy density would no longer be directly proportional to temperature, which as previously claimed helped explain why vibrational energies of polyatomic gases are frozen out as $T \to 0$. Furthermore, the listed latent heats of vaporizations are all corrected for 298 K but our concern remains their values, at their boiling point.

Table 11.1

Ideal Gas	Molecular latent heat of vaporization (J)	Boiling T (T_b)(K)	kT_b (J)	Ratio $l_{(l \to g)}/kT_b$
He	1.38×10^{-22}	4.2	5.80×10^{-23}	2.4
Ne	2.88×10^{-21}	27.1	3.74×10^{-22}	7.7
Ar	1.07×10^{-20}	87.3	1.21×10^{-21}	8.8
Kr	1.49×10^{-20}	119.4	1.65×10^{-21}	9.0
Xe	2.10×10^{-20}	165	2.27×10^{-21}	9.3

Reference:[6] Chemicool Periodic Table.

Due to their low boiling points leave helium and neon out of our analysis; does the above mean that in order for a liquid molecule to become an ideal gas that it needs to extract nine packets of accessible energy $k\overline{T}_b$? At first this seems troublesome, because there are only six neighbors each with a mean accessible energy along a given axis of $k\overline{T}_b$. Therefore, the answer requires some more contemplation.

At the time of writing this author can envision more than one possible explanation, but none actually feel 100% definitive. Begin with a hypothetical statement; the vaporization of noble molecules requires $kinematicnumber * k\overline{T}_b$. Based upon table 11.1, one might ponder that the number 9 is the best *kinematic number* to define the path for the latent heat of vaporization at T_b. [4]

Plausible Vaporization Paths for Kinematic Number 9

In the vapor state a molecule's mean kinetic energy is $3k\overline{T}_b/2$. In the liquid state its mean thermal energy is $3k\overline{T}_b$. The mean work required per molecule to displace the atmosphere is $W = Pv$, where v is gas's mean molecular volume. For the isobaric case, the energy of the gas, which displaces the atmosphere, is $3Pv/2 = 3k\overline{T}_b/2$. Can we now speculate upon $9k\overline{T}_b$ being required for an ideal liquid molecule to break the bonds with its neighbors plus perform the work ($W = P_g dV_g = P_{atm} dV_{atm}$) required to vaporize?

Scenario 1): At some instant, ideal molecule #7 of Fig. 11.5 would have to extract a mean thermal energy of $3k\overline{T}_b/2$ from each of its neighbors, allowing molecule #7 to vaporize. Perhaps, the six neighboring molecules pass all of their mean kinetic energy, onto vaporizing molecule #7, at some instant. All suggests that the mean kinetic energy from along all three orthogonal axis (total of $3k\overline{T}_b/2$) gets passed onto the vaporizing molecule, i.e. all the energy is given along one axis. It is mathematically eloquent: $6(3k\overline{T}_b/2) = 9k\overline{T}_b$, and molecule #7 retains its kinetic energy ($3k\overline{T}_b/2$). Still there remains the issue that herein the accessible energy only involves kinetic energy, i.e. being a multiple of $k\overline{T}/2$, rather than $k\overline{T}$, as is traditionally professed.

Scenario 2): At some instant, six packets of accessible energy ($k\overline{T}$) are extracted from the six neighbors, giving a total mean extracted energy of $6k\overline{T}_b$. Plus the vaporizing molecule's own thermal energy $3k\overline{T}_b$, certainly would give the mathematical required total thermal energy of $9k\overline{T}_b$. The problem remains that the vaporized molecule still possesses a mean energy of $3k\overline{T}_b/2$ after vaporization. Seemingly something feels amiss with this approach.

As clinical as both Scenario 1) & 2) may seem in explaining the kinematic number 9, they each possess apparent weaknesses, therefore other possible solutions or variations are sought. Another plausibility; consider that a given amount of kinetic energy is required to break the liquid's bonds. Once the bonds are broken then there would be another probability that the unbound liquid molecule attains kinetic energy from collisions with other liquid molecules enabling it to escape from the liquid rather than re-bonding with the surrounding liquid molecules, i.e. a multi-step process?

A multi-step process, Scenario 3): At some instant, six packets of accessible energy ($k\overline{T}$) are extracted from a vaporizing molecule's six neighbors, giving a total mean extracted energy of $6k\overline{T}_b$. Plus the vaporizing molecule's mean kinetic energy $3k\overline{T}_b/2$, allows it to break the six bonds and perform the necessary work (PdV). Note: Herein there is no potential energy associated with the vaporizing molecule because its bonds are broken. The reasoning being, there are six bonds and the potential energy associated with those bonds was considered to be part of the six surrounding molecule's accessible energy. The vaporizing molecule's then collides with other molecules, attaining another $3k\overline{T}_b/2$, which is its kinetic energy in the gaseous state.

Certainly, other paths do exist, as well as variations of the above paths, all warranting due consideration. What truly differentiates the logic presented herein from traditional theory is that this author suspects that the six neighboring molecules all collide with the vaporizing molecule at some instant and then bounce off along their orthogonal neighborly directions, thus breaking all six bonds required for vaporization. Then the question becomes when/how is the work done.

Vaporization Paths for Kinematic number 8.5

Perhaps 9 is not the best kinematic number, i.e. perhaps the kinematic number should be 8.5. Herein, at some instant, six packets of accessible energy ($k\overline{T}$) are extracted from a vaporizing molecule's six neighbors, giving a total mean extracted energy of $6k\overline{T}_b$. Plus the vaporizing molecule's mean kinetic energy $3k\overline{T}_b/2$, allows it to break the six bonds. Next the work (PdV) is extracted through more collisions, which equates to $k\overline{T}_b$. Giving a total required energy of: $6k\overline{T}_b + 3k\overline{T}_b/2 + k\overline{T}_b = 8.5k\overline{T}_b$. Herein the vaporizing molecule starts with a kinetic energy of $3k\overline{T}_b/2$, and it ends with the same kinetic energy, when in the gaseous state. So it either gets its kinetic energy in a gaseous state from its potential energy (?), or more from collisions with other molecules/walls after the liquid intermolecular bonds are broken.

When first envisioned this author felt that either 1), or 3), is the most likely path with kinetic number 9, as was published in Physics Essays[3] (and in part due to a reviewer's comments). However at the time of the final writing of this book this author started thinking again in terms of 8.5 being the better kinetic number. Of course if this is the case then even the ideal gases may be less than ideal. Furthermore I also found myself questioning just how many bonds are broken with a liquid molecule bonded to six neighbors and the final result still has six bound neighbors, it is just that the bound liquid molecules are now bound to molecules other than the one that has just vaporized. I still believe the answer should be six less bonds. Hopefully someone will deduce a more satisfactory path, or even somehow validate one of the above.

Remember, a molecule vibrates with frequencies in the order to 10^{13} times a second, thus in a second all neighboring molecules will possess, and exchange all sorts of packets of energy. It is only when all the criteria of total sufficient energy required to break the bonds is met, can vaporization occur. If boiling were some simple path-independent process then all the liquid molecules would vaporize at the instant that the energy requirements were attained, e.g. at the liquid's boiling point. Since a somewhat complex path exists, it helps to explain why boiling requires time, i.e. the right combination and permutation requires time to occur.

Kinematics Probability

The traditional probability defined by eqn 11.3, is based upon the concept that: $P'_{(E)} \rightarrow 100\%$, as $kT \rightarrow E$. When considering vaporization, eqn 11.4 was logistically weakened because $l_{(l \rightarrow g)} >> kT_b$. Reconsidering, the probability of molecule #7 becoming a gas; wouldn't it be better if it were rewritten in the following format?

$$P'_{(v)} = B\exp(-kT_b/kT) \qquad 11.6$$

Dividing both the nominator, and denominator, in the exponential by k, gives[3]:

$$P'_{(v)} = B\exp(-T_b/T) \qquad 11.7$$

Now if, $P'_{(v)} \rightarrow 100\%$, as $(T_b/T) \rightarrow 1$, when $B=1$, i.e. all liquid molecules would try to vaporize at once[3]. What prevents this is that B correlates to the likelihood of all six neighbors passing on their discrete energies at some instant plus the likelihood of the unbound vaporizing molecule actually escaping the liquid[3] plus whatever else is required to happen. Understandably eqn 11.7 doesn't describe some preferred path rather it just renders an improved probability based upon path dependent kinematics. Note: Clarification may be needed as to what extent B vs the normalization factor prevents all molecules from instantaneous vaporization

One may ponder why has eqn 11.4 been accepted for so long, as the correct probability function for boiling. Firstly, the thought that only one unidirectional packet of energy was required to break six different orthogonal orientated bonds feels troublesome enough in by itself. Certainly there can be nothing wrong with contemplating that each neighbor provides some, if not all, of the energy required to break its bond, plus do its share of the work, all of which is required in order for a given molecule to vaporize. Secondly, there are inherent dangers to using any such probability function, because it can provide false confidence, due to the normalization procedure, e.g. one can use the wrong exponential relation and still normalize their experimental data. This will be discussed in more depth in Chapter 12.

Latent Heat for Non-Ideal Substances

How valid is our conceptualization for the latent heat of vaporization for strongly bonded liquids vaporizing into non-ideal gases? Investigating, located at the end of this chapter is Table 11.2 (located at end of this chapter), which gives molar latent heats for various elements. Using molar latent heats doesn't alter the ratio, i.e.:

Ratio= (Molar latent heat of vaporization)/ $RT_b = l_{(l \rightarrow g)}) / kT_b$.

Table 11.2, shows that the majority of elements have ratios greater than: 9. An explanation may be required as to why the heaviest elements in Group 1 and 2, namely cesium and barium have ratios slightly below 9 i.e., ratios are 8.6 and 8.8 respectively. Perhaps 8.5 is the better kinematic number! Still why does the strongest diamagnetic element, bismuth, have a ratio of 6.9?

In all likelihood our analysis for ideal molecules will not be universally applicable. Different sized and shaped molecules will have differing number of neighbors. Each of which may require modeling. So although the number of six neighbors fits for the ideal it is not universal. Although with larger molecules one may have to consider the vibrational energies.

Reconsider water and its ratio of 13.1. Assuming $9k\overline{T}_b$ is required to both break the liquid bonds, and allow the vapor molecule to displace Earth's atmosphere, then can it be said that $13.1k\overline{T}_b - 9k\overline{T}_b = 4.1k\overline{T}_b$, represents the energy required to align/enclose a mole of water molecule's dipole moments in the vaporous state? This will require modeling and then again perhaps 8.5 is a better kinematic number.

Whatever the ultimate explanation becomes, there is some merit to this way of thinking. Unfortunately, the problem remains that things may not as readily discernible i.e. one needs to model any potential energy associated with a cloud of dipoles. It makes sense that for non-ideal gases, as the atmosphere is displaced, additional energy associated with electromagnetic bonding potentials, is required. Certainly this additional energy could cause the ratio for real gases, to be higher than the kinematic number.

Interestingly, for our dipole water molecule, vaporization requires a decreasing amount of energy as their mean molecular volume decreases i.e. latent heat of vaporization decreases with increasing pressure (Graph 1.11.1). As logical as that sounds, there still must also be an energy associated with the state of dipole moments initially wanting to repulse each other due to similar poles, i.e. an energy associated with the dipoles initial assembly, which increased the latent heat, hence caused the ratio higher

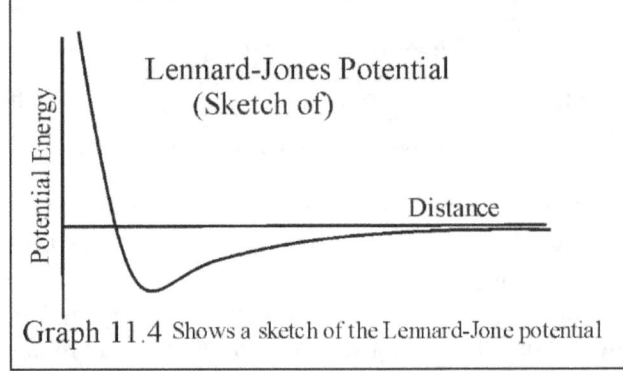

Graph 11.4 Shows a sketch of the Lennard-Jone potential

in the first place. After which the latent heat starts decreasing with pressure. Perhaps the inherent logic parallels the empirically known *Lennard-Jones potential,* as illustrated by Graph 11.4?

Certainly, we cannot boast 100% clarity at this time. Interestingly, in Appendix A.1 this author demonstrates that any decrease in latent heat with increasing pressure is of the scale one would expect if we were considering a cloud of point charges.

Why doesn't the gas molecules simply occupy a lesser volume so that all that is required is $9k\overline{T}_b$? Think back to Chapter 7. Walls vibrating molecules pump a mean kinetic energy of $k\overline{T}/2$, along each axis onto the gas molecules, ensuring gases obey Avogadro's hypothesis etc. Since the gas molecules are forced to occupy the specified volume (22.4 L, at 273°K, and 1 atm), insights arise into why energy associated with bonding potentials is required. Perhaps the energy associated with the bonding potential is extracted from the system right after the liquid molecule has vaporized. Thus for vaporization into a non-ideal gas, the extraction of the required latent heat from a system may be better explained as some multi-step process.

Reconsidering the Latent Heat of Vaporization

Of the energy required for vaporization, how much goes into doing work? The mean kinetic energy of a (sufficiently dilute) solitary monatomic gas molecule, at its boiling temperature is $3k\overline{T}_b/2$. This being the amount of energy associated with a solitary molecule that has displaced the Earth's atmosphere, whose mean volume is defined by Avogadro's hypothesis. However the actual work done onto the atmosphere is only $k\overline{T}_b$. Remember: Based upon kinetic theory only 2/3 of an ideal gas's kinetic energy change can be used for work.

At this point there is certainly more than one way that we can write the latent heat of vaporization. On a per molecule basis; in the terms of the kinematic number, plus the energy for dipole alignment ($u_?$) minus the bonding potential energy in the gaseous state (u_g), i.e.:

$$l_{(l \to g)} = 9kT_b + u_? - u_g \qquad 11.8\ (a)$$

$$l_{(l \to g)} = 8.5(kT_b) + u_? - u_g \quad 11.8\ (b)$$

Or do we just write it in terms of the bonding potential energy, and acknowledge that the bonding potential energy is generally a decreasing function for dipoles. I.e.

$$l_{(l \to g)} = 9kT_b + u_g \qquad 11.9\ (a)$$

$$l_{(l \to g)} = 8.5(kT_b) + u_g \qquad 11.9\ (b)$$

An extensive analysis when dealing with changes to bonding potentials of non-ideal gases will not be a simple task. Simple attractive, or, repulsive forces, should not be as difficult as dipoles, quadrapoles etc, in the gaseous state. See Appendix A.1.

The previous conjecture was entertaining and hopefully will inspire someone to deduce a more encompassing analysis. At this time all we can really do is say the latent heat of vaporization into energy ($3k\overline{T}_b/2$), and/or work ($k\overline{T}_b$), required to displace our atmosphere plus the energy required for changes in bonding energy: Δu. Accordingly:

$$l_{(l \to g)} = W_{atm} + \Delta u \qquad 11.10$$

The molar latent heat then becomes:

$$L_{(l \to g)} = W_{atm} + \Delta U \qquad 11.11$$

Trouton's Rule Revisted

As was discussed in Chapter 10: The concept that latent heat of vaporization can be thought of as a constant time boiling temperature is considered as Trouton's rule i.e. eqn 10.16: $L_{(l \to g)} = C''T_b = 10.5RT_b$.

It is an interesting that as a kinematic number, 10.5 could be readily explained in terms of the six neighbors plus vaporizing molecule all contributing $3kT/2$ to the vaporization process i.e. 7x3.5 = 10.5, which is similar to previously discussed scenario 1) . However, this is probably more of a coincidence than anything.

Latent Heat of Condensation

It was discussed that the latent heat of condensation does not involve the work term. Therefore, the molecular and molar latent heat of condensation would be as follows:

$$l_{(g \to l)} = -\Delta u \qquad 11.12$$

$$L_{(g \to l)} = -\Delta U \qquad 11.13$$

Boiling in Space

Consider the boiling of water in the space lab orbiting Earth. There is no gravity thus vaporization does not involve work (PdV) being passed onto the atmosphere. Therefore, boiling in space should require less energy than here on Earth.

Investigating further: The space lab is in an isometric (closed rigid) system hence the work (or potential work), as defined by $V_{(lab)}dP_{(lab)}$, should be required as the lab's pressure increases. As a potential to do work, it is not necessarily lost work. Even so let us play the game and say our first inclination might be to write:

$$L_{(l \to g)} = \Delta U + V_{(lab)}dP_{(lab)} \qquad 11.14$$

How accurate is eqn 11.14? Consider that a huge pot of water is boiled in the space lab, hence the lab's pressure increases. Eqn 11.4 is problematic because as the pressure inside the space lab increases significantly, then the mean molecular volume in the gaseous state must decrease. Therefore boiling in space actually requires an equation of the form:

$$L_{(l \to g)} = \Delta U + \int VdP_{(lab)} \qquad 11.15$$

If the gas inside of the space lab is ideal ($PV = C' = NkT$) then integrating:

$$L_{(l \to g)} = \Delta U + (NkT)\ln(P_i/P_f) \qquad 11.16$$

Based upon eqn 11.16, the latent heat required per molecule should increase as the pressure increases ($P_f \uparrow$). Note : Eqn 11.16 is isothermal hence ignores that this may also result in heat being added to the space lab's interior, due to increased molecular friction i.e. natural P-T relationship, or if preferred a viscous dissipation increase.

Imagine the far-fetched possibility that the space lab had some elaborate engineering that manages to keep the space lab isobaric. Then $P_f \approx P_i$ hence $ln(P_i/P_f) \approx 0$. And eqn 11.16 becomes:

$$l_{(l \to g)} = \Delta u \qquad 11.17$$

Do we now have an understanding as to why boiling in space requires significantly less energy than boiling on Earth's surface? Not exactly! Consider that the pot of water is one system and the rest of the isometric space lab is another. Then this boiling process involved mass transfer hence eqn 11.16 does NOT exactly apply because in the space lab the number of gaseous molecules increased during boiling i.e. $N_f > N_i$. Again if the space lab releases gas molecules through a valve in order to remain isobaric, then this should elevate the accuracy of eqn 11.16.

The realization is that eliminating gravity simply means that no work was actually required. Think of it this way, weight of the atmosphere on Earth means there is a downward force that expanding system's must overcome, and this force is non-existent in space.

Therefore boiling in the space lab is nothing more than an increase to both the thermal energy and the number of gaseous molecules in the space lab. If in its interior pressure and/or temperature were allowed to increase then the ability of the gas (within the lab) to do work will increase. Even so no work is actually done.

There has been research done in zero gravity boiling by Herman Merte[6] and others. They have found that due the lack of buoyancy and convection in weightless environments, the boiling tends to involve large often singular bubbles, with the bubble tending to stay near the heater, rather than form the cascade of smaller bubbles as normally witnessed here on Earth. It should be stated that such experiments were performed under the premises of traditional thought, rather than the understanding presented in this book. Note: This author plan on publishing a book on bubble nucleation in the near future.

Temperature Dependence for Latent Heat of Water

Let us now reconsider the latent heat of vaporization for water at temperatures near its boiling point. The approximate value for the molar latent heat of vaporization, at various temperatures is given in Table 11.3. Note: These values are approximations, visually taken from a graph published: Internet Wikipedia[7] with Graph 11.5 being a sketch of that graph for the latent heat of water versus temperature.

Table 11.3

T (°K)	Molar latent heat of vaporization (kJ/mol)
280	45
298	40.65
400	39
440	37
480	34.5

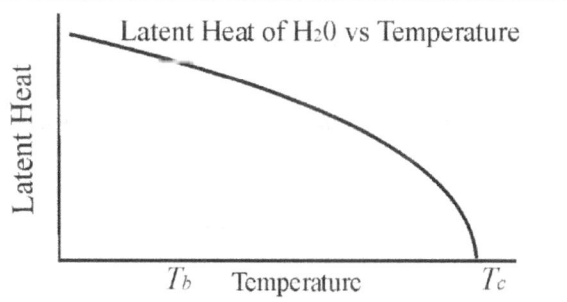

Graph 11.5   Sketch of how water's latent heat decreases with increasing temperature. Shown are the boiling temperature, T_b & critical temperature, T_c

For temperatures near the boiling point the molar latent heat decreases linearly. Using the data from Table 11.3, a rough approximate slope for changes in latent heat with temperature, for temperatures near the boiling point, is:

$(45-34.5)/(280-480) = -0.0525 \, KJ/mol \cdot K = -52.5 \, J/mol \cdot K$

Since the work done onto the atmosphere is proportional to the temperature, our expectation is that the above slope is due to the fact that the liquid's cohesive forces are approximately constant. I.e. as the temperature increases, the energy required to break a vaporizing liquid molecule's bonds should be

Kent W. Mayhew

decreasing. When considering vaporization paths, it was postulated that six packets of accessible energy (kT) from the six neighbors, and possibly also the vaporizing molecule's kinetic energy, was required to break the six bonds.

Therefore, as the water's temperature (T) increases the change in energy required to break those six bonds for a solitary molecule, should approximately be: $6k(T-T_b)$. Similarly, for a mole of molecules it would be approximated by: $6R(T-T_b) = -6R(T_b-T)$. Therefore the slope should approximately be: $-6R = -6 \times 8.31 = -50$ J/mol·K. Our roughly calculated slope of -50 J/mol·K is close to our approximated slope of: -52.5 J/mol·K. Although this was only a rough analysis, the requirement of six packets of accessible energy to break the six bonds seemingly fits with known data. Of course a more precise analysis is warranted.

Remember, the above analysis is based upon approximations, plus no consideration was given to changes in water vapor bonding, and system pressure (see: Appendix A.1). As previously stated for most pressure changes we can assume that the energy required to break liquid bonds is constant. However, when comparing to pressures associated with critical energy wherein pressure increases are an order of magnitude or more, we cannot make such assumptions.

Critical Temperature and the Kinematic Number 5

The critical temperature is the temperature above-which, no molecules can exist in the liquid state, irrelevant of the pressure. As was done for their boiling temperatures, we can calculate the molecular thermal energy (kT) of ideal gases molecules if they were in their liquid state, at their critical temperature, as shown in Table 11.4.

Exception being: He, we can see that the ratio of latent heat to mean molecular energy at the critical temperature, approximates 5. We shall call 5 the kinematic number for a liquid's critical temperature. Again we can attribute the lack of correlation for Helium (He) to the theoretical issues concerning gases nearing absolute zero.

Example calculation for a molecule of Argon:

$kTc = [1.38 \times 10^{-23}$ (J/K) $\times 151$ (K)$] = 2.08 \times 10^{-21}$ Joules
$l_{(l \to g)}) / kT_c =$ (Molecular latent heat of vaporization)/(kTc)
$= [1.07 \times 10^{-20}$ (J)/ 1.04×10^{-21} (J)$] = 5.1$

Table 11.4

Ideal Gas	Molecular latent heat of vaporization (J)	Critical T (T_c) (K)	kT_c (J)	Ratio: $l_{(l \to g)}) / kT_c$
He	1.38×10^{-22}	5.19	7.16×10^{-23}	1.9
Ne	2.88×10^{-21}	44.4	6.13×10^{-22}	4.7
Ar	1.07×10^{-20}	151	2.08×10^{-21}	5.1
Kr	1.49×10^{-20}	209	2.88×10^{-21}	5.2
Xe	2.10×10^{-20}	289.8	4.00×10^{-21}	5.3

Consider molecule #3 in Fig 11.5, which resides upon the liquid's surface. Molecule #3 only has five neighboring molecules from which it can extract thermal energy. Even so the prospect that molecule #3

140

will attain a mean thermal energy of kT_c from each of its five neighboring molecules, which is enough to break the liquid bonds plus do any required work, explains why, when the liquid is at, or above, its critical temperature, the likelihood of any molecules[4] existing in the liquid state must be zero. Hence explaining why no tensile layer can exist, when $T \geq T_c$. No tensile layer then no liquid state!

Critical Temperature and Non-Ideal Substances

What happens when we consider non-ideal substances and their critical temperatures? Dealing with the bonding potentials of high-density gases would become an extremely complex analysis. Even so, look at Table 11.5 (located on the last page of this chapter). It shows that for many elements the ratio of molar latent heat to mean energy of a mole of molecules at the critical temperature is close to 5. The exception being diatomic gases! Certainly, such diatomic gases would involve other considerations, such as the fact that not all bonds are broken, hence this will be left to others to contemplate. Note: Only a few elements were discussed because there seemingly is a lack of good data for the critical temperature. Moreover, many values given in literature are calculated hence questionable logic behind their calculation.

Multiple Paths

More than one path may exist for a given process, i.e. the probability of occurrence must be the summation of probabilities from all the possible paths. Each path may have its own temperature, wherein the probability along that path approaches unity, if and when a sufficient amount of time is given. Letting the numbers 1,2,3...N', signify different possible paths, then the total probability is:

$$P'_{(v)total} = P'_{(c)1} + P'_{(c)2} + P'_{(c)3}...P'_{(c)N''} \qquad 11.18$$

Obviously for the critical temperature there were well-defined paths at the lower temperature, that being the boiling point and a limiting path beyond which no liquid state can exist over any time frame. And certainly, other paths exist.

Closing Remarks

Interestingly, we provided insight into both the latent heat of vaporization, and the critical temperature, which is something that is traditionally lacking. What is important is that we introduced the need to discern the one, or more paths, for a given process to extract sufficient energy.

An indirect proof of concept will be given in Appendix A1, wherein we will demonstrate that the energy associated with changes to latent heat are of the order/scale/magnitude that one might expect from changes to electromagnetic potentials of a cloud of charged particles.

References:

1. Fundamentals of Statistical and Thermal Physics", F. Reif, McGraw-Hill, New York, 1965
2. "Statistical Physics", F. Reif, McGraw-Hill, New York, 1967
3. K. Mayhew, Phys. Essays **19**, vol 4, 604(2013)
4. Jozsef Garai, Fluid Phase Equilibria, 183, 89-92 (2009)
5. Pollack - fourth phase of water
6. Merte H And. Heat transfer 19, 181 (1973)
7. Chemicool Periodic Table. 4/4/2010 <http://www.chemcool.com.html>)
8. Wikipedia.org/wiki/enthalpy_of_vaporization (2011)
9. http://www. Wikipedia.org/wiki/vapor_pressure (2011)

Table 11.2: Latent heat of vaporization vs Boiling Temperature

Group	Atomic Number	Element	Exp. Molar Latent Heat of Vaporization (KJ/mol)	Absolute Boiling Temp. (K)	RT_b at Boiling Temp. (KJ/mol)	Ratio (Molar Latent Heat)/(RT_b)
TABLE 11.2						
Vaporization into ideal gases						
18	2	He	0.1	4.2	0.03	2.3
N	10	Ne	1.7	27.1	0.23	7.7
	18	Ar	6.4	87.3	0.73	8.9
	36	Kr	9.0	119.4	0.99	9.1
	54	Xe	12.6	167	1.39	9.1
	86	Rn	16.4	211	1.75	9.4
Vaporization: Non-ideal Gases						
1	3	Li	145.9	1620	13.46	10.8
	11	Na	97.0	1156	9.61	10.1
	19	K	79.9	1047	8.70	9.2
	39	Rb	72.2	961	7.99	9.0
	55	Cs	67.7	951	7.90	8.6
2	12	Mg	127.4	1380	11.47	11.1
	20	Ca	153.6	1757	14.60	10.5
	38	Sr	144.0	1656	13.76	10.5
	56	Ba	140.3	1913	15.90	8.8
3	21	Sc	332.7	3109	25.84	12.9
	39	Y	365.0	3618	30.07	12.1
	57	La	402.1	3737	31.05	12.9
4	40	Zr	573.0	4650	38.64	14.8
5	41	Nb	682.0	5200	43.21	15.8
	73	Ta	743.0	5698	47.35	15.7
6	74	W	824.0	5933	49.30	16.7
7	75	Re	715.0	5900	49.03	14.6
9	77	Ir	604.0	4800	39.89	15.1
11	29	Cu	300.4	2840	23.60	12.7
	47	Ag	258.0	2485	20.65	12.5
	79	Au	324.0	3129	26.00	12.5
12	80	Hg	151.0	630	5.24	28.8
13	13	Al	293.0	2740	22.77	12.9
	49	In	231.8	2273	18.89	12.3
	81	Tl	164.0	1730	14.38	11.4
14	14	Si	359.0	3538	29.40	12.2
	32	Ge	334.0	3106	25.81	12.9
	50	Sn	295.8	2543	21.13	14.0
	82	Pb	177.7	2013	16.73	10.6
15	83	Bi	104.8	1833	15.23	6.9

L	58	Ce	398.0	3716	30.88	12.9
	59	Pr	331.0	3793	31.52	10.5
	60	Nd	289.0	3347	27.81	10.4
	62	Sm	165.0	2067	17.18	9.6
	63	Eu	176.0	1802	14.97	11.8
	64	Gd	301.3	3546	29.47	10.2
	65	Tb	293.0	3503	29.11	10.1
	66	Dy	280.0	2840	23.60	11.9
	67	Ho	265.0	2973	24.71	10.7
	68	Er	280.0	3141	26.10	10.7
	70	Yb	159.0	1469	12.21	13.0
	71	Lu	414.0	3675	30.54	13.6

Most values for boiling temperature and experimental molar latent were taken from Jozsef Garai

paper: Reference: [4]Fluid Phase Equilibria, 183, 89-92 (2009)

Exception are the noble gases as indicated by ***, which calculated from
Chermicool Periodic Table 4/7/2010
http://www.chemicool.com/elements.html[28]

TABLE 11.5   Latent heat vs Critical Temperature

Element	Latent heat of vaporization	Critical Temperature	RT_c	Ratio: (Latent Heat)/RT_c
Ideal Gases	(kJ/mol)	(T_c) (°K)	(kJ/mol)	
He	0.08	5.19	0.04	1.9
Ne	1.73	44.4	0.37	4.7
Ar	6.45	151	1.25	5.1
Kr	9.03	209	1.74	5.2
Xe	12.64	289.8	2.41	5.2
Non-Ideal Gases				
Li	147.10	3220	26.76	5.5
F	6.62	144	1.20	5.5
Al	293.40	7850	65.23	4.5
Fe	349.60	8500	70.64	4.9
Cs	67.74	1938	16.10	4.2
Hg	59.23	1750.1	14.54	4.1
Au	334.40	7250	60.25	5.6
Substance				
Water (H_2O)	40.65	647	5.38	7.6
Ammonia (NH_3)	23.35	405.5	3.37	6.9
Diatomic Gases				
H	0.90	33.2	0.28	3.3
Ne	5.57	126	1.05	5.3

Critical Temperature was taken from website: [7] wikipedia.org/critical point_(thermodynamic) Jan 2011

Chapter 12: **Rates, and Vapor Pressure**

It was previously determined that the probability of a molecule vaporizing could be written strictly in terms of temperature: eqn 11.7: $P'_{(v)} = B\exp(-T_b/T)$. Eqn 11.7 was based upon the probability of a given molecule extracting enough energy from its six neighboring liquid molecules, in order to vaporize. There is no transgression in logic for the probability of a group of molecules vaporizing. Specifically, both the numerator and denominator within the exponential are simply multiplied by the number of molecules vaporizing.

Before we do, the symbol for the number of molecules, N needs to be addressed! Up to this point tradition has been followed by using "N" to signify the number of molecules. However, when dealing with rates at which molecules change states, there is the problem of the letter "N" also signifying the *normalization constant*. In order to avoid the headache of having N, signifying two different variables and/or constants in the same equation, herein the variable "X" will now signify the number of molecules, while N now becomes the *normalization factor*. Sorry for any confusion.

Consider an ensemble of X ideal molecules vaporizing through a liquid-gas interface. Assuming that the required energy per molecule is independent of the size of X, then the probability of the X molecules vaporizing becomes:

$$P'_{(v)} = B\exp(-XT_b/XT) \quad 12.1$$

Of course eqn 12.1 is simply a version of kinematic probability: eqn 11.7: $P'_{(v)} = B\exp(-T_b/T)$

Vaporization Rate at Tb

The subscripts "$_{(l \rightarrow g)}$" signifies evaporation/vaporization. The rate $[J_{(l \rightarrow g)}]$ at which molecules evaporate through the interface is a function of the probability $[P'_{(v)}]$ of vaporization, some normalization factor (N), as well as, the concentration ($[C]$) of the molecules in question. The vaporization rate, being:

$$J_{(l \rightarrow g)} - N[C]\exp(-T_b/T) \quad 12.2$$

For a pure liquid, the concentration is 100% i.e. unity: $[C] = 1$, therefore:

$$J_{(l \rightarrow g)} = N\exp(-T_b/T) \quad 12.3$$

For latent heat, the normalization factor, N, is considered to be constant, throughout some thickness, D, as is illustrated in Fig 12.1. A more precise analysis would consider that the closer a liquid molecule is to the surface, the more likely that it will vaporize.

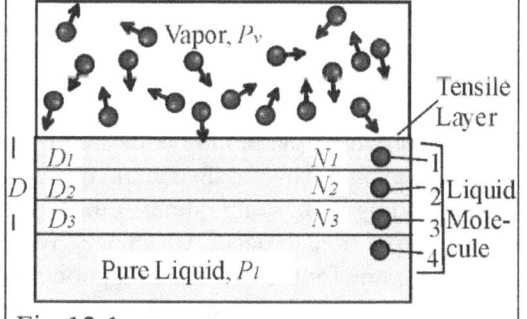

Fig 12.1 Shows a pure liquid with a vapor above it. D is the thickness beyond which no liquid molecules can vaporize.

Remember herein it is assumed that in order for vaporization from the bulk liquid, then a molecule that can extract energy from its six neighbors, would be the most likely one to do so. Hence, the likelihood of molecules vaporizing is going to be greatest for those liquid molecules located just below the tensile layer.

More precisely: The thickness D can be separated into successive layers, (D_1, D_2, D_3), as is shown in Fig 12.1. Each layer having their own value for N, e.g. N_1, N_2, N_3 etc. such that: $N_1 > N_2 > N_3$. Therefore, the rate of vaporization from each successive layer would progressively decrease, as the distance from the tensile layer increases. The total depth D is the limit, beyond which no molecules can escape the liquid.

For example: Let us say that there are 3 successive layers, from which liquid molecules can escape. Then:

$$J_{(l \to g)1} = N_1[C]\exp(-T_b/T) \qquad 12.4$$

$$J_{(l \to g)2} = N_2[C]\exp(-T_b/T) \qquad 12.5$$

$$J_{(l \to g)3} = N_3[C]\exp(-T_b/T) \qquad 12.6$$

The total rate at which liquid molecules become vaporous molecules would be:

$$J_{(l \to g)} = J_{(l \to g)1} + J_{(l \to g)2} + J_{(l \to g)3} \qquad 12.7$$

The more general form for the total vaporization rate through a liquid/gas interface would consider "i" successive layers, which would be written:

$$J_{(l \to g)} = \sum_i J_{(l \to g)i} \qquad 12.8$$

Eqn 12.3 can still be used to define the evaporation rate by contemplating that N is a constant over some prescribed depth (D), i.e., the value of N is the average: $N = \sum_i N/i$.

Normalization Factor (N)

The normalization factor can be thought of as helping define the rate of occurrence over a given time frame. To reiterate, as was discussed in the previous chapter: Imagine that the liquid is at its boiling point, hence from an energy perspective all the bulk liquid's molecules could vaporize. But this does not happen. Certainly, there is a probability that one or more neighbors have the required mean accessible energy, but the other neighbors may possess some other values, or simply have their net motion in the wrong direction. The combinations and permutation become endless. The fact that the vaporizing molecule will only break its bonds with the liquid when the sum of accessible energy from all of its neighbors, is enough, and directionally correct, as to break the liquid molecule's bonds, means that vaporization is never some instantaneous process. As was previously discussed; to what extend this is due to B vs N may be subject of debate, requiring more clarification in the future.

Furthermore, once the liquid bonds are broken, then the unbound liquid molecule may (depending upon the actual path) have to then attain more energy from its neighboring vibrating liquid molecules without rebinding to those liquid molecules, which involves further probabilities. All of this requires some duration of time to occur. Ultimately, the point is that in many situations the normalization factor gives a time frame for the event, based upon some series of other probable possibilities. If you prefer: The normalization factor (N) is system dependent variable, which puts the rate for a given system, into an observable time frame.

Pressure Dependence & Normalization Factor

As the pressure increases, the normalization factor for vaporization should decrease. Specifically, a pressure increase reduces the rate of vaporization. E.g., consider a liquid molecule located near the interface, which is constrained by a combination of the liquid's cohesive forces, and the pressure exerted upon the liquid. As the pressure increases, the physical constraint upon a given liquid molecule increases. Therefore, both the ability, and the depth (D) from which a given liquid molecule can readily escape from the bulk liquid, should decrease, as the pressure increases. Obviously, N is a function of P.

The above can be readily visualized. Consider, Fig 12.2 A: Illustrated are the average positions and cohesive forces acting upon a dozen liquid molecules located near a liquid-gas interface. In order for liquid molecule #11 to escape from some depth D through the interface, it collides with other molecules without rebinding, and perhaps certain general motions of the liquid molecules located above are needed, e.g. shown in Fig 12.2 B, molecules #1,2,5 and 6 are moving left while molecules #3,4,7 and 8 are moving right thus allowing liquid molecule #11 to readily escape from the liquid into the gas above. Furthermore, as $P\uparrow$, the likelihood of a given liquid molecule escaping through D, decreases i.e. the vaporization rate's normalization factor, N, decreases with increasing pressure.

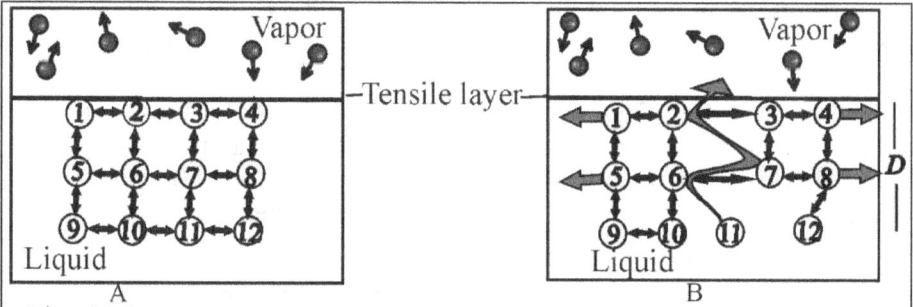

Fig. 12.2   Shows a dozen liquid molecules located near a tensile layer. In (A) the liquid's cohesive forces hold molecules together. In (B) molecule 11, which is located a distance D from the interface, then extracts sufficient energy to break its bonds and then escape through the interface. Possibly helping molecule 11 escape is the motion of molecules 1,2,5 & 6 to the left, while molecules 3,4,7, & 8 move right.

The above also improves our understanding as to why a liquid's boiling temperature increases with increasing pressure. Specifically, a pressure increase increases the physical constraints upon the liquid molecules, increasing the energy needed for vaporization.

Normalization Constant & Errors

If the *normalization factor* (N) is dependent upon all sorts of variables that define a system. Then, for the value of N in any two systems to be identical, both systems must be very similar. Strangely, scientists often wrongly visualize the normalization factor (N), as being part of some universally applicable constant. The reason resides in the method in which N is calculated. For example: We start out by saying that there exists some *normalization constant* (J_n), such that:

$$J_n = N[C] \qquad\qquad 12.9$$

The vaporization rate, as given by eqn 12.9, now becomes:

$$J_{(l \to g)} = J_n \exp(-T / T_b) \qquad 12.10$$

J_n can be empirically determined using graphically analysis. However, the act of determining J_n, is the act of treating J_n as some constant, thus J_n is the normalization constant, although it can only really be considered a constant for that particular system at those particular conditions. Again traditional conceptualization was based upon circle logic.

Consider the nature of the normalization constant that being, it can be obtained by plotting empirical data. Next consider that you are using an incorrect equation based upon an incorrect probability. Due to the power of the normalization process, your data can still be normalized. This explains how empirical data convinced researchers that eqn 1.12.1 was the correct probability function, when in fact eqn 1.12.7 is the more logical choice.

Normalization Constant Units

Before continuing let us take a step back and reconsider the equations for the vaporization rate, i.e. eqn 12.10. Since the vaporization rate is a number per unit surface area, then the units for: $J_n = N[C]$ must be per unit area $(1/m^2)$. Does this mean that the normalization factor (N) has units of per unit area?

Well that depends upon how the concentration $([C])$ is treated. Consider the concentration as a simple fraction, e.g. 90%=0.9, or say that the substance is pure and set: $[C] = 1$, and then the answer is yes. Conversely, if the concentration is expressed as some number per unit volume, then the answer is no. I.e.:

$$[C] = number/V \qquad 12.11$$

One could try to extend the thought process by considering that the interface has a surface area, A. Then, consider that liquid molecules can only escape through a distance D of liquid. Rather than writing the concentration, $[C]$, in terms of some number per unit volume, it is now written $[C]$ in terms of the same number but per unit area multiplied by the distance, D. Therefore, eqn 12.11 becomes:

$$[C] = number/(AD) \qquad 12.12$$

Vapor Pressure

Vapor pressure is the pressure exerted by the liquid molecules that possess enough energy to vaporize, and are located close enough to the interface etc. In reality the term is a tad fictitious because molecules in the liquid state do not actually exert pressure in the same manner as gas/vapor does. No matter this traditional conceptualization improves our understanding of a liquid-vapor interface.

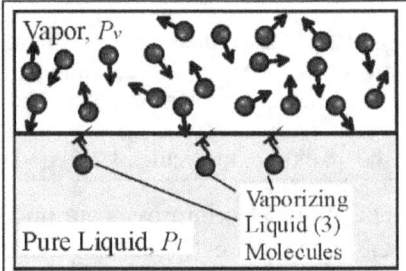

Fig 12.3 Shows a pure liquid with a vapor above it. Equilibrium exists between the number (3) of vaporous molecules condensing and the number (3) of liquid molecules vaporizing.

Let "Y" represent the number of vaporous molecules striking the tensile layer and then condensing into the liquid state. Let "X" represent the number of liquid molecules vaporizing. In Fig. 12.3, the system is drawn in equilibrium. E.g., at any instant, 3 vapor molecules are condensing, while 3 liquid molecules are vaporizing. Since: $X = Y$, therefore equilibrium exists. If $X > Y$, then the volume of the liquid would gradually decrease with time. Conversely, if $X < Y$, then the volume of the liquid would gradually increase with time.

In terms of vapor pressure:

1) If $X = Y$, then the liquid's vapor pressure equals the pressure of vapor that is located above the liquid.
2) If $X > Y$, then the liquid's vapor pressure is greater than the pressure of vapor that is located above the liquid.
3) If $X < Y$, then the liquid's vapor pressure is less than the pressure of vapor that is located above the liquid.

If the system is rigid and closed, and $X > Y$, then the pressure of vapor above the liquid will increase with time, as more liquid molecules vaporize, until the condition $X = Y$, is attained. Conversely, if $X < Y$, then the pressure of vapor above the liquid will decrease with time, as more vaporous molecules condense until the condition $X = Y$, is attained.

Furthermore, the number of molecules vaporizing per unit time increases as the surface area, A, increases. Similarly, the number of molecules condensing will increase. Thus, the simple act of increasing the interface's surface area, A, does not change the equilibrium condition of $X=Y$, although it changes the magnitude of both. However, starting off with the non-equilibrium condition of either, $X > Y$, or $X < Y$, and then increasing the surface area, A, will alter the time required to attain equilibrium.

Vapor Pressure Via Rate Equation

To approximate the number of molecules vaporizing (X) per unit time, start by realizing that each unit of surface area gives a vaporization rate that can be approx. by eqn 12.2: $J_{(l \to g)} = N[C]\exp(-T_b/T)$. Therefore, the determination of X per unit time through the surface area, A, becomes:

$$X = AJ_{(l \to g)} \approx AN[C]\exp(-T_b/T) \qquad 12.13$$

Vapor pressure can be considered in terms of, how much pressure is exerted by the vapors above the liquid under the equilibrium condition: $X=Y$. For equilibrium, eqn 12.13 can be rewritten in terms of the number of molecules condensing (Y):

$$Y = AJ_{(g \to l)} = AJ_{(l \to g)} \approx AN[C]\exp(-T_b/T) \qquad 12.14$$

Realizing that each molecule in the vaporous state that impacts an interface exerts a force upon that interface. Therefore, the total force exerted upon the interface's surface area (A) must be a function of the number of vapor molecules striking it, which is: Y. More specifically, if each vapor molecule exerts an average force, as given by: $\langle F_i \rangle$, then the total force (FT) exerted upon the interface would be:

$$F_T = \langle F_i \rangle Y = \langle F_i \rangle AN[C]\exp(-T_b/T) \qquad 12.15$$

Since: $P=F/A$, therefore the vapor must exert a pressure (P_v), upon the interface:

$$P_v = F_T / A = \langle F_i \rangle N[C]\exp(-T_b/T) \qquad 12.16$$

For a pure liquid, the concentration is set to unity: $[C]=1$, and eqn 12.16 becomes:

$$P_v = \langle F_i \rangle N \exp(-T_b/T) \qquad 12.17$$

Eqn 12.17 states: The pressure exerted by a vapor equals the rate of vapor molecules impacting the interface, times the mean molecular force exerted. Herein it is assumed that all molecules that impact the interface do condense, which may not be true.

Taking this a step further, remember that the normalization constant (N) has units of: $1/m^2$. Remember: $\langle F_i \rangle N = n'\langle P_i \rangle$, where n' is considered as some constant such that: $N = n'/A$. Note: A is area. In which case eqn 12.17 becomes:

$$P_v = n'\langle P_i \rangle \exp(-T_b/T) \qquad 12.18$$

Consider: $n'\langle P_i \rangle = C'P_b$, wherein C' is some constant and P_b is the boiling pressure.

$$P_v = C''P_b\exp(-T_b/T) \quad 12.19$$

We must not forget that when we derived 12.17, we stated that not all molecules that hit the interface necessarily would condense. So at this point all we can say is that C' must be determined from experiment, or if you prefer: $C'P_b$ can be considered as a constant. Either way once the constant is determined, we can use the *vapor pressure equation* to estimate the liquid's vapor pressure at various temperatures.

Traditional Approach: Clausius-Clapeyron Equation

A traditionally accepted mathematic model for vapor pressure as a function of temperature is the Clausius-Clapeyron equation[1], which can be written:

$$P_v = P_0 \exp[-L_{(l \to g)}/RT] \quad 12.20$$

where P_v = Vapor pressure, P_0 = Constant, $L_{(l \to g)} = Xl_{(l \to g)} = \Delta H_{vap}$ = Latent heat for a mole of molecules, and R = gas constant = 8.31 J/mol·K

Comparing eqn 12.19 to eqn 12.20: Beyond writing the constant in a slightly different manner, the real difference is what is written in the exponential. Which is also how we differentiated our new probability for vaporization (eqn 1.12.7), from the traditional probability (eqn 1.12.4). Again, the Clausius-Clapeyron equation suffers the same traditional detriment i.e. $L_{(l \to g)} >>> RT$, for all substances.

The astute reader should realize that if the traditional probability was used, followed by the same logic in going from eqn 12.13 to 12.19, then our final result would have been an equation very similar to the Clausius-Clapeyron.

The *vapor pressure equation* (eqn 12.19) can be rewritten in terms of a mole of molecules by using the ideal gas law at the boiling point: $T_b = P_b V_b / R$. Substituting into eqn 12.19, then in terms of P-V space, the vapor pressure becomes:

$$P_v = C' P_b \exp(-V_b P_b / RT) \quad 12.21$$

For eqn 12.21 to equal eqn 12.20 then: $L_{(l \to g)} = P_b V_b$, which is incorrect. The fact that the Clausius-Clapeyron equation has been used throughout the 20th century providing reasonable results reinforces the issues associated with using exponential equations, i.e. Boltzmann factor too often allows the wrong exponential to be wrongly normalized to empirical finding. In other words again circular logic is reinforced by math and the calculation of so-called constants, irrelevant to whether it be calculated by plotting empirical data or, other means.

Mixtures

Consider a mixture of two liquids, whose densities are similar so that neither liquid floats nor sinks. For such a mixture of liquids, each liquid's vaporization needs to be treated separately. Fig 12.4 illustrates a binomial liquid where the concentration of dark liquid molecules near the interface is 60%, while the concentration of light liquid molecules near the interface is 40%. No matter which liquid molecule is considered, the mean accessible energy, as defined by kT is going to be the same. However, the energy required as latent heat, may be different for each separate liquid. Furthermore, the fact that this is a mixture may alter the energy requirements for vaporization, since the bonding to neighboring molecules may vary.

As an approximation: Consider the normalization factor to be some constant over some depth, then using eqn 12.2 gives; the rate at which the light liquid molecules vaporize through the interface is:

$$J_{(l \to g)(light)} = N_{(light)}(0.40) \exp(-T/T_{b(light)}) \quad 12.22$$

Similarly, the rate at which the dark liquid molecules vaporize through the interface is:

$$J_{(l \to g)(dark)} = N_{(dark)}(0.60) \exp(-T/T_{b(dark)}) \quad 12.23$$

Understandably, if both normalization factors, and boiling temperatures, were the same for both the dark and light liquid molecules, then the proportions of light and dark molecules in the vaporous state would be the identical to their concentrations in the liquid state. For our example, 40% light molecules and 60% dark molecules, constituting the vaporous state, as is shown in Fig 12.4.

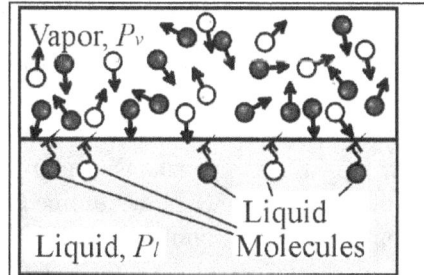

Fig 12.4 Shows a binomial liquid with a vapor above it. Equilibrium exists between the number (5) of vaporous molecules becoming liquid and the number (5) of liquid molecules becoming vaporous.

If the normalization factors were the same for both the dark and light liquid molecules, while their boiling points were not, then the expectation is that the contribution to the total vapor pressure would increase for those molecules, whose boiling point was lowest. Contributions to the total vapor pressure would be defined by the ratio of their evaporation rates, i.e. the ratio of light to dark molecules in the vaporous state would be:

$$Light : dark = [N_{(ligh6)}(0.40)\exp(-T/T_{b(light)})]/[N_{(dark)}(0.60)\exp(-T/T_{b(dark)})] \qquad 12.24$$

Eqn 12.24 gives the ratio of their vapor pressures, which traditionally was done using the Clausius-Clapeyron equation.

Vapor Pressure Equation

Another equation used for obtaining the vapor pressure is the Antoine equation, which is sometimes traditionally derived from the Clausius-Clapeyron equation. Again, instead of following tradition let us now derive an equation similar to the Antoine equation.

Starting with eqn 12.19: $P_v = C'P_b\exp(-T_b/T)$. Taking the log of both sides gives:

$$\log(P_v/C'P_b) = -T_b/T \qquad 12.25$$

Applying a law of logarithms, eqn 12.25 becomes:

$$\log(P_v) - \log(C'P_b) = -T_b/T \qquad 12.26$$

Which can be rewritten as:

$$\log(P_v) = \log(C'P_b) - T_b/T \qquad 12.27$$

We can say that the pressure at the boiling point is some constant (A) for a particular substance, therefore: $\log(C'P_b) = A$. Therefore, eqn 12.27 becomes:

$$\log(P_v) = A - T_b/T \qquad 12.28$$

Consider that the boiling point is a constant for each substance at 1 atm pressure: I.e. $T_b = B$, where: "B" is some constant. Therefore eqn 12.28 becomes:

$$\log(P_v) = A - B/T \qquad 12.29$$

The above resembles the two-coefficient version of the *Antoine equation*. It becomes interesting to ponder the exact validity of the Antoine equation, as certainly nobody has previously suggested that the constant B is the boiling temperature, or even related to it.

Rewriting eqn 12.29 as:

Kent W. Mayhew

$$B/T = A - \log(P_v) \qquad 12.30$$

In terms of temperature we can rewrite the two-coefficient version as:

$$T = B/[A - \log(P_v)] \qquad 12.31$$

Now the values of constants A, and B, are to be calculated from experimental data. It turns out that the two-coefficient version of the Antoine equation is not particularly accurate. Accordingly, a more commonly used equation is:

$$T = B/[A - \log(P_v)] \text{-C} \qquad 12.32$$

Even the above three-coefficient version of the Antoine equation is very limited, e.g. only valid for pressures with 10% of standard[7]. Even so, it is commonly used with the values for the constants A, B, and C, for various substances published in literature.

One may ponder why researchers have the need for the three-coefficient version of the equation. Well if you remember back to our discussion of the normalization factor (N) and how it must be a variable, which depends upon the system's parameters, e.g. pressure. You see the problem in developing a universal equation, when the reality is that N is actually a variable, which was foolishly treated as some constant. Of course herein this fictitious constant (N) was transformed into another constant (C').

Boiling Point vs Pressure

The problem with the fictitious constants is such that we cannot readily use any of the predetermined equations to calculate the changes to boiling points with pressure. For example we could start off with eqn 12.28 and move the variables so that it gave an equation that equated to the boiling temperature, e.g. arrive at an equation like: $T_b = (PV/R)[A - \log(P_v)]$. Which implies that the boiling temperature decreases with increasing pressure, which is NOT VALID as the opposite is true.

This can be taken a step further by saying: If the rate of molecules evaporating is simply a function of the probability, as defined by eqn 1.12.7, then combined with the fact that the latent heat of vaporization for water decreases with increasing pressure, one might be inclined to believe that the number of water molecules evaporating through the interface will increase with increasing pressure. Again this is not the case.

Water's boiling temperature increases with pressure, e.g. Table 12.1.

Table 12.1: Boiling Temperature of water versus Pressure

P_{sys} (bar)	0.8	1.0	1.15	2.8	3.5	4.0
T_b (°C)	93.5	99.6	116.9	131.2	138.9	143.6
W_l (kj/kg)	2274.1	2257.9	2210.8	2170.1	2147.4	2133.0

As previously discussed the pressure increase increases the force of constraint upon the liquid molecules hence higher temperatures are required for vaporization. This shows how strongly the fictitious normalization constant is influenced by pressure. Note: The boiling temperature of water as a function of pressure is often approximated by a limited three-coefficient version of the Antoine equation: $T_b = [1730.63/(8.07131-\log_{10}P)]- 233.426$. Wherein: T_b is in degrees Celsius, and P is in Torrs.

Closing Remarks

In this chapter, our new understanding was applied to vaporization arriving at similar but certainly different equations to those traditionally beheld. Since our approach was based upon fundamental principles, it certainly warrants consideration as a replacement method of analysis, as well as new equations with possible improved insights concerning the vapor pressures of liquids at various temperatures and pressures.

Reference:

1, http://www. Wikipedia.org/wiki/vapor_pressure (2011)

Chapter 13: **Solubility**

Solubility of Gases

Dissolved gas molecules in a liquid are shown in Fig. 13.1. Work is required for a dissolved gas molecule to change from an aqueous to a gaseous state. For an open system, the work required is related to the isobaric volume increase, i.e. increase Earth's atmosphere's potential. Remember, for a closed rigid system, the work results in an isometric pressure increase.

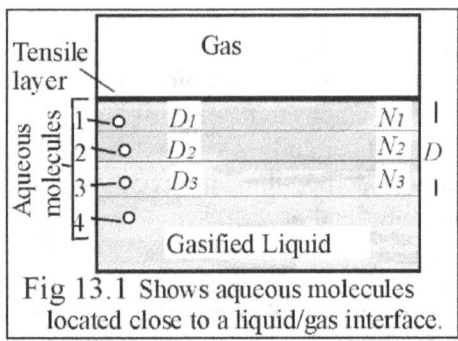

Fig 13.1 Shows aqueous molecules located close to a liquid/gas interface.

Now ponder: Should an aqueous molecule be considered as a liquid, or a gas? As a liquid there would be some bonding with the neighboring liquid molecules, thus it would have a mean kinetic energy defined by $3k\overline{T}$. As a gas, there should be no bonding, in which case the aqueous molecule should possess a mean energy of $3k\overline{T}/2$.

Assuming no bonding, then the aqueous molecule behaves as a gas. Let "$e_{(a \to g)}$" be the energy required for evaporation, that being the energy required for a molecule to go from an aqueous to a gaseous state. Then: $e_{(a \to g)}$ involves any work that is done. Note as was the case for latent heat of vaporization, in equilibrium no work is done, where equilibrium means: $N_{evaporating} = N_{condensing}$.

If the aqueous molecule does work when it becomes gaseous, then as it increases Earth's atmosphere's potential and/or thermal energy. In which case the liquid's temperature decreases i.e. the aqueous molecule extracts thermal energy from its surroundings, cooling it down by an infinitesimal amount (may be too small to measure). Firstly, an aqueous molecule is above its boiling temperature, thus there is no energy is required to break bonds. Secondly, the aqueous molecule has a mean energy of $3kT/2$, which is enough energy to displace the Earth's atmosphere. Furthermore, it would readily extract back any lost thermal energy through collisions with its surroundings.

Gaseous to Aqueous: No Work

When a gas molecule becomes aqueous then the Earth's atmosphere is now displaced downwards. Hence, part of the Earth's atmosphere must experience a change from potential into kinetic energy. Again no actual work is done, thus there should be no energy change as the gaseous molecule becomes aqueous. This assumes that there is no intermolecular bonding.

Rate of Aqueous Becoming Gaseous

Since, the energy requirements are small then the probability of an aqueous molecule extracting enough energy to become gaseous should be:

$$P'_{(a)} = B\exp(-\beta e_{(a \to g)}) = 1 \qquad 13.1$$

where the subscript "(a)" stands for aqueous and B is the proportionality constant.

Since the probability is unity, then the rate at which aqueous molecules become gaseous ($J_{(a \to g)}$) would be based upon eqn 12.2: $J_{(l \to g)} = N[C]\exp(-T_b/T)$, i.e.:

$$J_{(a \to g)} = N[C] \qquad 13.2$$

The normalization factor (N) times the concentration ($[C]$) basically defines how many aqueous molecules are close enough to the interface, and moving in the right direction, with enough energy to escape from the liquid. Again, the normalization factor could be considered as a constant over some distance, D. A more exacting analysis would break the thickness (D) into successive layers: D_1, D_2, D_3, as shown in Fig 13.1. Each layer would have its own value for: N, namely: N_1, N_2, N_3 etc., such that: $N_1 > N_2 > N_3$. Therefore, the rate of evaporation from each layer would progressively decrease, as the distance from the liquid's surface increases. The thickness (D), then defines the limit, beyond which no dissolved gases can escape the liquid. Either way, the value of β and $e_{(a \to g)}$ are considered as constant throughout the thickness (D). The general form for the equation defining the rate from the i_{th} layer becomes:

$$J_{(a \to g)} = N_i[C] \qquad 13.3$$

The total rate ($J_{(a \to g)}$) for aqueous molecules becoming gaseous would be the summation from the various layers, i.e.:

$$J_{(a \to g)} = \sum_i J_{(a \to g)i} \qquad 13.4$$

Eqn 13.4 can be rewritten as:

$$J_{(a \to g)} = \sum_i N_i[C] \qquad 13.5$$

Table 13.1: Solubility versus Temperature

Temperature (K)	Solubility : Carbon Dioxide (CO_2)	Solubility: Oxygen (O_2)	Solubility :Nitrogen (N_2)
288.15	2.095×10^{-5}	2.299×10^{-4}	1.386×10^{-5}
293.15	1.1618×10^{-5}	1.1645×10^{-4}	1.274×10^{-5}
298.15	1.774×10^{-5}	1.671×10^{-4}	1.183×10^{-5}
303.15	1.657×10^{-5}	1.457×10^{-4}	1.178×10^{-5}
308.15	1.562×10^{-5}	1.288×10^{-4}	1.047×10^{-5}

Table 13.1 gives the accepted solubility of 3 common gases in water as a function of temperature[1]. Clearly, the solubility of these gases decreases with increasing temperature. Based purely upon equation 13.5, it remains awkward to explain how the solubility of dissolved gas becomes a function of temperature, unless the normalization factor (N_i) is also a function of temperature.

Fig 13.2 illustrates a dissolved gas molecule, located near interface, which is surrounded by liquid molecules. If the aqueous molecule can squeeze by liquid molecules #6,7,2 and then it will become gaseous. Surmising that the normalization factor (N) is a function of temperature then an explanation for dependence could be based upon two factors.

Firstly, a liquid's intermolecular cohesive force decreases

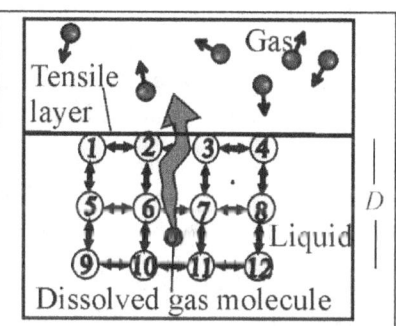

Fig. 13.2   Shows a dozen liquid molecules located near a tensile layer, surrounding a dissolved gas molecule. The liquid's cohesive forces prevent the dissolved gas from escaping.

with increasing temperature. Since the liquid's cohesive forces act as a force of constraint, a temperature increase should allow the aqueous molecule to increase its mean molecular volume. Remember, an aqueous molecule is considered to reside between neighboring cohesively bonded liquid molecules.

Secondly, the aqueous molecule's mean kinetic energy increases, as the temperature increases, thus increasing the vigor with which an aqueous molecule bounces off of its surrounding liquid molecules. Thus also increases the aqueous molecule's *mean effective volume*.

Due to the mean molecule volume increase, the aqueous molecule's scattering cross-section will increase, thus increasing the likelihood of the aqueous molecule interacting with the interface and becoming gaseous, as is shown in Fig 13.2.

Henry's Law

Henry's law defines a gas's solubility as a function of pressure (*P*), molar solubility (*S*) (mole/liter) and Henry's constant (k_H) i.e.:

$$S = k_H P \qquad\qquad 13.6$$

Henry's law traditionally states that the liquid's dissolved gas concentration ([*C*]) is proportional to the overlying gas's pressure. Any increase in [*C*] causes an increase in the rate, at which dissolved gases leave the liquid, as defined by eqn 13.2: $J_{(a \to g)} = NC$. This results in a new equilibrium.

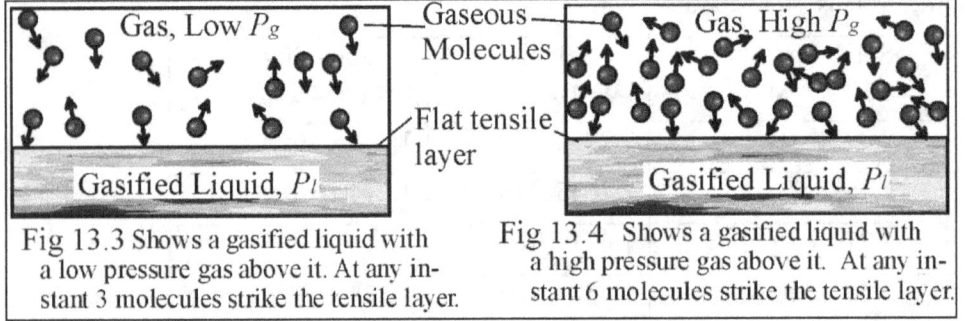

Fig 13.3 Shows a gasified liquid with a low pressure gas above it. At any instant 3 molecules strike the tensile layer.

Fig 13.4 Shows a gasified liquid with a high pressure gas above it. At any instant 6 molecules strike the tensile layer.

Investigating Henry's law further; fig 13.3 shows a gasified liquid below a gas, where at any instant there are 3 gaseous molecules striking the liquid's surface.

Fig 13.4 shows an increase to the pressure of the gas resulting in an increase to the number of gaseous molecules striking the interface (flux increase), i.e. from 3 to 6. Herein the addition of more gaseous molecules into an isometric system increased the gas's pressure. Similar results can be obtained by decreasing the volume occupied by a given number of gaseous molecules.

A gas's pressure can also be increased by increasing the gas's temperature, thus increasing the gas's mean molecular momentum. Herein, the mean force that each gas molecule exerts upon the liquid, increases, rather than the flux of gas molecules hitting the surface.

No matter how the pressure is increased, the result per unit time becomes: The number of gaseous molecules that impacts the liquid's tensile surface with enough momentum to become aqueous is directly proportional to the pressure. Herein, pressure is a measure of the *effective concentration* of gas molecules that interact with the liquid. The inherent problem with is that it only tells half of the story. Specifically, it ignores how pressure also acts a *force of constraint* upon the aqueous molecules in the liquid.

Reconsider the aqueous molecule located between liquid molecules, as was illustrated in Fig 13.2. Imagine that two forces constrain the aqueous molecule's motions and/or mean volume. The forces being:

1) the system's pressure. 2) The liquid's intermolecular cohesive forces. As previously discussed, constricting an aqueous molecule's motions and/or volume will affect the rate at which aqueous molecules escape the liquid.

Envisioning that the aqueous molecule's mean volume, is a function of pressure, the so too is its scattering cross-section. The implication becomes that the normalization constant (N) is also a function of pressure. This can conceptualized, as a combination of a decrease in the depth, D, from which an aqueous molecule can escape the liquid, therefore the normalization factor (N) decreases with increasing depth. The net result being; the rate, at which aqueous molecules escape the liquid, decreases as the pressure (depth) increases.

There is a need to separate pressure as a force of constraint from pressure as a measure of effective concentration. In order to understand just consider the following. Fig 13.5 shows a gas in equilibrium with a gasified liquid. What would happen if the liquid were enclosed with a membrane, as is shown in Fig 13.6, and then the pressure within the liquid was increased, while the overlying gas's pressure remained constant? Based solely upon the traditional interpretation of Henry's law, the answer would wrongly be nothing. Specifically, since the pressure of the gas above the liquid is the same in both Fig 13.5 and Fig 13.6, then the rate of gas molecules becoming aqueous molecules will be the same for both! Herein, pressure is a measure of the effective concentration, hence number of vapor molecules that interact with the liquid's surface!

However, the rate of aqueous molecules escaping the liquid will be lower in the pressurized liquid i.e. Fig 13.6. Herein pressure is a measure of the force of constraint upon the system. Ultimately, once equilibrium is re-attained then the dissolved gas concentration will be highest in the pressurized liquid, i.e. liquid under pressure. Accordingly:

(Dissolved gas concentration in Fig 13.6) > (Dissolved gas concentration in Fig 13.5).

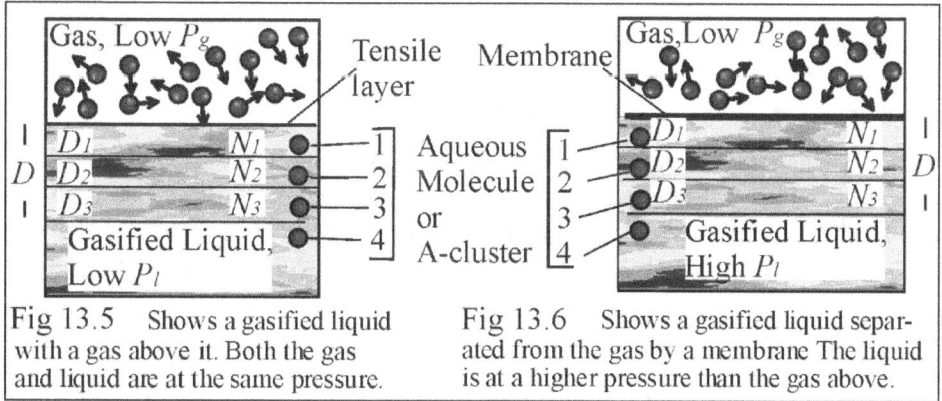

Fig 13.5   Shows a gasified liquid with a gas above it. Both the gas and liquid are at the same pressure.

Fig 13.6   Shows a gasified liquid separated from the gas by a membrane The liquid is at a higher pressure than the gas above.

In order to increase the aqueous concentration in a liquid one can either:

    1)   Increase the pressure of the gas above the liquid (effective concentration).

Or

    2)   Increase the pressure applied to the liquid (pressure as a force of constraint).

The reality becomes that Henry's law should be broken down into two parts, thus taking into account the above two separate components.

Gas Mixture

If dealing with a gaseous mixture, then P in eqn 13.6: $S = k_H P$, becomes the partial pressure of the gas in question. The total pressure (P_t) exerted is given by Dalton's law of partial pressures:

$$P_t = P_1 + P_2 + P_3 + ... + P_n \qquad 13.7$$

where P_t = total pressure. Subscripts "1 through n" represent each partial pressure for each component of the gas mixture. Each partial pressure signifies a unique type of gas, with a unique Henry's constant. Thus each type of gas has to be dealt with independently.

As far as the rate at which gaseous molecules condense into the liquid, the only concern is the partial pressure of that particular gas, i.e. as a measure of the effective concentration. Conversely, the rate at which dissolved gas molecules vaporize, concerns the total pressure, i.e. pressure as a force of constraint.

Solubility of Solids

When a solid dissolves in a liquid, work generally is not a significant component of any energy requirements for the solubility of solids. Fig 13.1 is a sketch for the solubility of several solids as a function of temperature. For most substances, the hotter a liquid's temperature is, then the greater the solubility of a given solid within that liquid will be. Interestingly, there exists a minority of solutes whose solubility decreases with increasing temperature, e.g. sodium sulfate: $T > 32.4$ °C.

The increase in solubility of solids as a function of temperature is traditionally poorly explained in terms of entropy. This author's preference would be one based upon energy requirements i.e. the increase in a liquid's temperature results in an increase in energy available to break ionic bonds.

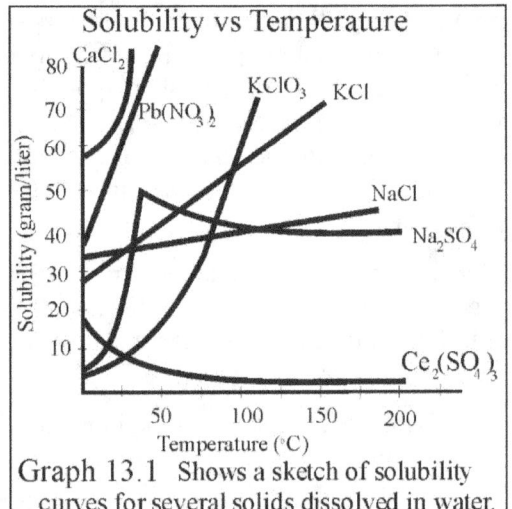

Graph 13.1 Shows a sketch of solubility curves for several solids dissolved in water.

Such an explanation may include the requirement of a specified frequency from the spectrum of thermal energy, or the spectrum itself. Either way, the required energy generally should be small in comparison to the surrounding liquid's thermal energy, hence may not be readily measured. This may explain why energy changes are not associated with solubility, e.g. the equation for the dissolving of copper chloride:

$$CuCl_2(s) \Leftrightarrow Cu^{2+}(aq) + 2Cl^-(aq)$$

Of course, there will be a myriad of other relevant conditions such as the concentrations of various ions in the water, and/or, the water's ph, all influencing the solid's solubility.

Closing Remarks

In this chapter, the solubility of gases was explained without the woeful need of considering entropy, as is traditionally done. As for the solubility of solids, there is a need for some more thoughts, and insights.

Chapter 14: **Physical Chemistry**

Physical chemistry demonstrates how a poorly conceived foundation results in an over-complication of a science. It remains troublesome that so many are so fully indoctrinated. Theory has been massaged so that it seemingly validates empirical data, although so much of what is claimed remains awkward to fully envision. Moreover, how does one now convert all that data, backing centuries of misinterpretation into something more readily understood? Estimated recovery time; generations.

There will be those who will argue that this author is not a chemist thus should leave well enough alone. To which, I fully agree. But I equally feel that the rethink has to start somewhere, so here I begrudgingly will attempt a start.

Traditional Thoughts on Entropy

Traditional interpretations: One often starts with this total differential:

$$\partial S = (\partial S / \partial \varepsilon)d\varepsilon + (\partial S / \partial v)dV + \sum\nolimits_{i=1}(\partial S / \partial N_i)dN \qquad \text{Trad A)}$$

where N_i is the number of particles involved in the reaction.

Applying the equations of state, Trad A) becomes the preferred reactions equation:

$$\partial S = d\varepsilon / T + (P / T)dV + \sum\nolimits_{i=1}(\mu_i / T)dN_i \qquad \text{Trad B)}$$

All the above are extensive variables then integrating gives for isothermal entropy:

$$S = \varepsilon / T + PV / T - \sum\nolimits_{i-1} u_i N_i / T \qquad \text{Trad C)}$$

where $\sum u_i N_i$ is taken to be the chemical potential.

Issues with the above include:

1) There is not a proper universal understanding for entropy, yet it remains so fundamental.
2)   Many chemical reactions are limited to isothermal processes i.e. heat bath experiments.
3) Our understanding of internal energy is not much better than entropy's.
4) A better understanding the differences between a system's ability to do work (PV) vs its total energy ($3PV/2$ for monatomic gas) e.g. enthalpy vs energy.

Reaction Rates

Chemical reactions generally when molecules/atoms are in thermal contact. The implication being that the molecules/atoms should be capable of movement, hence, for the most part reactions naturally occur when one or more of the reactants are either in the liquid, or, gaseous state. Of course solids can be crushed, and mixed, thus enabling reactions to occur.

Higher temperatures results in higher mean energetics. Hence for gases this generally means increased velocities. Furthermore, for polyatomic gases, solids and liquids, higher temperatures mean increased stronger molecular vibrations. Accordingly temperature increases generally result in an increased energy exchanged per collision, thus increasing reaction rates.

Chemical Reaction

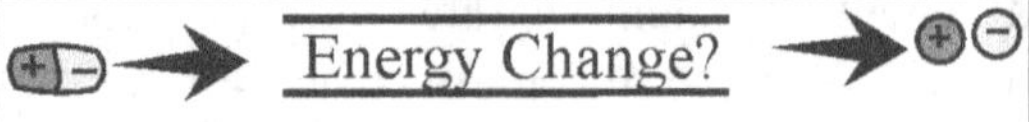

Fig 14.1 illustrates a simple chemical reaction; two charged atoms/ions on the L.H.S. combine to form a single polar molecule, as is shown on the R.H.S.

Fig. 14.1 Shows a simple chemical reaction.

Let "U" be the bonding energy, which includes the intermolecular and/or intramolecular bonding energy. If a system of N_i molecules undergoes a chemical reaction then the change in the system's chemical potential ($\sum u_i N_i$) becomes:

$$\sum_{i=1} \mu_i N_i = \Delta U \qquad 14.1$$

The change in bonding potential should include changes to both the intramolecular bonding energy and/or the intermolecular potential/bonding energy.

Enthalpy of Reaction

Imagine that a reaction occurs resulting in a volume increase, which displaces our atmosphere. Next consider the enthalpy relation: $H = \varepsilon + PV$. For a chemical reaction the internal energy (ε) is considered to be based solely upon bonding energy/potential (U), in which case the enthalpy relation becomes:

$$H = U + PV \qquad 14.2$$

Differentiating eqn 14.2 gives the change to the enthalpy, hence:

$$dH = dU + d(PV) \approx dU + VdP + PdV \qquad 14.3$$

Understand that $d(PV) \approx VdP + PdV$ is actually only a valid approximation for infinitesimally small change. Less obvious is that eqn 14.3 assumes the reaction to be isothermal! For an isobaric reaction, eqn 14.3 becomes:

$$dH = dU + PdV \qquad 14.4 \text{ (a)}$$

As defined by eqn 14.4 (a), the enthalpy change is the change to chemical potential (dU) plus work (PdV) that is done onto the surrounding atmosphere during an isobaric isothermal reaction.

An Argument against Tradition

As stated in Chapter 1; The internal energy (ε) is accepted as being the summation of all microscopic energies, i.e. that being system's total energy. Consider eqn 1.8 ; $H = \varepsilon + PV$. If the pressure and volume are a direct result of those very microscopic internal energies then the enthalpy relation is arguably illogical since the mechanical parameters (PV) are added to the total internal energy in order to give the systems total enthalpy! So what is enthalpy *total energy* plus PV of the system? No way!

Ignoring the above problematic reality now reconsider eqn 14.4 (a). It states that the only change to a system's internal energy is the change to its bonding potentials (dU), which sounds fine until you ask what about eqn 14.2? It lacks any real validity because to say that the system's total energy is solely based upon their chemical potential (U) also is non-sensible.

However, recognizing that work is done external to the system, therefore expanding systems tend to do work onto our atmosphere ($W = P_{atm}dV$) hence P in eqn 14.4 is actually the atmospheric pressure (P_{atm}). Now we can properly rewrite eqn 14.4 (a) as.

$$dH = dU + P_{atm}dV \qquad\qquad 14.4$$

As previously discussed: dV can be considered as the expanding system's volume increase, however since the expanding system is a subsystem of the atmosphere, then dV is equally the atmosphere's volume increase. Eqn 14.4 has a certain logic that neither eqn 14.2 nor 14.4 (a) nor eqn 1.8 possess.

Since expanding systems tend to do work onto the atmosphere, therefore insulated expanding system experience a temperature decrease. Hence isothermal eqn 14.4 is really limited to idealistic expanding system surrounded by heat baths, where thermal energy that is freely given into the expanding system. Accordingly, eqn 14.4 still remains conceptually limited.

Heat of Reaction

"*Heat of reaction*" signifies the amount of thermal energy exchanged between a system and its surroundings during an isothermal chemical reaction. For an isolated system that exchanges neither energy, nor matter, with its surroundings a reaction results in a systematic energy change. The classification of reactions depends upon whether eqn 14.4 is positive or negative, e.g.:

1) An *exothermic* chemical reaction releases energy onto its surroundings: $dH < 0$, hence the reaction can occur spontaneously, if and when the reactants are in thermal contact.

2) An *endothermic* chemical reaction requires energy onto its surroundings: $dH > 0$, hence energy is required for the reaction to occur, when the reactants are in thermal contact.

In practice one generally calculates how much energy would be required to restore the system to its initial state based upon by its temperature change multiplied by its heat capacity i.e. one of eqn 1.28; $c_{p'} = (1/m)(dQ/dT)_P$, thru eqn 1.31: $C_v = (1/n)(dQ/dT)_v$.

Discussion Concerning a System's Energy and What We See of a Reaction

What is the energy contained within a system? As previously stated, it is not actually defined by the enthalpy relation: $H = \varepsilon + PV$. I.e. a monatomic ideal gas's total energy is kinetic (E_T): $E_T = 3PV/2$. Accordingly, a system's total thermal energy (E_I) is defined in terms of its isometric molar heat capacity (C_v), temperature (T), and number of moles (n) could be written in terms of eqn 5.6: $E_T \approx nC_vT$.

Remember eqn 5.6 assumes that the heat capacity is constant through all temperature regimes and, this may not necessarily be the case i.e., eqn 5.6 is an approximation, which is valid for most temperatures that we readily experience. Differentiating eqn 5.6 determines an isometric system's total thermal energy change:

$$dE_T = nC_v dT \qquad 14.5$$

Is it safe to assume that change to the bonding potential (dU) has no temperature dependence? If the bonding potential is solely related to the mean distance (r) between bound molecules then there should be little temperature dependence for condensed matter and/or isometric gases. Accordingly the total energy (E_{total}= binding potential plus thermal energy) could be approximated in terms of the isometric heat capacity (C_v), i.e.:

$$E_{Total} = U + nC_v T \qquad 14.6$$

Differentiating eqn 14.6 gives the change to total energy:

$$dE_{Total} = dU + nC_v dT \quad 14.7$$

Exothermic Reaction

Imagine that infrared phonons/photons are released by an exothermic reaction in a fully insulated isometric system, as illustrated in Fig. 14.2. Such phonons/photons should be absorbed by the surrounding condensed matter thus raising the system's temperature.

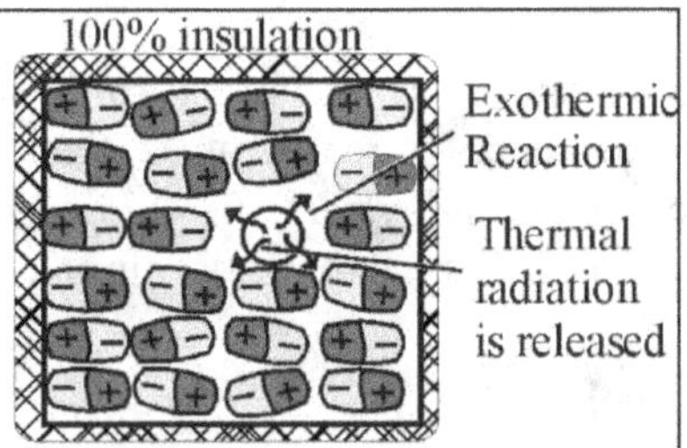

Fig. 14.2   Shows an exothermic reaction, emitting thermal radiation which gets absorbed thus increasing the system's temperature.

However, the total energy of the system does not change. Specifically, the lowering (change) of chemical potential energy simply resulted in the system's temperature increase. For this isometric reaction:

$$dU = -nC_v dT \quad \text{and} \quad dE_{Total} = 0 \qquad 14.8$$

In the above exothermic case, the change in chemical potential energy is negative ($dU < 0$), hence the system's temperature increase is positive as defined by:

$$dT = -dU / nC_v = |dU| / nC_v \qquad 14.9$$

Next imagine the case of a system of a dilute gas, wherein the system's walls absorb the energy from an exothermic reaction. Again, the system's temperature (including walls) should increase, assuming that the temperature change is actually measureable. Accordingly, for isolated rigid systems exothermic reactions result in an increase to the system's temperature i.e. $dT \uparrow$, $dU \downarrow$, $dE_{total} = 0$!

If the system is not insulated and is in thermal contact with a heat bath/sink then the above processes may appear isothermal. For this seemingly-isothermal exothermic reaction: $dU < 0$,

$dT = 0$, and $dE_{Total} < 0$. Remember the surrounding atmosphere can be considered as being an isothermal heat bath/sink for most quasi-static processes.

Endothermic Reaction

Now consider an endothermic reaction in an insulated closed rigid system. Since the energy required for the reaction comes from the thermal phonons/photons contained within system then: $dU > 0$, and: $dT < 0$ (assuming that it is measurable), such that: $dE_{Total} = 0$. Again equations 14.8 and 14.9 hold true. Accordingly, for isolated rigid systems, endothermic reactions results in a system's temperature decrease i.e. $dT \downarrow$, $dU \uparrow$, $dE_{total} = 0$!

If the above system is not insulated and is in thermal contact with a heat bath/sink then the process may appear isothermal. For this case of an endothermic reaction: $dU > 0$, $dT = 0$, and $dE_{Total} > 0$. Again the surrounding atmosphere maybe an isothermal heat bath/sink for most quasi-static processes.

If a chemical reaction releases high-energy, short-wavelength phonons/photons, of the kind not readily absorbed by condensed matter, then the expectation is that these phonons/photons may pass through the system's matter (condensed and/or gaseous). Such high-frequency phonons/photons may also interact with matter in other ways, e.g. be absorbed by another reaction, or strike a high-energy photon detector, allowing us to measure it.

Consideration of Work

It must be emphasized that whether the reaction is exothermic or endothermic, any expanding system that does work onto their surroundings such as upwardly displacing of our atmosphere (lost work), then that process is irreversible. Which is to say that the reverse reaction will not involve the same magnitude of energy, as the initial reaction had?

Accordingly: If the forward endothermic reaction involves work, i.e. the heat of reaction is defined by isobaric enthalpy change (dH or ΔH) then this differs from reverse reaction, i.e. the heat of reaction is now the negative change in bonding potential (dU or ΔU) rather than the negative of enthalpy! I.e.:

Forward heat of endothermic reaction that does work onto the atmosphere: $\Delta H = \Delta U + P\Delta V$
Reverse heat of the above endothermic reaction: $-\Delta U$

The above is really no different than when the differences between isometric and isobaric heat capacities of gases was discussed back in Chapter 5.

Conversely: If the forward exothermic reaction involves work, i.e. the heat of reaction is defined by isobaric enthalpy change (dH or ΔH) then this also differs from reverse reaction, i.e. the negative change in bonding potential (dU or ΔU) plus any work that is done! I.e.:

Forward heat of exothermic reaction that does work onto the atmosphere: $\Delta H = -|\Delta U| + P\Delta V$
Reverse heat for the above exothermic reaction = ΔU

Traditional Assertion of Work into a System

It is of interest that in traditional physical chemistry you may read about *pressure-volume work* (*PdV*) being done onto the surrounding atmosphere but there seems to be little understanding that this means the process is irreversible. Remember that in terms of the expanding system (chemical reaction) that the work is negative (*W=-PdV*), i.e. a loss of energy by expanding system, while in terms of the surrounding atmosphere this work is positive (*W=+PdV*), i.e. a potential energy gain.

To improve our understanding, consider the following gaseous reaction of carbon monoxide and oxygen being the reactants with the product being carbon dioxide:

$$2CO(g) + O_2(g) \rightarrow 2CO_2(g)$$

Assume the gases are ideal then the volume of the products must be less than the volume of the reactants. Hence, if performed in an isobaric calorimeter at constant temperature of 298 K, one attains an enthalpy change: $\Delta H = -566$ kJ.

The traditional argument is based upon the ideal gas law and writing the work done onto the system as; $P\Delta V = RT(n_f - n_i)$. And since two is the number of moles of products (n_f=2), while three is the number of moles of reactants (n_i=3), then $P\Delta V = -2.5$ kJ. Accordingly, based upon traditional eqn 14.4 (A): $\Delta H = \Delta U + P\Delta V$, then; $\Delta U = -(566 - 2.5) = -563.5$ kJ. Note in these considerations, the vibrational energy (in carbon dioxide) is seemingly forgotten or perhaps it is considered as part of the bonding energy. This will be addressed in the following section.

The above wrong assertion that the system's volume decreased means that work is done onto the system/reaction by surrounding atmosphere is unpalatable. The reality is that no work is ever done into a shrinking system by the surrounding atmosphere! Rather the surrounding atmosphere would simply experience a change of some of its potential energy into kinetic energy, due to such a volume decrease!

So how do we resolve the above issues? One must understand that the change in bonding potential (ΔU) is generally an isometric consideration, in which case the pressure decrease is due to the reduction in number of moles of gas. If this occurs in an insulated system, then there should be a temperature decrease due to any natural *P-T* system relationships, as discussed in Chapter 7!

Seemingly the above reaction shoes product was carbon dioxide warrabnts a rethink. I will leave it up to those who deal with such matter to properly fill in all of the numbers as I remain unsure as to what exactly and/or how the measurements were taken. However I will now present an approach that at least makes more sense to this author.

Reaction & Isothermal Change to the Number of Gaseous Molecules

Investigating the previous discussed formation of carbon dioxide by starting with the reverse reaction that does work.

Consider that State 1 is a sufficiently-dilute diatomic gas and that becomes a dilute monatomic gas, State 2, while all enclosed by walls. Herein, the number of gaseous molecules doubles from N to $2N$. If the process is isobaric, then the volume occupied by State 2, will become twice that occupied by State 1, as is illustrated in Figures 14.3 and 14.4. Next consider that the systems are surrounded by a heat bath hence they remain both isothermal and isobaric but not isometric.

In Chapter 2, the total translational & rotational plus vibrational energy associated with a contained gas was given by eqn 2.19: $E_T \cong NkT(n''+1/2)$. As a reminder: The validity of eqn 2.19 is really limited to smaller polyatomic gases i.e. $n''>4$ as they generally would be a rough approximation at best for larger polyatomic molecules, as was discussed back in Chapter 5.

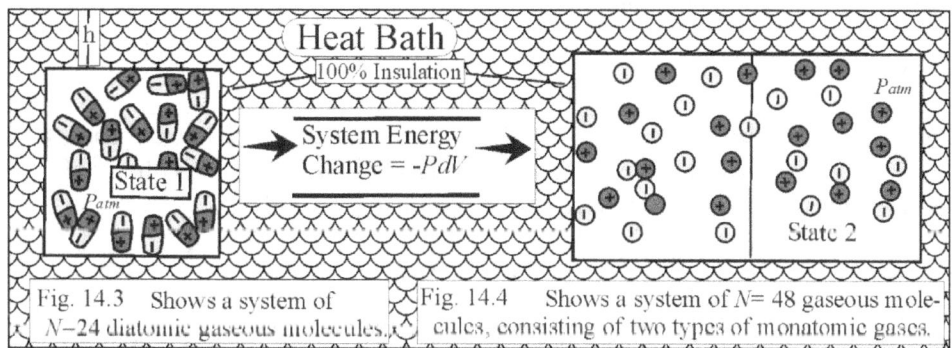

Fig. 14.3   Shows a system of N–24 diatomic gaseous molecules.

Fig. 14.4   Shows a system of N= 48 gaseous molecules, consisting of two types of monatomic gases.

For the products: N monatomic ($n''=1$) gas molecules, which have no vibrational energies, then the gas' translational & rotational energy, as defined by eqn 2.19 is: $3NkT/2$. But there are $2N$ monatomic gas molecules, hence the total kinetic energy of the products is:

$$E_{T(prod)} = 3(2N)kT/2 = 3NkT \qquad 14.10$$

For the reactants: Diatomic gases ($n''=2$) consisting of N gas molecules, the gas' translational & rotational plus vibrational energy is also defined by eqn 2.19. Hence the total energy of the reactants was:

$$E_{T(react)} = 5NkT/2 \qquad 14.11$$

Therefore, for a reaction wherein the reactants are diatomic gas, while the products are monatomic gas molecules, then based purely upon kinematics (translational & rotational plus vibrational energy), the required energy for an isobaric isothermal process is:

$$E_{T(prod)} - E_{T(react)} = 3NkT - 5NkT/2 = NkT/2 \qquad 14.12$$

What about work? The isobaric volume increase means that isothermal work ($W_{atm} = P_{atm}dV$) is done when the process involves the displacement of Earth's atmosphere. The work required to displace the weight of the overlying portion of the heat bath can be ignored if the heat bath is shallow enough i.e. $h \approx 0$.

A sufficiently-dilute contained gas can obey both Avogadro's hypothesis and the ideal gas law. For such a gas the volume change equals the original volume of State 1, i.e.: $dV = V_{(State1)}$, then the work that was done by the expanding closed system of gas becomes:

$$W_{atm} = P_{atm}dV = NkT \quad 14.13$$

Therefore, the energy required for both work done and changes to kinematics, would be obtained by adding eqn 14.13 to eqn 14.12, which gives the total required kinematic energy as.

$$E_{T(required)} = NkT + NkT/2 = 3NkT/2 \quad 14.14$$

The actual total energy change required in going from a diatomic to monatomic gas, would also have to consider changes to bonding potential (U) i.e. change to potential energy change of a cloud of charged gas plus/minus intermolecular bonding:

$$E_{T(required)} = 3NkT/2 \pm [U_{(monatomic)} - U_{(diatomic)}] \quad 14.15$$

Conversely, if the product was a diatomic gas while the reactants were two monatomic gases, then no work would be done: $W_{atm} = 0$ because the volume of products is less than the volume of reactants. Therefore, the total energy change associated with that reaction should be:

$$E_{T(diatomic)} - E_{T(monatomic)} = \pm [U_{(diaatomic)} - U_{(monatomic)}] - NkT/2 \quad 14.16$$

If there are no changes to chemical potential energy, then the only energy change is due to kinematics, in which case the energy change is:

$$E_{T(prod)} - E_{T(react)} = E_{T(diatomic)} - E_{T(monatomic)} = -NkT/2 \quad 14.17$$

And the negative sign means energy is released i.e. would be spontaneous. Remember the reason that the magnitude of the reverse reaction does not equal the original reaction is that no work was done in the reverse reaction. Of course we could go on and on. These basic principles that can be applied to other polyatomic gases!

Non-Isothermal Reactions

Now consider the same reaction and expansion but instead of the heat bath, the experimental system is fully insulated, as is illustrated in Figures 14.5 and 14.6.

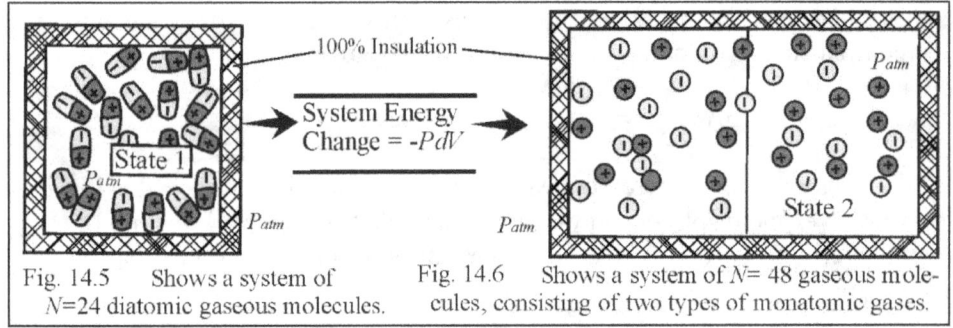

Fig. 14.5   Shows a system of $N=24$ diatomic gaseous molecules.

Fig. 14.6   Shows a system of $N= 48$ gaseous molecules, consisting of two types of monatomic gases.

Again the kinematic energy of the monatomic gases (product) is defined by: $E_{T(prod)} = 3NkT$, while the diatomic gas's (reactants) kinematic energy is defined by: $E_{T(react)} = 5NkT/2$. Therefore the change in kinematics is given by eqn 14.12: $E_{T(prod)} - E_{T(react)} = NkT/2$. Or is it? Eqn 14.12 assumes that the process is isothermal but as the system expands it does work, hence its temperature decreases.

If the reaction is considered to only involves kinematics then:

$$E_{T(react)} = E_{T(prod)} + P_{atm}dV \qquad 14.18$$

Substituting in gives:

$$5NkT_i/2 = 3NkT_f - P_{atm}dV \qquad 14.19$$

Assume that the system's initial temperature equals that of the surroundings, then based upon the ideal gas law: $P_{atm}dV = P_{atm}(V_f - V_i) = NkT_i$. Therefore eqn 14.19 becomes:

$$5NkT_i/2 = 3NkT_f - NkT_i \qquad 14.20$$

Based solely upon kinematic energy one might be inclined to write:

$$3NkT_f = 5NkT_i/2 - NkT_i = 3NkT_i/2 \qquad 14.21$$

Solving gives:

$$T_f = T_i/2 \qquad 14.22$$

Remember eqn 14.22 is an interesting result but it does NOT consider the energy associated with change to chemical potential energy (U). If the potential energy were considered then eqn 14.21 could be rewritten:

$$U_f + 3NkT_f = U_i + 3NkT_i/2 \qquad 14.23$$

Certainly the temperature change is not as readily determined as it was for the case of no chemical potential energy change.

Note: If the reaction was initially spontaneous, then there may be a point where the temperature has declined so much that there is no longer sufficient thermal energy to drive the reaction. This differentiates from the same process in contact with a heat bath where thermal energy is freely extracted from the heat bath. Remember often the surrounding atmosphere may enable a reaction to continue by maintaining the isothermal nature of that process, e.g. for many quasi-static reactions.

Another Discussion

Traditionally in chemistry one often talks about the energy of reaction versus randomness. Of course randomness is not a particularly scientific term i.e. it lacks clarity. No matter, in the above, it was discussed how a reaction that involves volume increase requires work while its reverse reaction does not. Hopefully you will take this heart, realizing that concepts involving any direct association of randomness with energy are based upon ill-conceived traditional conscripts, similar to those beholden onto entropy.

Endothermic Reaction and Work by Heating

An endothermic reaction often requires energy from an external source. Fig 14.7 shows System 1 in State 1 being heated, thus providing a total thermal energy ($Q_1 = dE_{T1}$). The result: System 1 is transformed into State 2, wherein a volume increase has occurred hence work (W) is done as shown in Fig.14.8.

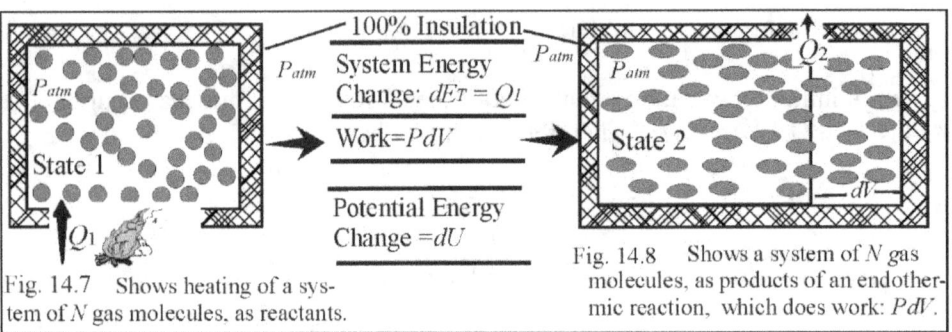

Fig. 14.7 Shows heating of a system of N gas molecules, as reactants.

Fig. 14.8 Shows a system of N gas molecules, as products of an endothermic reaction, which does work: PdV.

Instead of thinking in terms of changes to kinematics, the view in terms of heat capacities will be contemplated. Consider that the temperature of State 2 is higher than that of State 1, therefore the thermal energy required for the temperature increase is: $E_T = nC_p dT$. Start by assuming that no energy leaves State 2: $Q_2 = 0$. Since the isobaric heat capacity (C_p) includes the work done onto the atmosphere, the increase in the energy in going from State 1 to State 2 is:

$$Q_1 = dE_{Total1} = nC_p dT + dU \qquad 14.24$$

The isometric heat capacity (C_v) does not include the work done/lost, i.e. $W = P_{atm}dV_{atm}$ onto our atmosphere. Therefore in terms of isometric heat capacity eqn 14.24 can be rewritten as:

$$Q_1 = nC_v dT + P_{atm}dV_{atm} + dU \qquad 14.25$$

If energy is extracted from State 2: $Q_2 > 0$, then eqn 14.24 becomes:

$$Q_1 = nC_p dT + dU + Q_2 \qquad 14.26$$

And the maximum amount of energy that can possibly be extracted from State 2 becomes:

$$Q_2 = Q_1 - nC_p dT - dU \qquad 14.27$$

If this were a continuous process then the preference may be to completely rewrite the above in differential form. I.e. instead of the total heat (Q), derivatives (dQ) would be used, thus:

$$dQ_2 = dQ_1 - nC_p dT - dU \qquad 14.28$$

Helmholtz free energy

As was briefly discussed in Chapter 9, free energy is basically the energy that can be extracted from a system, hence is available to do work, or to be used as energy by another system. Substituting $dF_{(free)}$ for dQ_2 then based upon eqn 14.28, the maximum amount of energy that can be extracted becomes:

$$dF_{(free)} = dQ_1 - nC_p dT - dU \qquad 14.29$$

Of course this extracted energy (dF or dQ_2) could be used for work in which case dF could be replaced with W_{out} if the efficiency was 100%. Ultimately, $dF_{(free)}$ is a generality for the maximum amount of energy that can be extracted from a system.

In English the free energy as described by eqn 14.29 is for an energy input (dQ_1) minus the energy used in heating the system ($nC_p dT$) minus the energy required by an endothermic chemical reaction (dU).

Next consider the case of no heat input ($dQ_1 = 0$) hence eqn 14.29 becomes:

$$dF_{(free)} = -nC_p dT - dU \qquad 14.30$$

If the reaction releases energy then: $dU < 0$. Now the maximum amount of energy that can be extracted from this exothermic reaction becomes:

$$dF_{(free)} = dU - nC_p dT \qquad 14.31$$

Which can be rewritten in terms of molar isometric heat capacity (C_v) and work done onto the atmosphere [$P_{(atm)} dV$] as:

$$dF_{(free)} = dU - nC_v dT - P_{atm} dV \qquad 14.32$$

Eqn 14.32 mimics the traditionally accepted change to Helmholtz free energy (dF) for a chemical reaction. If there were no chemical reaction ($dU = 0$) then the maximum amount of energy/work that could be extracted becomes:

$$dF_{(free)} = -nC_v dT - P_{atm} dV \qquad 14.33$$

Compare eqn 14.33 to a traditionally accepted equation for changes to Helmholtz free energy (dF) i.e. eqn 15.13: $dF = -SdT - PdV$. Seemingly, now entropy (S) equals the number of moles time isometric molar heat capacity (nC_v). It seems that Helmholtz free energy is right but arguably for the wrong reasons. Note this will be discussed again in the next Chapter 15:

For exothermic chemical reaction Helmholtz free energy can be written as: $dF = -SdT - PdV + dU$, you also may see the change in bonding potential written as: $dU = d(N\mu)$, where μ is the molecular bonding potential. Note chemists often use dA instead of dF which is preferred by physicists for Helmholtz free energy.

Endothermic Reaction and Work

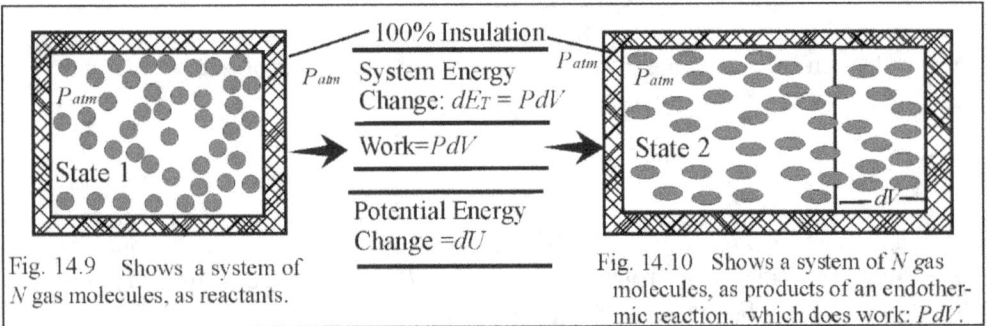

Fig. 14.9 Shows a system of N gas molecules, as reactants.

Fig. 14.10 Shows a system of N gas molecules, as products of an endothermic reaction, which does work: PdV.

Next consider that the energy for an endothermic reaction comes from with the thermal energy within a system: $dU > 0$, and: $dT < 0$ and that work is done onto the surrounding atmosphere. Now the system does work, i.e. an isobaric volume increase as is illustrated in Figures 14.9 and 14.10. Therefore:

$$dU + P_{atm}dV = -nC_vdT \qquad 14.34$$

And since the work is done externally to the system then:

$$dE_{Total} = d\varepsilon = P_{atm}dV \qquad 14.35$$

Eqn 14.35 describes that the change to a system's energy (A.K.A. internal energy) is actually the lost work that is done onto the surrounding atmosphere.

Reconsider that the energy required for an endothermic reaction comes from with the thermal energy contained within a system: $dU > 0$, and: $dT < 0$, but now the system experiences an isometric pressure increase as is illustrated in Figures 14.11 and 14.12.

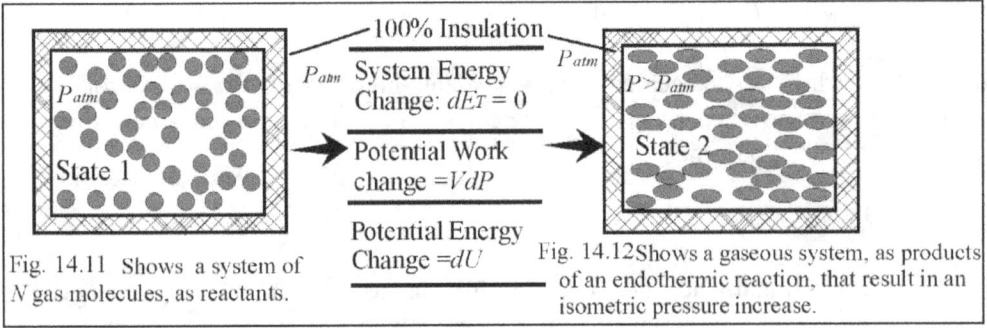

Fig. 14.11 Shows a system of N gas molecules, as reactants.

Fig. 14.12 Shows a gaseous system, as products of an endothermic reaction, that result in an isometric pressure increase.

There is no actual work done onto/into the rigid closed & insulated system! But there is a pressure increase that results in an increase in the system's potential to do work. As with any pressure increase there should be an associated temperature increase unless thermal energy is allowed to radiate away into the surroundings. Can we now write?

$$dU - VdP = -nC_vdT \text{ And } dE_{Total} = 0 \qquad 14.36$$

The problems with eqn 14.36 are:

1) The increase in potential to do work (VdP) is not the same as the change in the system's internal energy i.e. as defined by kinetic theory; an isometric ideal monatomic gas' energy change is: $3VdP/2$.

2) Does VdP actually address the closed & insulated system's temperature increase or is there more to its functionality with temperature? And is this functionality state dependent?

Accept that some thought followed by clarity is required. If these issues associated with eqn 14.36 are conveniently ignored, then the energy that could be theoretically extracted ($G_{extract}$) becomes:

$$G_{extract} = dU - VdP + nC_vdT \qquad 14.37$$

Now look at the similarities between Gibbs-Duhem relation ($G = Nd\mu - VdP + SdT$) and eqn 14.37. Again if $S=nC_v$ then eqn 14.37 corresponds to the Gibbs-Duhem relation. Certainly there is room for more contemplation and assertion by others.

Addressing the above problematic issue 1) for gases, one might be inclined to rewrite eqn 14.36 as:

$$G_{extract} = dU - 3VdP/2 + nC_vdT \qquad 14.38$$

Certainly eqn 14.38 does present some food for thought, but even if accepted one must bear in mind it is limited to gases that behave ideally, i.e. are sufficiently dilute and are not large polyatomic molecules and are in system's with walls i.e. most experimental apparatus. Of course if eqn 14.38 has any real validity then one might consider the problematic issue 2) along with 1).

What then about a temperature change associated with natural P-T system relations? More food for thought. What if VdP in eqn 14.37 actually relates to the temperature change associated with the natural P-T relationship. In other words it correlated to dT, perhaps by the ideal gas law for isometric systems i.e. $VdP=NkdT$.

Again at this point this author does not claim to hold all the answers, hence is hoping for insights by others.

Traditional Assertions of Enthalpy

It has been discussed how problematic the simple enthalpy relation can be. It must be emphasized that seemingly one never actually considers a system's actual enthalpy as defined by eqn 14.2. Rather we only consider the system's enthalpy change as defined by eqn 14.4 (A): $dH = dU + PdV$. In hindsight this alone should have set off alarms! As discussed in fact the only way to think would be in terms of eqn 14.4: $dH = dU + P_{atm}dV$

Certainly, limiting ones thought to enthalpy change diverted our attention from the fact that the enthalpy relation makes little sense, except for when change that includes any work being done onto the surrounding atmosphere. The issues go deeper because enthalpy change was allowed become such a dominate aspect for all chemical analysis even though it actual interpretation was fraught in deceptive logic, i.e. the so-called *standard enthalpy of reaction* (ΔH^o) is actually in practice the **enthalpy change of reaction**

Just consider Hess's law, which fundamentally states that enthalpy's are additive. It is not that the enthalpies are additive, rather it is a case of the enthalpy changes can be additive! It is easy to understand that the chemical potential changes (ΔU) in some multi-step reaction are additive but so too are energy losses attributable to the volume increases. Accordingly, if in one step of a reaction there is a volume increase, while at the same time the next step involves another volume increase then those volume increases are additive, i.e. the work done onto the surroundings is additive.

Similarly, if in one step of a reaction involves a volume increase, while another step involves a volume decrease then those volume may or may not be additive. It may depend upon whether the volume increase actually does work onto the atmosphere before the volume decrease occurs, or if the increase and decrease all happen in unison. Even so, if the net result of any step is an isobaric volume increase, then work is done onto surrounding atmosphere (*PdV*).

But there is never work done into the system/reaction by the surrounding atmosphere. Net system volume decreases, simply allows for the surrounding atmosphere to transform some of its potential energy into kinetic energy, which is actually equivalent to heat.

In the isometric systems where reactions result in a pressure change thus altering the natural *P-T* relationship within the system e.g. lower pressure in a system can lend itself to a temperature decrease. Conversely, if the pressure increases then higher system pressure can lend itself to a temperature increase.

Note: The standard state for enthalpy (change?) (ΔH^o) is 1 bar pressure (10^5 Pa), at some defined temperature which is taken to be for gases, *T*=298.15 k (25° C) unless otherwise stated.

It should be further understood that enthalpy changes can generally be experimentally analyzed, hence are readily measured when there is a volume increase. However the reverse experiment is not always readily accomplished. Specifically when there is a volume increase then what one measures is the enthalpy change. However when there is a net volume decrease then the measurement may lack experimental clarity, even though the result should simply be the change in bonding potential e.g. no work is done.

Cohesive Forces Changes in Liquids

Pressure change in gaseous state is readily envisioned but what about chemical reactions in the liquid state? Before answering the cohesive forces in the liquid state should also be reconsidered.

Chemical reactions often occur in the incompressible liquid state, e.g. an isometric system. Consider a liquid system of atoms reacting and becoming a system of dipole molecules. Specifically, State 1 consists of 24 positive charged atoms, and 24 negative charged atoms, as illustrated in Fig 14.13. While as shown in Fig 14.14, State 2 consists of 24 polar molecules.

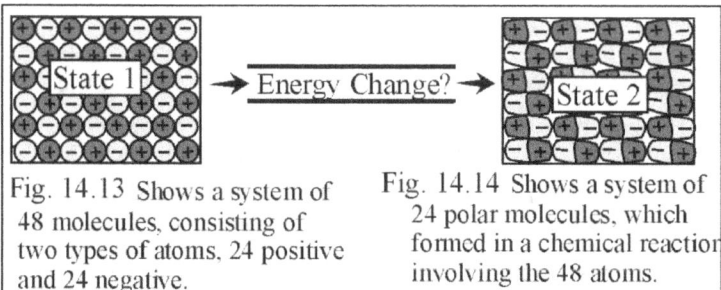

Fig. 14.13 Shows a system of 48 molecules, consisting of two types of atoms, 24 positive and 24 negative.

Fig. 14.14 Shows a system of 24 polar molecules, which formed in a chemical reaction involving the 48 atoms.

The intermolecular bonding energy in State 1 is due to the attraction between oppositely charged atoms/molecules. The bonding energy associated with State 2 is due to two factors. Firstly: The attraction between the newly formed molecule's dipole moments e.g. an intermolecular potential energy. Secondly: The *molecular bonding energy* when the molecule formed from a pair of atoms e.g. intramolecular energy. The two energies could be separated or left so that the bonding energy (U) is due to both.

Interestingly, changes in intermolecular potential energy (dU), should result in changes in the liquid's cohesive forces, which may actually alter the pressure felt within the liquid! This concept confounds traditional thermodynamics and will be discussed in the second book of this series (assuming it gets published) wherein tensile layers, and their associated cohesive forces are re-analyzed. Therein, it will be discussed why pressure changes due to changes in a liquid's cohesive forces do occur but would <u>not</u> be readily measurable with any pressure probe.

So although not readily measurable, the changes in intermolecular forces should result in a pressure change within that liquid, in which case the energy change associated with the isometric pressure change becomes: VdP. Again this may or may not, result in a temperature change depending in part, upon how the experiment is designed.

Consider that a system of liquid molecules is State 1, as illustrated in Fig. 14.15. Imagine that an isometric reaction occurs, resulting in a change to the intermolecular cohesive forces, e.g. State 2, as illustrated in Fig 14.16. Furthermore consider that there is no attraction between the walls and contained liquid, i.e. if water then hydrophobic walls.

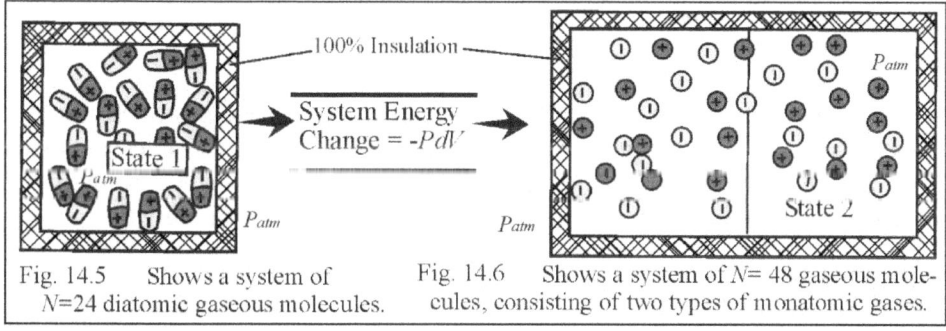

Fig. 14.5 Shows a system of N=24 diatomic gaseous molecules.

Fig. 14.6 Shows a system of N= 48 gaseous molecules, consisting of two types of monatomic gases.

Accepting that an isometric pressure increase is work into the liquid then again we have eqn 14.36: $dU - VdP = -nC_vdT$ And $dE_{Total} = 0$, which could be rewritten as:

$$dU = -C_v dT + VdP \qquad 14.39$$

Eqn 14.41 implies a correlation between intermolecular bonding change, and the isometric liquid pressure change. Traditional thermodynamics considers changes to Gibbs free energy (dG) as eqn 15.17: $dG = -SdT + VdP$. Although our analysis, and reasoning, is very different to that which is currently accepted, you should see the resemblance to eqn 14.39. It is intriguing but not necessarily fulfilling. And no this author is not saying that on a per mole basis: $S = C_y$, nor that: $dU = dG$. Rather this author is saying that the science needs a rethink. Note: More on the traditional derivation for Gibbs free energy will be given in Chapter 15.

Do these equations fail to acknowledge that a natural temperature –pressure (*P-T*) relation exists? Or is it actually part of the temperature decrease? Moreover, does the increase in potential (VdP) to do work in a liquid actually represent the energy increase within that liquid, becomes a question that warrants consideration. Furthermore, do we treat a liquid similar to a gas or can we simply apply the Gibb-Duham relation (eqn 14.37: $G_{extract} = dU - VdP + nC_v dT$) to liquids?

Remember: It must be understood that VdP can represent an increase in a gaseous system's potential to do work but that is not to say any work is actually done! Specifically work is always done external to a system and any isometric pressure increase may be nothing more than an increase in the ability/potential to do work, and this is often based upon a comparison to the surrounding's pressure, which may or may not actually represent true energy change within the system. As previously stated: In both liquids and gases VdP may also have an association with temperature change.

Traditional chemistry

Change to Gibbs free energy is accepted as the definitive factor in determining whether, or not, a given reaction occurs within a system. Specifically, if: $dG < 0$, then a chemical reaction is considered *exergonic*, i.e. it releases energy, therefore it should occur spontaneously, when all the reactants are in thermal contact. Conversely, when: $dG > 0$ then the chemical reaction is considered *endergonic*, i.e. it requires energy to proceed, when all the reactants are in thermal contact.

Chemists prefer to use the following formulation for changes to Gibbs free energy when deciding whether a reaction is spontaneous or not, that being:

$$\Delta G = \Delta H - T\Delta S \qquad 14.40$$

As previously discussed, change to enthalpy only makes sense when the system does work onto its surroundings. What does this actually mean to eqn 14.40? A chemist will go to his/her tables and read off the change in enthalpy and the calculated/tabulated entropy for a given substance. Since they are reading the change in enthalpy rather than enthalpy, they really are not using eqn 14.40. Seemingly they are actually using:

$$\Delta G = \Delta(\Delta H) - T\Delta S \qquad 14.41$$

So are we taking the change in the change of enthalpy minus the absolute temperature multiplied by the change in entropy. Now for enthalpy change: $\Delta H = \Delta \varepsilon + P\Delta V$. If the only internal energy change is the chemical potential, i.e. $\Delta \varepsilon = \Delta U$ then: $\Delta H = \Delta U + P\Delta V$. Rewriting eqn 14.41, now gives:

$$\Delta G = \Delta(\Delta U + P\Delta V) - T\Delta S = \Delta\Delta U + \Delta(P\Delta V) - T\Delta S \qquad 14.42$$

Translating eqn 14.42 into English; the change in Gibbs free energy equals the change of the change in chemical potential plus the work done onto the surroundings, minus temperature multiplied by the entropy change. One must be careful when dealing with such things.

Assuming that: $\Delta(\Delta U) = \Delta U$, then what about $\Delta(P\Delta V)$. Most likely it is the changes to any work that is done. There are issues. Firstly, once the work is done onto the surroundings, it generally is lost, i.e. work tends to be irreversible. Secondly, an ideal gas whose energy is for the most part kinetic energy will be more efficient at doing work than other gases. Certainly, if gases are part of the reaction then we would have to consider any changes in chemical potential plus any work which is done. And then compare this to the thermal energy that can be readily extracted.

Like entropy, no one claims to know what changes to Gibbs free energy really means, except to say that it determines spontaneous from non-spontaneous reactions. Entropy whatever its true guise may be, retains similarity to heat capacity but seemingly entropy has been transformed into something that also makes Gibbs free energy function in real world applications. In other words, the reason Gibbs free energy works as a function is that entropy has been twisted and prodded/massaged to make it so. No matter what the final outcome is, the traditional understanding of Gibb's free energy maybe in a shambles, although it seemingly works, possibly for the some of the wrong reasons!

Up to this point it has become obvious that traditional thermodynamics is in need of an overhaul, although correlations between our new perspective and traditional assertions can be seen. Let us reinvestigate some basics concerning reactions just so that there are no misunderstandings.

Logarithmic Functionality

Many consider that entropy change and its association to statistical physics, is a reason for the use of natural logarithmic functions. Certainly probabilities are logarithmic and traditional consideration of entropy is based upon such probability functionality. This all sounds grand until one realizes that no one really knows what entropy means. Moreover if you adhere to what is discussed in this book then it is really a form of circular logic. One must then ask; how does one explain this logarithmic functionality that seemingly drives so much of physical chemistry?

In Chapter 4, the rates of heat transfer were considered. Although heat flows from cold to hot, as well as from hot to cold, the net flow is always from hot to cold. This led to eqn 4.43: $dq_1 / dt = C In(T_1 / T_{heat})$

For an exothermic chemical reaction the rate at which heat is transferred into the surrounding will certainly adhere to similar logic. In so far as eqn 4.43 helps explain a need for using natural logs, it is not the only reason. Earlier in Chapter 11 the probability for latent heat was discussed, as well as its rates in Chapters 12 and 13. Certainly similar rational apply to chemical reactions.

Plausible Considerations for Probability of Reaction

Imagine that a chemical reaction that needs to extract some of the thermal energy within a system. In terms of change in chemical potential energy and Boltzmann's exponential ($e^{-\beta E}$), as is used to define the probability of reaction, then:

$$P'_{(r)} = B\exp(-\Delta u / kT) \qquad 14.43\ (a)$$

$$P'_{(r)} = B\exp(-\Delta U / RT) \qquad 14.43\ (b)$$

Eqn 14.43 is fundamental to accepted considerations for chemical reactions. Furthermore it implies that as the mean thermal energy that can passed along a given direction approaches the change in chemical potential, then the probability of reaction approaches unity i.e. as $kT \to \Delta u$, $P'_{(r)} \to 100\%$.

An implication of eqn 14.43 is that the required energy involves a spectrum of thermal energy (phonons or photons) as defined by kT and this energy is extracted from only one neighboring molecule, which may or may not be the case for changes to some solitary bond. Certainly for latent heat of vaporization (Chapters 10&11) it was deduced that six bonds with the six neighbors are broken, hence our new kinematics based probability for vaporization [eqn 11.7: $P'_{(v)} = B\exp(-T_b/T)$] is required. Certainly, all the plausible issues concerning latent heat of vaporization have not been fully resolved. Even so, the realization now exists for the need of path determination before one can properly write a probability.

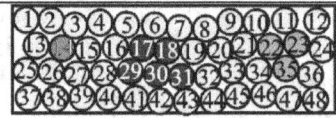

Fig. 14.17 Shows a system of N=48 molecules. If molecules were simple spheres then molecule 14 has six nearest neighbours. Now ask what how many neighbours does a group of reactants have, for example group 17,18,29,30,31 has? Or group 22,23,35 has?

What if the chemical bond required a specified frequency to break, then would eqn 14.43 actually still be the best equation? Although the odds of the required specified frequency's existence may increase with increasing temperature, in all likelihood the answer becomes no. Certainly a somewhat more complex analysis may be required. Remember even if the wrong exponent were used i.e. wrong $\Delta U / kT$, often the probability can be normalized so as to match empirical findings. Hence knowing the path is paramount!

Reconsider that a spectrum with a total thermal energy is required then the following needs to be asked: How many neighbors are there from whom required energy can be extracted? For reactions in the liquid state; if the reaction involved change to a solitary bond then does the energy required come from a solitary neighbor, or does it involve the energy from its six neighbors from along the three dimensions i.e. molecule #14 in Fig 14.17 is at some instant in physical contact with molecules #3, #13, #15 and #26 plus two neighbors perpendicular to the page? This is likely frivolous if the change is only to one bond and all that is required is the thermal energy from one neighbor, i.e. eqn 14.43 applies!

Consider that molecules #22,#23 and #35 in Fig 14.13 are the reactants for a reaction. The chemical reaction no longer involves one bond between two atoms, so what happens to the probability? Next consider that molecules #17, #18, #29,#30, and #31 in Fig 14.17 are five reactant molecules in a reaction. Accordingly, dependent upon how you model the reaction the number of neighbors could reach two dozen but how many actually contribute their energy to the reaction is open for debate. And of course the number of neighbors involved in a reaction will vary when comparing reactions in the gaseous, versus liquid, or solid, or mixed states. Luckily most reactions are between two molecules/atoms, so the complexity should generally be manageable.

Do not forget: If the energy requirement is contained within the exponential part of the probability, then the so-called constant (B) contains the odds/likelihood of the neighbors all acting at some instant, as was previously discussed in Chapters 11 and 12. I.e. B is just a constant for that particular process in that particular apparatus at those particular conditions.

Another issue to consider is the states of reactants and/or products i.e. liquid, solid and/or gaseous states and how energy can be exchanged in these various states. Certainly the total thermal energy per molecule will be different in the gaseous state and then there is the issue of any changes to the gaseous potential energy e.g. sees discussion concerning potential of a cloud of gas in Appendix A1.

Still another possible issue being; is the intramolecular vibrational energy included as part of the energy accessible for a reaction? Yet another issue being; if the number of molecules in reactants does not equal the number in products then what is the number of neighbors? In general the expectation is that the probability of energy extraction is for a one-step process, but is this always the case? And certainly more valid questions will remain open for discussion. One's model for paths hence approach, hence how the probability is written, may vary depending upon an array of factors. Perhaps the probability will require a kinematic number approach as was previously discussed for latent heats in Chapter 11.

Spontaneous Reactions

Based upon eqn 14.43, $P'_{(r)} = B\exp(-\Delta u / kT)$ if the reaction involves the extraction of thermal energy (kT) from only one neighbor then the reaction is spontaneous when; $kT \geq \Delta u$.

Reactions that do work

Next consider that the energy of reaction includes change to bonding potential (ΔU) plus work that is done in the process of the reaction ($P\Delta V$) i.e. real enthalpy change (ΔH). It has to be said that a precise analysis would include as part of it, the change in bonding potential, and any bonding /repulsion/attraction potentials in the gaseous state assuming that a gas is a product of the reaction. When gases are products of a reaction then the probability should be written in terms of enthalpy change per molecule (Δh).

If the reaction involved the exchange of thermal energy from one neighbor to break one bond then our expected relations for probability should take the following form:

$$P'_{(r)} = B\exp(-\Delta h / kT) \qquad 14.44 \text{ (a)}$$

$$P'_{(r)} = B\exp(-\Delta H / RT) \qquad 14.44 \text{ (b)}$$

Gibbs free energy and Probability

If the chemical reaction results in an internal pressure change within the system, and the probability of reaction has temperature dependence then the probability may take the following form.

$$P'_{(r)} = B\exp(-\Delta g / kT) \qquad 14.45 \text{ (a)}$$

$$P'_{(r)} = B\exp(-\Delta G / RT) \qquad 14.45 \text{ (b)}$$

Rather than be completely wrong this author would prefer to leave it to those in the field to determine some new clarity concerning Gibbs free energy before discussing eqn 14.45.

A Discussion

The reality; if one accepts that the atmosphere has mass and its upward displacement requires work (lost). Then physical chemistry needs a complete overhaul. And this includes changes to our understanding of entropy and what the values in chemistry tables really mean.

The real tragedy remains that although theoretically cumbersome, Gibbs free energy and entropy have all been obtained/deduced/calculated for most substances through a combination of experiments and calculations all most likely based upon gross misunderstandings. And never forget the previously discussed issues with exponentially based probabilities, especially those based upon circular logic. Specifically, when one uses the wrong function in an exponential, and then applies it to a data set. Thenthe act of normalization may empower them to firmly believe in what they have done, albeit it is completely wrong. In other words normalization remains both the strength and weakness of exponential probabilities.

Closing Remarks: Physical Chemistry

We have only skimmed the surface concerning physical chemistry and how this author envisions its relation to thermodynamics could/should be perceived. A major thing to come out of all this is the realization that physical chemistry does require a rethink.

We also arrived at better possible understandings/limitations of both enthalpy and entropy. Entropy being related to heat capacity but unlike heat capacity, which has a system wide application, entropy may be better envisioned in terms of how a reaction extracts a system's thermal energy enabling chemical reactions. Specifically, entropy in chemistry tables is a combination of empirically measured and calculated values all made to render changes to Gibbs free energy into a function that separates spontaneous from non-spontaneous reactions. A separation that perhaps should be based upon the readily accessible/ extractable thermal energy versus the energy required for a given reaction.

Gibbs free energy may have to remain a part of the sciences for the time being, although its true original identity seemingly lacks solid constructive logic. Similarly, entropy arguably remains the thermodynamic parameter with too many plausible inexact interpretations, i.e. it still requires clarity if not a complete rethink. And of course the sciences may eventually become simplified by abandoning entropy and/or Gibbs free energy and/or Helmholtz free energy altogether.

More exacting models on how various reactants become products should be in our future once we learn to properly model paths determining how this thermal energy is exchanged and the associated path dependent kinematic numbers. Whatever the final outcome, reactions involving high frequency energy may require separate consideration from those involving thermal energies. Importantly, a new perspective was presented, without emphasizing entropy, as is traditionally done. Hopefully some will embrace the concepts discussed herein, and then write more expansive books concerning physical chemistry, books that are both simple and eloquent, while enshrined in constructive logic. And then maybe this author will begin to actually fully understand physical chemistry.   Make me proud, and call me a simpleton!

Chapter 15: **The Differential Shuffle**

In Chapter 14 it was discussed that physical chemistry should be simplified. Importantly it was demonstrated that there may be other interpretations for Gibbs free energy, which is a foundation of modern chemistry. Similarly in Chapters 9 and 14, other simpler interpretations for Helmholtz free energy were provided.It remains this author's belief that these new interpretations not only allow for a simpler understanding of the sciences but they may also help convince the indoctrinated to actually open their minds. Moreover, the fundamental reason for the following bizarre differential shuffle remains traditional thermodynamics wrongly perceived need to maintain the so-called second law as a postulate.

Traditional Analysis

Tradition starts off with the isothermal ($dT = 0$), and isobaric ($dP = 0$) changes to the so-called equation of state[1], A.K.A. eqn 1.14.

$$TdS = d\varepsilon + PdV \qquad 15.1$$

One might ask why start with eqn 15.1? The primary reason is the so-called second law of thermodynamics concerns isothermal entropy change. Therefore if one wants to write the science (as Planck pointed out) around the second law then this is a starting point that does not challenge its validity. Remember: This text was written to clearly demonstrate that thermodynamics simplifies once you remove the second law as a universal postulate. Continuing with tradition: Based upon eqn 15.1:

$$d\varepsilon = TdS - PdV \qquad 15.2$$

Eqn 15.2 is often traditionally referred to as the combined first and second law. When transforming either 15.1 or 15.2, most texts[1pg162] will employ the following:

$$PdV = d(PV) - VdP \qquad 15.3$$

Again a more precise analysis would realize:

$$d(PV) = PdV + VdP + dPdV \qquad 15.4$$

For infinitesimal changes ($\Delta \rightarrow 0$): $dPdV <<< PdV$, or VdP, then changes as described in eqn 15.3 approximates changes as described in eqn 15.4. It must be said, that for some processes infinitesimal change is not be the case! Remember: Infinitesimal change allows for the illogical traditional considerations for reversibility, as was discussed in Chapters 4 and 9!!

So-called: Helmholtz Free Energy

Continuing with the traditional: Combining eqn 15.2 with eqn 15.3 gives:

$$d\varepsilon = TdS - d(PV) + VdP \qquad 15.5$$

Collecting the terms, then eqn 15.5 can be rewritten.

$$d(\varepsilon + PV) = TdS + VdP \qquad 15.6$$

Traditional thermodynamics defines "enthalpy" as:

$$H = \varepsilon + PV \qquad 15.7$$

Note: As previously discussed the above enthalpy relation is only truly valid for ε being the system's total energy and PdV being the work done onto the system's surroundings!

Traditional thermodynamics rewrites eqn 15.6 as the "enthalpy relation":

$$dH = TdS + VdP \qquad 15.8$$

Applying similar logic, traditional rewrites: TdS in the following manner:

$$TdS = d(TS) - SdT \qquad 15.9$$

Therefore, eqn 15.8 becomes:

$$dH = d(TS) - SdT + VdP \qquad 15.10$$

Which can be rewritten as:

$$d(\varepsilon - TS) = -SdT + VdP \qquad 15.11$$

"Helmholtz free energy" as[7]: $F = \varepsilon - TS \qquad 15.12$

Consequentially, eqn 15.11 can be rewritten as:

$$dF = -SdT - PdV \qquad 15.13$$

Seemingly, Helmholtz free energy change is for changes in temperature (T) and volume (V). It was demonstrated in Chapter 14 that based upon eqn 14.33: $dF = -nC_v dT - P_{atm}dV$, that an understanding of Helmholtz free energy based upon constructive logic was readily determinable.

Understandably traditionalists may counter that their attaining eqn 15.13 through their differential shuffle makes perfect sense. This author remains unconvinced that it is actually based upon constructive logic, rather than an illogical shuffling of variables/differentials until one arrives at something that can be claimed to match empirical findings. For the case of an exothermic reaction, then as discussed in Chapter 14 Helmholtz free energy can be rewritten as:

$$dF = dU - SdT - PdV \qquad 15.14$$

Interestingly, Dunning- Davis[2] discuss how eqn 15.14 is often used as the introductory equation for relating statistical thermodynamics to classical thermodynamics. Although actually valid for the wrong reasons, the traditional faulty logic, hence limitations of eqn 15.14 applicability should now be readily apparent to you the reader.

So-called: Gibbs Free Energy

Again, traditional starts off with eqn 15.1: $TdS = d\varepsilon + PdV$. Applying the transformations for $d(TS)$ and $d(PV)$, as given by eqn 15.3 and eqn 15.9 respectively, gives:

$$d\varepsilon = d(TS) - SdT - d(PV) + VdP \qquad 15.14$$

Eqn 15.14 can be rewritten as:

$$d(\varepsilon - TS + PV) = -SdT + VdP \qquad 15.15$$

Gibbs free energy as[1]: $G \equiv \varepsilon - TS + PV \qquad 15.16$

By inserting 15.16 into eqn 15.15, traditional thermodynamics obtains[7]:

$$dG = -SdT + VdP \qquad 15.17$$

Changes to Gibbs free energy (eqn 15.17) applies to processes that are both isometric ($dV = 0$) and isentropic ($dS = 0$?): Isentropic's real meaning depends upon one's interpretation of entropy (S), Traditionalist may defend their differential shuffle in attaining eqn 15.17. Again this author will argue that a simpler similar understanding based upon constructive logic is readily determined i.e. eqn 14.37:

$G_{extract} = dU - VdP + nC_vdT$

Problematic Traditional Thought

Traditional thermodynamics is unique in its use of differentials! It starts with a part: PdV, from which the whole: $d(PV)$ is then subtracted, obtaining the other parts: VdP. Certainly logical dictates that if one started off with the whole: $d(PV)$, one could then deduce the parts: PdV & VdP.

As previously stated, the reason that eqn 15.1 was beheld with such relevance is that it allowed for a connection between the so-called second law and lost work e.g. Carnot cycle (Chapter 9). The equating of work to entropy was part of the mental progression to Clausius's understanding that ST defines energy. Furthermore, lost work meant that the Carnot engine could not return to its original state without an influx of energy, leading to Lord Kelvin's declaration concerning the second law of thermodynamics. It all would be so humorous if it were not for the fact that the second law and entropy, both have taken on a demigod status, and the 150 yrs of indoctrination that has followed.

Although lacking clarity entropy (S) was construed so that its relation to both volume (V) and internal energy (ε) explained empirical data, i.e. entropy became a mathematical contrivance that it is. The indoctrination of the differential shuffle is all embedded in circular logic. The fact that statistical mechanics is accepted as the inarguable proof behind thermodynamics, speaks more of the power of statistics, than anything else. And of course the equating of Boltzmann's constant (k) so that it explains empirical data here on Earth just reinforced what was claimed. Again circular logic i.e. equating equations to empirical data and then claiming that this is some kind of proof is NOT constructive logic!

Moreover, in using eqn 15.1 as a foundation thermodynamics has failed to equally treat pressure, and volume, as parameters of relevance. The ramifications to processes wherein both pressure, and volume increase are profound, e.g. bubble nucleation. Also consider a cosmological black hole that being some constant volume horizon wherein the pressure increases, as more and more matter enters it. No wonder paradoxes arose when the so-called second law was applied to black holes.

Closing Remarks

Although the similar theoretical results were attained, our new perspective completely alters our understanding, and hopefully for the better. Certainly considering: $W=dU+ d(PV)$ has allowed this author to be the first to successfully calculate the energy required to nucleate a bubble[3]. Since writing that paper this author realized that there are fundamental issues with the traditional thermodynamic relations as described in this chapter.Thermodynamics can be considered a lesson in the use of differential equation. However, simply moving variables around, without instilled logic in each and every step can led to an abuse of differential equations. Once more two plausible explanations for the same empirical findings are presented for both Helmholtz and Gibbs free energy. Which is right? Based upon Occam's razor this author will argue for the explanation based upon our new perspective is the better explanation e.g. as was presented in Chapter 14. Although not necessarily politically correct, one might ascertain that traditional thermodynamics is fool's gold based upon sledge hammer contrived mathematics.

References:

1) "Fundamentals of Statistical and Thermal Physics", F. Reif, McGraw-Hill, New York, 1965
2) Dunning-Davis, Jeremy,& Lawrence, Norma, Richard "Truth in Paradigm" Stand Dead Publications 2017
3) K. Mayhew, "Energetics of Nucleation" Phys. Essays **17**, vol 4, 476 (2004)

Chapter 16: **Entropy and those Indignant Laws**

Perspectives do change! In the 18th century heat caloric theory dominated with a mass-less substance called *phlogiston* was considered to be part of combustible bodies resulting in the conservation of phlogiston being an early thermodynamic law. Our comprehension changed, starting in 1798, when Count Rumford (1753-1814) professed, *"heat is merely a form of motion of the particles in a body"*[1pg 134]. The count's radical statement was not accepted until well into the 19th century.

Interestingly, Robert Mayer argued against expressing heat/energy in terms of the increased motions of the smallest parts of matter[2], which remain the basis of thermodynamics into the 21st century. Statistical thermodynamics wrongly limited temperature to molecular kinematics, wherein entropy and the second law, reigned supreme in a probability founded universe. Interestingly, in 1894 Planck wrote[2]: *"it is completely unfounded, simply to assume that changes in nature always proceed in the direction of lesser to greater probability"*.

Certainly this book has demonstrated that another perspective exists, one that is based upon constructive logic rather than mathematical axioms. Now reconsider entropy and it accompanying laws.

Zeroth Law of Thermodynamics

The Zeroth law of thermodynamics is based upon "thermal equilibrium", i.e. the equality of temperature: *"If two systems are in thermal equilibrium with a third system, they must be in thermal equilibrium with each other*[3pg 282]: The Zeroth law stands as it allows us to use a thermometer to measure a temperature and then compare that measurement to other systems.

First Law of Thermodynamics

The traditional statement: *"An equilibrium macrostate of a system can be characterized by a quantity E (called its internal energy), which has the property that for an isolated system: E= constant. If the system is allowed to interact and thus go from one macrostate to another, the resulting change in E can be written in the form:*

$$\Delta E = W + \Delta Q \qquad 16.1$$

where W is the macroscopic work done on the system as a result of changes of the external parameters of the system. The quantity ΔQ is called the heat absorbed by the system"[3pg282].

Conversely, if the system does work onto its surroundings then the first law becomes:

$$\Delta Q = \Delta E + W \qquad 16.2$$

This essence of the first law is agreed upon. However the law as written has issues:

1) Consider the statement: *"W is the macroscopic work done on the system"*. Work done onto a system (eqn 16.1) often changes the system's internal energy, which may or may not equate to the system's ability and/or potential to do work increases.
1) Clarity is obtained by specifying that work is done to the system's surroundings (eqn 16.2), whether the surrounding be our atmosphere or our atmosphere and a mechanically driven device.
2) What exactly is our understanding of internal energy (E or ε). Any definition must acknowledge that the macro energy is a summation of the micro energies. Rather than enter such a quandary eqn 16.1 would be better served if E was considered as the system's total energy.
3) The notion that work can only use 2/3 of a closed system of gas's kinetic energy increases. Certainly the traditional understanding may render one to exclaim that eqn 16.1 is compare apples

with oranges and equates it to fruit. Clarity is attained by realizing that not all the kinetic energy of a gaseous system can be used for work, hence the total thermal energy of a system doing work tends to increase. This also means that useful systems going through cyclic steps tend to get hotter. A natural inefficiency experienced by most useful processes, thus irreversible.

4) Work is often not reversible but the difference in writing of eqn 16.1 vs eqn 16.2 is not expressed.

The true purpose of the first law of thermodynamics is to affirm that energy is conserved! To this end, there is total agreement with traditional thermodynamics. It should be noted that depending upon the author and the application you might see eqns 16.1 and/or 16.2 written in various different formats. Rather than getting hung up with all the parameters, perhaps, the first law should simply read: *The energy change within a system must equal the total influx of energy into a system minus the efflux of energy from that system* i.e. *energy can be converted from one form to another but it cannot be created, or destroyed.*

Either way this author envisions that better clarity would be attained if the term internal energy was replaced with total energy. Throughout this text, both entropy and the second law have been treated in a circumlocutory manner. Why? Because they invite unnecessary over-complications!!!!

Entropy, Boltzmann's Insights, Statistical

Investigating: Part of Ludwig Boltzmann brilliance was his insight in the concept of the number of accessible energy states (Ω). Although the following equation is written on Boltzmann's tombstone, Planck is attributed to its final writing, which equates it to entropy (S):[3]

$$S = kIn(\Omega) \qquad 16.3$$

Eqn 16.3 forms the basis of Boltzmann's entropy (S) which has become a foundation for statistical thermodynamic. An entropy increase can then be written as:

$$dS = d[kIn(\Omega)] = kIn(\Omega_f / \Omega_i) \quad 16.4$$

An often used variation of eqn 16.4 rewrites entropy change in terms of the volume increase of a system, i.e.:

$$dS = d[kIn(\Omega)] = kIn(V_f / V_i) \quad 16.5$$

Furthermore if two systems are put into thermal contact then, the number of microstates is still obtained by multiplying the number of microstates for each system, i.e.:

$$\Omega_{final} = (\Omega_1)(\Omega_2) \qquad 16.6$$

And the new final total entropy (S_{final}) still remains as large as possible, that being:

$$S_{final} = k(\log \Omega_1 + \log \Omega_2) \qquad 16.7$$

Randomness: The 20th century Assertion

Equations 16.3 through 16.7 demonstrate the powers of the math. However, their application often lacks constructive logic i.e. the 20th century assertion that entropy is a measure of a system's molecular randomness. Sadly as ascribed by Ben Arien[4] randomness remains in the eye of the beholder i.e. show a dozen people various pictures of systems and ask them to put them in order of their randomness you might get twelve different answers hence randomness is not a particularly scientific term.

Even so the act of increasing volume (eqn 16.5) was accepted as increasing the above mathematical result that defines the number of possible states that molecules can access. Sadly eqn 16.5 lacks meaning until both the surroundings and forces of expansion are clearly defined. Okay it sounds like repetition but it not. The lack of clarity allows for the erroneous association between energy that maybe lost and an increase to the number of energy states in an expanding system. Are we to believe that all expanding systems are irreversible? Of course not! If the increase to the number of possible energy states is due to heating, and the system expands hence does work onto the surrounding atmosphere, then the process is irreversible! But this new understanding is based upon the simple principle of lost work and not eqn 16.5.

Furthermore, eqn 16.5 assumes the number of possible energy states is directly related to the mean molecular volume e.g. the way physical chemists tend to view things. However considering that changes to the randomness or even mean molecular volume, in itself signifies an energy change again puts the cart ahead of the horse. The reality is when energy is added then:

a) Closed isometric/isochloric systems tend to experience a pressure increase.
b) Closed isobaric systems tend to experience volume increases, hence do work on their surroundings, e.g. when surroundings are our atmosphere then: $W_{lost} = P_{atm}dV$.

Interestingly, as a system's volume increases then molecular dispersal will cause the molecules to become evenly distributed which is the maximum state of molecular randomness. Is entropy needed? The answer is again an emphatic NO! Moreover, given a sufficient amount of time then dispersion will cause maximum randomness, allowing us to say that in terms of molecular volume (v):

1) With a given set of constraints, any molecule with a given energy will attempt to attain a maximum mean molecular volume (v).
2) As we add energy to certain systems (especially gases), the molecules within that system, often will attain a higher mean molecular volume (v).
3) In any expanding system, the mean molecular volume (v) will continuously increase.

Nothing is extraordinary about the above three common sense statements. Yet similar statements concerning traditional entropy are shamefully deemed profound! Entropy may remain some wondrous mathematical tool allowing statistical thermodynamics to relate to what is often witnessed. But at the end of the day one cannot forget most useful expanding systems displace the Earth's atmosphere against gravity. And this explains lost work irrelevant of a system's randomness!

If it wasn't for gravity then the natural state would be disorder hence. Interestingly, we now realize that Boltzmann's constant (k) correlates to Earth's gravity through the enthalpy equation. Perhaps from this point on science will accept that Boltzmann's constant allows for a correlation between gravity here on Earth and temperature.

Second Law of Thermodynamics & Entropy

Consider the following traditional statements[3pg 283]: "*An equilibrium macrostate of a system can be characterized by a quantity S (called entropy), which has the following properties*:

i)   In any infinitesimal quasi-static process in which a system absorbs heat, its entropy changes by an amount:

$$dS = dQ/T \qquad 16.8$$

Where T is a parameter characteristic of the macrostate of the system and is called its absolute temperature

ii)   *In any process in which a thermally isolated system changes from one macrostate to another, its entropy tends to increase, e.g.*

$$\Delta S \geq 0 \quad " \qquad 16.9$$

Throughout this text, it was understood that unless a system is truly isolated (insolated and surrounded by a vacuum) then a volume increase often requires the displacement its surroundings. And if its surroundings has mass that is displaced against a gravitational field, then this requires work e.g. a potential energy increase of the Earth's atmosphere ($P_{atm}dV$), i.e. "lost work", which explains Maxwell's demon, hence dethrones the second law. Accordingly, the second law is not required to explain why perpetual motion machines remain idealistic devices. All this acknowledges that <u>isolated systems</u> are not natural systems here on mother Earth.

Consider, one of the better descriptions for the second law of thermodynamics was found on the internet, written by Matt McIrvin[5]: "*A physical system will, if isolated (that is, if energy cannot get in or out), tend toward the available macroscopic state in which the number of possible microscopic states is largest.*" In other words, an isolated system will tend towards an equilibrium macrostate, which is as random as possible. If one realizes that dispersal of a gas due to intermolecular collisions will do the very same thing, then why call it entropy, i.e. why not call it dispersal? Okay dispersal as a term is not palatable when applied to Clausius's equation. But still dispersal is what Matt is describes although it would not be an ideal term from a statistical/quantum understanding.

As for mechanical devices, friction creates heat, resulting in energy loss i.e. dissipated energy. Neither entropy, nor the second law is required to realize that such heat radiated into the atmosphere is not recoverable. Plain and simply put, perpetual motion is not possible! Again there is no need for either entropy or the second law to explain why, irrelevant of the process!

Second law: Confusion in Heating

Eqn 16.8, is sometimes considered as the differential equivalent to the Clausius equation ($T\Delta S = \Delta Q$), which traditionally is considered as applicable to any process where there is an isothermal transfer of heat. Certainly as a mathematical contrivance entropy became a tool applicable to empirical research. As great as that sounds, logic screams that any input of thermal energy (Q or ΔQ) generally should result in a system temperature increase, even if that increase is too small to measure! And the tradition of claiming infinitesimals to alter that logic is ****. Okay let us investigate this in more detail.

As an exercise combine, equations 16.3 and 16.8 to obtain the following isothermal relation:

$$d[kIn(\Omega)] = dE/T \qquad or \qquad dE = Td[kIn(\Omega)] \quad 16.10$$

Common sense dictates that the number of possible energy states should increase as a system's thermal energy increases. Strangely eqn 16.10 is an isothermal relation that is often traditionally applied to almost all systems where thermal energy is added. Theoretically this can happen if the work done (W) by the system equals the thermal energy input into that system.

For example, consider that a system experiences an isobaric volume increase due to an input of thermal energy (E_{in}), such that:

$$E_{in} = W = P_{atm}dV \qquad\qquad 16.11$$

Now the energy input (E_{in}) equals work done (W) by the expanding system onto its surroundings i.e. the upward displacement of our atmosphere (lost work $= P_{atm}dV$). Accordingly, there is no real system energy change (forgoing any changes to the system's blackbody energy), thus the system remains isothermal although it has expanded. Ask yourself, did the number of accessible energy states actually increase?

If the number of accessible energy states is strictly a function of the system's total energy, then the answer would be no! Conversely, if the number of accessible energy states is a function of volume, then the answer becomes yes! Which answer do you choose? Explain yourself!

Investigating the above scenario further; traditional thermodynamics wrongly associates the "lost work ($P_{atm}dV$)" with an entropy increase, i.e. the system has greater volume hence the molecules are seemingly more random. At the risk of continuously repetition, the traditional perspective is one of wearing blinders, because the reality is that the work was lost due to the upward displacement of the surrounding atmosphere! Again, this is the basis of the traditional association of entropy with volume. Strangely, traditionalists may answer yes, even though there is no change to the system's total energy.

Next consider the simply addition of thermal energy into an isometric isobaric system, such as adding piece of hot iron into a cup of water. If the piece of iron is large enough and hot enough, and the thermometer is accurate enough, then there will be a measurable temperature increase within that cup. Since there is a temperature increase, then this process cannot be isothermal. Hence you CANNOT simply apply isothermal relation eqn 16.8, as is often wrongly done.

Traditionally the application of eqn 16.8 absconds from the above logic by employing infinitesimal arguments, which is nothing less than losing sight of reality. For example one might argue that if the piece of iron was sufficiently small and/or the cup was sufficiently large, then your thermometer would not be able to measure a temperature increase, hence eqn 16.8 applies. Although this allows for the traditional treatment of entropy to be some magical isothermal heat capacity, it is really a question of scale rather than logic. Just because the temperature increase in the cup is immeasurable, does not mean it did not happen. Specifically, if your thermometer was more accurate, or the cup of water was smaller, then you would be able to measure a temperature change! Hence you couldn't use eqn 16.8.

Now ask: Did increasing the system's thermal energy result in an increase to the number of possible energy states? If the number of accessible energy states is strictly a function of volume, then the answer is no, because the water in the cup did not expand. Conversely, if the number of accessible energy states is a function of the system's total energy, then the answer is yes. In this author's way of thinking the number of accessible energy states cannot remain a function of both volume and energy, as is traditionally taught when traditional entropy based logic is applied to all types of energy transfer. Perhaps if the concept that randomness related to energy, then the illogical consideration of volume being associated with number of accessible states can be thrown out. The ramifications to the sciences maybe immense especially in subjects like physical chemistry.

Again, the above does not diminish Boltzmann's great mathematical skills, rather it adds context to statistical thermodynamics illustrating part of the traditional fundamental failure. Certainly, eqn 16.3 can remain a foundation of statistical thermodynamics, but there must be significant changes based upon our new understandings of Boltzmann's so-called constant (k) and what the number of accessible states (Ω).

Planck's Consideration and the Poor Path of Logic

Previously, it was discussed that Planck had reservations concerning probability theory but then is later attributed with writing its most fundamental eqn 16.3 (change in attitude?). In his 1917 treatise, Planck[6] does briefly discuss that work is done onto the surrounding atmosphere but he fails to understand that this is lost work. Certainly if he did the 20th century mayhem may have been averted.

The wall of illogical deductions predates Planck. It was either Kelvin or Clausius, who showed that heat exchanges are a path independent exact differential[7], hence reinforcing eqn 16.8. The issue with eqn 16.8 does not concern it being an exact differential but how it is equally applied to work and energy and then applied to all states of matter. It completely ignores the fact that the reason thermal energy associated with matter is a simple function of temperature is because of the thermal radiation from our Sun i.e. our Sun's radiation of thermal energy density can be approximated by the Rayleigh-Jeans equation: $\rho_R = a'T$.

The idea that entropy arose due to deductions based upon second law[7] clearly shows that traditional thermo is based upon mathematical conjecture rather than constructive logic. As was discussed in preface, the early consideration of the second law can be attributed to either of the following accepted statements[7]:

1) Kelvin: "It is impossible to transform an amount of heat completely in cyclic process in the absence of other effects"[7]
2) Clausius "It is impossible for heat to be transferred by a cyclic process from a body to one warmer that itself without producing other changes at the same time"[7]

Envisioning systems of large number of particles formulated statistical thermodynamics[7], with the second law, as previously described, becoming the core postulate (false?). Interestingly, the micro-canonical ensemble [variables (ε,N,V)] performs best for isothermal systems otherwise it becomes an approximation. In other words, statistical thermodynamics is based upon idealized systems surrounded by heat bath, while the mother of all heat baths, e.g. our Earth's atmosphere, oceans, have their thermal energy density approximated by a linear function of temperature. We now begin to understand how and why traditional thermodynamics became so logically derailed. Note other ensembles are also used i.e. canonical ensemble [variables (N,V,T)].

One may ask why it wasn't realize that in cyclic processes, i.e. Carnot cycle, there are steps where a system's entropy decreases just as there steps where it increases. Although blurred, the logic that isothermal entropy never decreases became enshrined dogma. Herein entropy is related to randomness hence ill-perceived randomness. Again if only Planck had realized what lost work is. But there is more, as the various mathematical ensembles are based upon elastic collisions, which are in reality, rare. Equally perverse was the previously discussed issue of considering the addition of heat in terms of isothermal entropy increase, which is really nothing short of some poor understanding concerning the reality of thermometers. Equally, painting work and all system energy with the same entropy brush may have been nothing short of fool's gold.

Circular logic

It has to be emphasized that traditional thermodynamics is often based upon circular rather than constructive logic. Think about it, you design a math to explain empirical data for say lost work, and then you exclaim that the empirical data proves your mathematical based theory and then universal assumptions are rendered. This author cannot help but think that this in part explains how Boltzmann designed his brilliant mathematical conceptualizations (basis of statistical thermodynamics), and determined his constant (k) all so that it equates to lost work here on Earth: ($W_{lost} = P_{atm}dV$).

Of course after enough repetition the poorly construed yet brilliant science becomes reinforced in our minds, in part because statistical thermodynamics now seemingly now explains what we witness. Ultimately it becomes so-believed inarguable all-encompassing proof, when in actuality it very inception was that it was equated to what was empirically known. This demonstrates circular logic at its best.

Kent W. Mayhew

Second law and Other

Is the universe's entropy really increasing?[8] That depends upon entropy's definition. An expanding universe implies a volume, hence randomness increasing. So the answer is seemingly yes! But what about all those clumps of matter that gravity assembles? Perhaps not! And then, the other cosmology extreme black holes, wherein the second law paradox used to exist. Sure one can circumnavigate the second law's complicated path of logic i.e. a black hole becomes an isometric horizon into which matter and energy seemingly disappear[9].

What about information? If we are talking black hole thermodynamics, who cares? Ben-Naim[4] discusses that Shannon may have been better off to calling it information rather than entropy, which would have blindsided any misunderstandings between thermodynamics and information theory. Interestingly, Ben-Naim[4pg7] demonstrates something this author has been saying for years: That entropy is a mathematical contrivance. How important of one remains in the eye of the beholder.

Interestingly, Dunning-Davis[7] discussed that information entropy is just a similar mathematical construct to thermodynamic entropy, all based upon probabilities (P') and constant (K') i.e:

$$S = -K' \sum P' \log(P') \qquad 16.12$$

Just because they own the same mathematical construct does not mean that they have the same profound meanings. Moreover questioning the fundamentals of information entropy means that the similar questions apply to the essence of thermodynamic entropy[7].

Carnot was probably the originator however the innovators for the modern second law concepts arguably are Rudolph Julius Emanuel Clausius (1822-1881) and/or Lord William Thompson Kelvin (1824-1907). Eqn 16.3 implies; if its entropy always increased, then the universe would eventually reach a state of uniform temperature and maximum entropy from which it would be impossible to extract any work. Lord Kelvin named this point of finality the "*Heat Death of the Universe*". How correct was Lord Kelvin? Note: As briefly discussed in preface Clausius somewhat abandoned this by 1876.

Ultimately, if our universe as a whole is expanding then it must be cooling down. However this has little to do with entropy and everything to do with the dispersal of both matter and energy. Moreover, if one envisions our universe obtaining some uniform cold temperature, then he/she has seemingly forgotten about gravity. Specifically, gravity tends to pull matter together creating hot spots throughout our universe. And so long as matter is gathered into clumps by gravity, then both the pressures and the temperatures associated with those clumps will remain elevated. Remember condensed matter and polyatomic gases adsorb thermal energy thus significantly increasing its density.

Numerous other challenges to the second law are well documented[10]. Interestingly, D. Sheehan[11] rightfully states: "*The second law of thermodynamics is an empirical law. It has no fully satisfactory theoretical proof. This being the case, its absolute validity depends upon its continued experimental verification in all thermodynamic regimes.*" Realizing that the second law and entropy were in part construed then formulated and then equated to "lost work"[12,13,14] renders any concept that they are empirically validated somewhat superficial.

Empathy may be given to scientists focusing upon a heat engine, and thinking solely in terms of internal molecular motion and probabilities. Sadly their focus led to a loss of reality. Of course the man on the moon simplifies our perspective by witnessing the atmosphere's volume increase.

Finally, there are those who believe that the second law prevents net thermal energy from being transferred from a cold system to a hotter system. In terms of our new perspective, logic simply dictates that although heat flows both ways, the net flow is always from hot to cold.

The Third Law of Thermodynamics

The traditional statement: *The entropy of a system has the limiting property that as:*

$$T \rightarrow 0 \ then \ S \rightarrow S_o$$

where S_o is a constant independent of the structure of the system [3pg284]:

The reason that the third law exists is due to eqn 16.3: $dS = dQ/T$, and the implication of absolute zero. More precisely as $T \rightarrow 0$, what happens to dS? Without the third law, then in traditional thermodynamics entropy change as defined by eqn 16.3, would approach infinity as $T \rightarrow 0$.

With the realization that entropy, no longer holds the theoretical reigns of thermodynamics! Hence the third law can join the second law. In terms of our new perspective, as temperature approaches zero, then the thermal energy density approaches zero. Importantly, as the temperature approaches absolute zero, then the thermal energy density would no longer be linearly proportional to temperature! Remember the thermal energy density differs for different material at the same temperature. And accept that vacuum's now have a temperature, something that was lost in the sledge hammer mathematics! And that is that!

The Combined Law of Thermodynamics

Traditional thermodynamics sometimes like to combine the first law with the second law and obtain what is sometimes referred to as the "combined law of thermodynamics":

$$dE - TdS + PdV \leq 0 \qquad 16.13$$

In terms of our new perspective, there are inherent problems with eqn 16.13, as have been discussed.

Entropy: It is One, or the Other
Entropy is either:
> 1) Something that when multiplied by temperature change gives changes to a system's ability to perform work in which case: hence : $TS = \varepsilon + PV$ is valid
> 2) Something that when multiplied by temperature change gives changes to a system's energy in which case: $dS = dQ/T$ is valid, and entropy becomes the heat capacity for inhomogeneous systems
> 3) A sort of heat capacity that determines the efficiency of Helmholtz free energy processes
> 4) A sort of heat capacity that enables Gibbs free energy to function i.e. a kinematic number times accessible thermal energy, that being a representation of the mean accessible energy from all the neighbors at a given T
> 5) A count of the number of microstates i.e. eqn 16.3

Even as a mathematical contrivance, entropy now needs to become one or the other, but it cannot remain everything to everyone. Certainly if 5) gives entropy its clarity, then entropy remains a mathematical construct based upon probabilities. Dunning[/], points out the probability based entropy has little resemblance to entropy as defined in classical sense fundamentally based upon 1).

I take matters further realizing that changes in a system's ability to do work and changes to its energy are not always one, especially when related to a system's temperature (change or otherwise). Accordingly, I presented five possibilities, trusting that more may exist.

Interestingly, Shufeng-Zhang[17] rightfully points out that Q is not a single valued function of T. Furthermore, this is entropy's fundamental error, which brings into question Clausius's entropy as seemingly was backed by Boltzmann's entropy. Of course our comprehension of this gross error extends further than this but at least others have been thinking about the impossibility of traditional thermodynamics.

New Realizations

Thermodynamics primarily concerns how systems of matter interact with thermal energy. It was written based upon two fundamental laws. The first law being energy is conserved and the so-called second law based upon the traditional assertion that isothermal entropy never decreases. In order to maintain the second law's glory, traditional thermodynamics has become a science based upon entropy change, all of which is now arguably a complication of the simple.

Our new perspective focuses upon energy change, with the understanding that work is an irreversible process, and that this has no real basis in entropy change! In other words, thermodynamics could be simplified by omitting and/or reconsidering entropy in all of it guises. Certainly, unnecessary complications arose because as a mathematical contrivance, entropy has too often been used explain so many of the sciences ailments hence becoming the science's cornerstone. Yet, strangely to this day entropy lacks the clarity of having a precise discernible meaning!

We now understand that the reason that energy relations tend to be directly proportional to temperature is because the Sun's rays controls the thermal energy density within our massive heat baths/sinks (oceans, atmosphere and/or our planet). And this renders the thermal energy density of systems into linear functions of temperature! This occurs as systems try to attain thermal equilibrium with these massive heat baths/sinks, which in part explains Clausius's consideration that something times temperature equates to energy. Initially there was nothing wrong with calling that something entropy! It is really the universality with which entropy multiplied by temperature is applied, that makes one ponder its sanity.

The complication arose partly because a system's thermal energy density results in linear functionality of temperature for both the system's thermal/blackbody radiation and the kinematics of its contents. Rather than understanding this, the scientists concentrated upon the kinematics of matter and then developed statistical thermodynamics based upon circular logic. Furthermore, the circular logic embraced a kinetic theory that put the cart in front of the horse i.e. it started with billiard ball physics wherein both momentum and kinetic energy were both strangely conserved. Herein we now realize that condensed matter and/or large polyatomic gases pump their kinematics onto smaller sufficiently dilute gases.

Moreover, there also was the whole unnecessary yet ridiculous conceptualization that all intermolecular collisions are elastic. Certainly this traditional belief was beheld due to some futile attempt to simplify the science. All it really accomplished was to promote the absurd science. Reality is probably better served by treating most intermolecular collisions as being inelastic, thus explaining P-T relations, molecular dissipation etc.

To further exasperate the problem, there are the issues associated with the traditional misunderstandings concerning lost work by expanding systems. Certainly Planck and his peers understood that work is often done onto the atmosphere. However they failed to realize that such work is lost work into our atmosphere ($P_{atm}dV$), which ultimately lends itself to the heating of the atmosphere (global warming?). Instead of realizing what lost work is, the science has clung onto the false postulate known as the second law. This lent itself to the crazy 20th century notion that a system's increase in randomness (?) explained why energy is lost hence served as circular reinforcement of the so-called second law.

It is interesting that today some seemingly understand that work is done onto the surrounding but few are willing to make the next necessary mental leap. So today the science stands transfixed in macabre thought, that there is still energy associated with an expanding system's randomness, and yet the work done is onto the surroundings. In part this is due to the hideous refusal to state the obvious that the surrounding tends to be our atmosphere!

Furthermore, work is a form of energy but a change to a system's ability to perform work does not equal its energy change, as can readily be witnessed by comparing the ideal gas law to a gas's kinetic energy. Equally perverse is the traditional indoctrination of limiting work to isobaric processes, and how unnecessarily complicated this made things when considering processes where both pressure and volume increased i.e. bubble nucleation.

Latent heat and Lost work

When contemplating latent heat, it was realized that the latent heat of vaporization should be greater than that of condensation by the amount of work lost during vaporization. The point being that it takes work to upwardly displace the weight of our atmosphere, but the converse does not hold for lost work. Specifically, when the atmosphere is displaced downwards, the only net effect is a transformation of potential energy into kinetic energy by any downwardly displaced atmospheric gas molecules. Obviously, the energy that was lost in vaporization cannot be magically found in condensation. Again neither entropy nor the second law is required to explain what we witness.

Other Realizations

We realized that in order to fully understand then blackbody/thermal radiation needs reconsideration. Thermal energy being the spectrum of photons that can be readily absorbed by condensed matter, resulting in vibrational energy within that matter. We concluded, if we were to insert a thermometer into the vacuum then a temperature reading would be obtained, hence the vacuum has a temperature. This is something not traditionally accepted wherein temperature is strictly associated with molecular kinematics.

When contemplating condensed matter, we followed tradition, basing the thermal energy upon the vibrational energy within condensed matter. An interesting evolution in thought occurred when we contemplated the kinetic energy of a gas, and realized that Avogadro's hypothesis was due to the energy exchanges with the molecules that constitute a system's walls. This also influenced how we view the ideal gas law and kinetic theory as a whole.

The next great insight that was developed was the concept that a liquid's boiling temperature is based upon a vaporizing molecule plus its six neighbors, all possessing enough accessible energy for the vaporizing molecule to break its intermolecular bonds and then escape the liquid, resulting in an explanation for the latent heat of vaporization. The same logic was extrapolated and applied to the critical temperature. Of course this all meant that we had to rewrite the probability function, and the ensuing rate equations. Interestingly, in Appendix A.1, a plausible proof to this logic will be ascertained.

The next insight was more of an evolution than revolution arriving at a new understanding for Henry's law. And then our analysis of vaporization rates led to a series of equations that were similar to heralded Clausius-Clapeyron equation.

Traditional Teaching

Entropy based thermodynamics has become enshrined early in most scientist's psyche. We were taught probability functions, and about the number of possible energy states within a system. And taught a probability based universe, which still may be the case but how these are written requires new thought. Never forget that statistical analysis is merely an eloquent mathematical language. Be humble because too often, learning heartens our benevolence thus encouraging faulty circular logic!

Certainly learning about entropy, probabilities and their result, has provided remarkable insights. Einstein realized that the entropy of radiation has the same form as that of a gas[15,16]. It sounds grand, but it also may just be a mathematical consequence to the fact that the thermal radiation density from our Sun is proportional to temperature, for most of the temperature regimes that we witness here on Earth.

For any given problem there is often more than one solution, each based upon a chosen perspective. This author's distain arises when one claims that there exists only one language to explain things, and then arrogantly states that it has to be. Only when something can be explained several different ways, in more than one language, in its simplistic context can we take such an arrogant stance.If we want language to distinguish mankind, whether it is mere words, or complex mathematics, we must be able to visualize that of which we speak. Without visualization, both language and math is worthless. This author remains unconvinced, that all, which is traditionally professed could be readily visualized.

The uncertainty principle is not some enshrined in either classical thermodynamics, nor in the new perspective presented herein. Actually uncertainty is a mathematical result of the application of statistical mechanics to thermodynamics[7]. It is this author's belief that when statistical mechanics is rewritten and then properly applied onto our new perspective, not only will it seemingly support this perspective but so to will any mathematical deduced uncertainty.

One may ask: Why rewrite statistical thermodynamics? Because it was developed with the principle that all molecular collisions are elastic! We have shown that empirical findings actual back molecular collisions being inelastic, therefore a complete rethink of statistical followed by quantum physics may be required. Hopefully others see this as an opportunity rather than insult to their humanity!

Closing Remarks: Revolution or Evolution

Clifford Truesdell[18]: "*I hesitate to use the term first law and second law, because there are almost as many first laws as there are thermodynamicists, and I have been told by these people for so many years that I disobey their laws that now I prefer to exult in my criminal status and non condemning names to the concrete mathematical axioms I wish to use in my outlaw studies of heat and temperature. The term entropy seems superfluous, also, since it suggests nothing to ordinary persons and only intense headaches to those who have studied thermodynamics but have not given in and joined the professionals.*"

Yes traditional theory seemingly explained so much, all based upon its sledge hammer driven mathematical structure. However, not realizing that volume increases signify work lost in displacing our atmosphere was missing the obvious. And insisiting all molecular collisions are elastic was floppish.A sad reality concerning our new perspective remains. As happened with phlogiston, it may take a few generations (if at all) before mankind accepts what thermodynamics is. Whether or not, this actually turns out to be the case is purely up to you the reader. If you still insist upon traditional entropy based thermodynamics, so be it. But please arrive at one all-encompassing interpretation of entropy, so us free thinkers can argue against its virtues. Furthermore at least accept that we did not simply bedevil our past. Rather we demonstrated that much of what has been claimed in the guise of entropy and the second law, can equally be explained using fundamental principles and constructive perspective.

The ultimate question that the independent reader, should so is: Apply Occam's razor! If you answer the new perspective presented herein is the simpler explanation, then you will be open to this providing answers, where none existed before. This is especially true in systems wherein both the volume and pressure change, i.e. subjects like nucleation, and cosmology endear themselves to this new perspective.

Alexander Unzicker[19] points out that in "the structure of scientific revolutions" Thomas Kunn considers that *"Long periods of "normal science" dominated by the paradigm of time, are interrupted by revolutions in which a large part of the existing knowledge becomes obsolete and a new paradigm, often simpler and more insightful is established"*. Is this book revolution or garble? You decide.

And thank you kindly, for reading what I wrote. You are the best! Are you ready for the future book on tensile layer, nucleation and DCI? It was written long before this book and will hopefully be coming soon. Never forget; ***What a tangled web we weave when at first we fail to perceive!*** *KWM*

References:

1) Juan Campanario, Brian Martin "Journal of Scientific Exploration" Vol 18 No 3 pg421-438 2004
2) Caroline Haartmann 21st century Science & Technology pp 19 Fall-winter 2008
3) Fundamentals of Statistical and Thermal Physics", F. Reif, McGraw-Hill, New York, 1965
4) "A Farewell to Entropy: Statistical Thermodynamics Based on Information". Arieh Ben-Naim World Scientific Publishing Co., London, New York 2011
5) Internet: A concise introduction to elementary statistical mechanics, or: where does the Boltzmann factor come from, by Matt McIrvin (2012)
6) Planck, Max "Treatise on Thermodynamics" Third edition, London, Logmans, Green and co., 1917
7) Dunning-Davis, Jeremy,& Lawrence, Norma, Richard "Truth in Paradigm" Stand Dead Publications 2017
8) Mayhew, K.W. "Improving our thermodynamic perspective", Phys. Essays **24**, vol 3,338 (2011)
9) Mayhew K.W. " Energetics of Nucleation", Phys. Essays 17,vol 4 (2004) 476
10) "Challenges to the second law of thermodynamics" Vladislav Capek and Daniel P. Sheehan, Springer Press, Netherlands 2005
11) D. Sheehan, J. Sci. Exploration, **12** (2), 303 (1998)
12) Mayhew,K.W. "Second Law and lost work" Phys. Essays **28**, 1,152-155 (2015)
13) Mayhew,K.W. "Entropy: An ill-conceived mathematical contrivance" Phys. Essays **28**, 3,352-357 (2015)
14) Mayhew, K.W. "A new perspective for kinetic theory and heat capacity", *Progress in Physics* Vol. 13 (**4**) 2017 pg 166-173
15) Norton, John D. "Atoms entropy quanta: Einstein's miraculous argument of 1905" His. & Phil.of Modern Phys. 37,(2006) pg 71-100
16) Irons F.E. "Reappraising Einstein's 1905 application of thermodynamics and statistics to radiation" Eur. J of Phys. 25 (2004) 269-277
17) Shufeng Zhang "Entropy: A concept that is not a physical quantity" Phy Essays **25**(2) 172-176 2012
18) Clifford Truesdell: www.eoht/info/page/clifford&trues
19) A.Unzicker, "On the Origin of the Constants *c* and *h*" or http://vixra.org/pdf/1508.0170v1.pdf

Appendix

Appendix A

Is an analysis by this author, one which employs accepted empirical data to show that latent heat changes are on the scale that one would expect for changes in potential energy of a cloud of dipoles! This is important as it backs our assertions concerning latent heat in Chapters 10 and 11.

Appendix B

For the most part, this appendix gives traditional interpretations. Its purpose is to expand concepts that were only briefly detailed in various chapters of this text, all of which rewrites previously accepted understandings. It also contains some more of this author's thoughts as a continuation of accepted conceptualizations.

Appendix A.1: **Latent Heat vs Pressure for Water**

Abstract: In Chapters 10 & 11, it was discussed that the latent heat of vaporization for water decreased with increasing pressure, as illustrated by Graph A.1.1. Furthermore, in Chapter 11, it was concluded that the latent heat of vaporization for ideal substances could be explained in terms of a kinematic number, possibly 9 or 8.5, which helps explain the vaporization path.

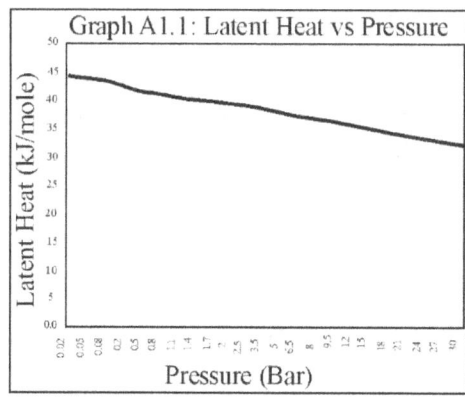

Graph A.1.1: Latent Heat vs Pressure

We shall show that the latent heat vs pressure is of the order that we might expect for changes to a cloud of dipoles with pressure. As for explaining why the laent heat for a real substance is greater than the kinematic number, well that would take some more modeling and analysis. However based upon the fact that changes to latent heat is now relatable to a cloud of dipoles, then that remains the most likely reason that ideal substances adhere to the kinematic number while real gases do not.

Theory

Graph A.1.1 was obtained by plotting data from the steam tables: Table A.2 Water molecules are polar molecules, which we will consider as dipoles. The non-ideal nature of water is evident by looking at how the steam's density changes with pressure. As the pressure is doubled, the steam's volume does not simply decrease by a factor of 2. For example, consider the change from 1 to 2 bar pressure, as given in Table A.2. If water vapor was an ideal gas, then its molecular volume would decrease by 50% as the pressure doubled, however the actual volume decrease is 47.8%.

In Fig.A.1.1, water's liquid state is modeled as having a strong alignment of the dipole moments in the liquid state, and a much weaker alignment in the vaporous state. Due to the fact that water in incompressible, when analyzing how the latent heat changes with pressure, the energy of molecular bonding in the liquid state will be considered as being independent of pressure. Therefore, changes in bonding energy (ΔU) are only due to changes of the bonding energy in the vaporous state.

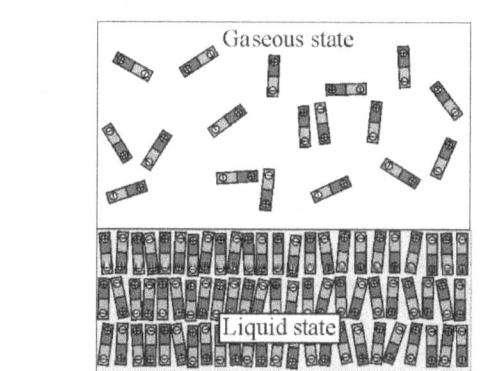

Gaseous state

Liquid state

Fig. A.1.1 Shows dipole molecules having strong alignment of their dipole moments in the liquid state versus a weaker alignment of dipole moments in the gaseous state.

As the pressure is increased, the mean molecular volume in the vaporous state will decrease therefore, the density of dipole moments must increase. There would be a certain bonding between the vaporous polar molecules, which increases as the density of their dipole moments increases. Our expectation would be that the change in molecular bonding (ΔU) between the liquid and vaporous state must decrease, as the pressure increases, with the pure liquid state representing the highest degree of molecular bonding.

At this point we really need to consider, how to model the energy change associated with the change in density of dipole moments. Note: The electrical force (F_e) between two charges q_1 and q_2 can be expressed as: $F_e \alpha q_1 q_2 / r^2$. Similarly, the magnetic force (F_m) between two poles p_1 and p_2 can be expressed as: $F_m \alpha p_1 p_2 / r^2$.

Unfortunately this author could not find an analysis for the change in potential of a cloud of dipoles. Certainly, calculations for a cloud of dipoles would be more exacting. However, at this time, we will have to make the best approximation possible using the traditional derivation for a cloud of charged particles. This author must emphasize that the following is to be taken as an approximation, and I sincerely hope that someone devises a better analogy/analysis for a cloud of dipoles.

Cloud of Similarly Charged Particles

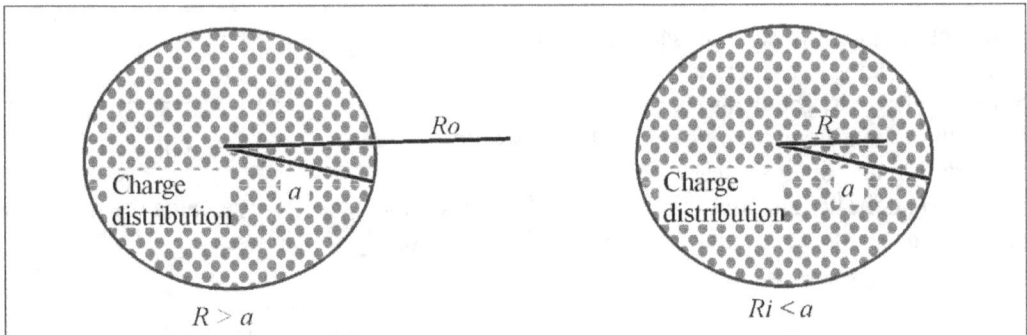

Fig A.1.2 Shows a spherical charge distribution of radius "a". On the L.H.S the electric field is measured at some radius (Ro) outside of the charge distribution. On the R.H.S the electric field is measured at some radius (Ri) located inside of the charge distribution.

Consider a spherical cloud of charged particles of radius a. Each particle has a similar charge q. The charge per unit volume is known as the charge density and is given by: ρ . The condition of spherical symmetry means that ρ only depends upon the distance of a point (at R) from the center and not upon the direction. Find the expression for the electric field (E') both outside and inside of the charge distribution:

For a point outside of a (i.e. $R > a$): Applying Gauss's law to a spherical Gaussian surface of radius R in Fig. A.1.1 leads to[1]:

$$E' = Q / 4\pi\varepsilon_0 R^2 \qquad\qquad A.1.1$$

where ε_0 = permittivity constant=8.85 x 10^{-12} F/m, and $Q = \sum_i q_i$, represents the total charge.

It turns out that for points outside the spherically symmetric charge distribution, the electric field (E') has the same value that it would have for a point charge q, where $q=Q$.

For a point inside of the charge distribution: $R < a$, Gauss's law gives:

$$\varepsilon_0 \oint E' * dS = \varepsilon_0 E'(4\pi R^2) = Q' \quad A.1.2$$

Or

$$E' = Q' / 4\pi\varepsilon_0 R^2 \qquad\qquad A.1.3$$

Q' is simply the part of Q that is contained within the sphere of radius R. The part of Q that lies outside of radius R makes no contribution to the electric field at the radius R.

For a uniform charge distribution, we can express Q' as:

$$Q' = Q[(4\pi r^3/3)/(4\pi a^3/3)] \qquad \text{A.1.4}$$

Which becomes:

$$Q' = Q(R/a)^3 \qquad \text{A.1.5}$$

The expression for the electric field becomes:

$$E' = QR/4\pi\varepsilon_0 a^3 \qquad \text{A.1.6}$$

In order to understand how the electric field varies between the inside and outside of a spherical cloud of radius a, of charged particle, we give the following Graph A.1.3 sketched from "The Electromagnetic Problem

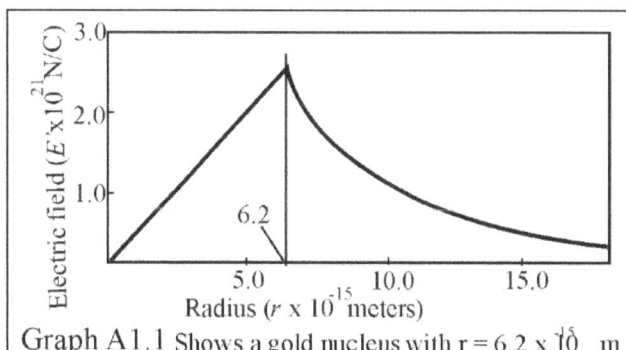

Graph A1.1 Shows a gold nucleus with $r = 6.2 \times 10^{-15}$ m and a total charge of: $q = Zc = (79)1.6 \times 10^{-19} = 1.3 \times 10^{-17}$ C

Solver"[1pg116], which shows the electric field both inside and outside of a gold nucleus. We can see that at its surface ($R = 6.2 \times 10^{-15}$ m) the electric field has a maximum value. At this maximum value eqn A.1.6 gives the same value as eqn A.1.3, which is expected when, $a = R$. Calculating the potential energy of a sphere of like charged particles, then we would start off with eqn A.1.1: $E' = Q/4\pi\varepsilon_0 R^2$.

In terms of charge density (ρ) we can write the total charge (Q) as:

$$Q = (4\pi a^3/3)\rho \qquad \text{A.1.7}$$

where "a" is the radius of the cloud of charged particles.

For a point located at some distance r, which is outside of the charged sphere of radius a, we can substitute eqn A.1.7 into eqn A.1.1, and write the electric field as:

$$E' = \rho a^3/3\varepsilon_0 R^2 \qquad \text{A.1.8}$$

We can write the potential (Φ) as given by the electric field: eqn A.1.8, as:

$$\Phi = \rho a^3/3\varepsilon_0 R \qquad \text{A.1.9}$$

If we consider that $R = a$ then have:

$$\Phi = \rho a^2/3\varepsilon_0 = \rho R^2/3\varepsilon_0 \qquad \text{A.1.10}$$

The energy of a sphere of charged particles of radius $a=R$ is:

$$U_e = 1/2 \int_V \rho(R)\Phi(r)dV \qquad \text{A.1.11}$$

Realizing that for a sphere of radius r: $dV = 4\pi r^2 dr$, then eqn A.1.11 becomes:

$$U_e = 1/2 \int_r \rho(R)\Phi(R)4\pi R^2 dR \qquad \text{A.1.12}$$

Substituting into eqn A.1.12, we obtain:

$$U_e = 1/2 \int_r \rho(\rho R^2/3\varepsilon_0)4\pi R^2 dR \qquad \text{A.1.13}$$

From eqn A.1.7 we realize that: $\rho = 3Q/4\pi R^3$ therefore, eqn A.1.13 becomes:

$$U_e = 1/2 \int_r (3Q/4\pi R^3)^2 (R^2/3\varepsilon_0) 4\pi R^2 dR \qquad \text{A.1.14}$$

Collecting the terms and moving the constants out of the integrand we obtain:

$$U_e = (3Q^2/8\pi\varepsilon_0) \int_r (1/R^2) dR \qquad \text{A.1.15}$$

For latent heat calculations: One needs to know how the energy of a spherical cloud of charged particles changes with radius (R). I.e. change in energy of a spherical cloud (eqn A.1.15), when R changes from an initial value (R_i) to a final value (R_f) is[1]:

$$dU_e/dR = (3Q^2/8\pi\varepsilon_0)(1/R_f{}^2 - 1/R_i{}^2) \qquad \text{A.1.16}$$

Our Analysis

Based upon the above standard theory, eqn A.1.16 is to be applied to a cloud of similar charged particles. The work (potential energy) required creating a cloud, of similarly charged particles increases as the mean distance between molecules decreases. Another way of viewing this is to consider that the charged particles are initially separated by an infinite distance. Since electromagnetic repulsion between the particles exists, then it requires energy to bring those charged particles closer together.

The corollary to the above being: In order to separate a cloud of oppositely charged particles, by some infinite distance requires energy (work). For the case of two similar versus two opposite charges, the magnitude of required energy is the same for both, therefore, as an approximation we shall assume: The energy required to separate a cloud of oppositely charged particles to an infinite distance, equals the energy required to form a cloud of similarly charged particles, from particles that were originally at an infinite distance.

Now consider that polar water vapor molecules align themselves due to the attraction of their opposite dipoles, as was illustrated in Fig. A.1.1. An approximation for dipoles could be that the bonding energy of vaporous dipole molecules decreases as the radius of molecule separation increases.

Let us consider a simplified version of eqn A.1.16, where: C'' is some constant and write:

$$dU/dr = C''(1/R_f{}^2 - 1/R_i{}^2) \qquad \text{A.1.17}$$

Instead of a cloud of radius R, we want to consider r to be the radius of the molecular volume (v) in the vaporous state, which we will call "intermolecular radius". We expect that the binding energy between vaporous dipole molecules increases as the intermolecular radius (r) decreases, in which case we would consider that eqn A.1.1 applies to the changes in bonding energy of a cloud of dipoles.

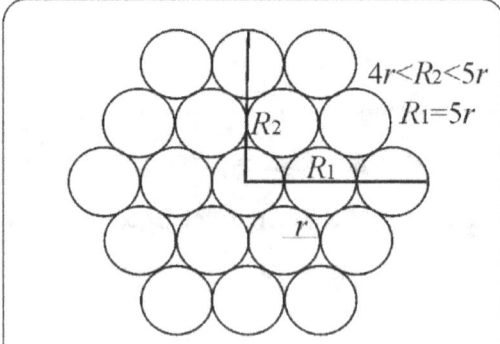

Fig. A.1.3 Shows 18 intermolecular volumes of radius r, closely packed.

We need to rewrite eqn A.1.17 in terms of per molecule in order to correlate with our analysis. Consider, the 18 closely packed gaseous molecules, as is shown in Fig A.1.3. Consider each gas molecule to approximately occupy a volume r, that being its intermolecular volume. We can see that the radius (R)

of total volume, occupied by the 18 molecules is between R_1 and R_2. According, we can say that: $16r < R^2 < 25r$.

The point for the 18 gaseous molecules becomes; each occupies a volume as defined by their intermolecular radius (r). On a per molecule basis eqn A.1.17 can be rewritten as:

$$du / dr = (1/18)dU / dR = C''[1/ R_f{}^2 - 1/ R_i{}^2] \qquad A.1.18$$

Realize that: $16r < R^2 < 25r$. According, for simplicity/sanity consider that: $R^2 \approx 18$. Please, keep in mind that our analysis is an approximation at best. Eqn A.1.18 can now be rewritten as:

$$du / dr = (1/18)dU / dR \approx C''[1/(18r_f{}^2) - 1/(18r_i{}^2)] \qquad A.1.19$$

Which can be rewritten as:

$$du / dr \approx C''(1/ r_f{}^2 - 1/ r_i{}^2) \qquad A.1.20$$

Although not precisely correct, eqn A.1.20 implies that eqn A.1.17, which for a cloud of molecules can be approximated with eqn A.1.20, which is on a per molecule basis. Our goal is to determine how the empirically found latent heat data in Table A.2 correlates to eqn A.1.20!

Letting the subscript "g" and "l" respectively represents the gaseous and liquid. Start by considering that ΔU is simply the energy required to break the bond when going from a liquid to a gaseous state. Therefore, we are interested in the energy required to separate all the molecules some infinite distance of separation. For a system of molecules, we shall utilize eqn 10.4: $L_{(l \to g)} = U_g - U_l + P_g V_g$, which can be rewritten in terms of the bonding energy (U) as:

$$U_g - U_l = -(L_{(l \to g)} - P_g V_g) \qquad A.1.21$$

We want to know how the bonding energy changes, with the intermolecular radius, i.e.: dU'/ dr. Where; $U' = U_g - U_l$. Although, it seems like a mess, the change in bonding energy with intermolecular radius can be rewritten, as:

$$dU'/ dr = d(U_g - U_l)/ dr = -d(L_{(l \to g)} - P_g V_g)/ dr$$

Let the subscripts "i" and "f" respectively represent the initial and final, state. Expanding the above eqn A.1.22 into a more general equation, gives:

$$[(U_{gf} - U_{lf}) - (U_{gi} - U_{li})]/(r_{gf} - r_{gi}) = -[L_{(l \to g)f} - P_{gf}V_{gf} - (L_{(l \to g)i} - P_{gi}V_{gi})]/(r_{gf} - r_{gi}) \qquad A.1.23$$

Due to water being incompressible, we assume that the liquid's molecular bonds remain constant, as we change the pressure. Therefore:

$$U_{lf} - U_{li} \qquad A.1.24$$

Based upon eqn A.1.24, we can simplify the L.H.S. of eqn A.1.23, and write:

$$(U_{gf} - U_{gi})/(r_{gf} - r_{gi}) = -[L_{(l \to g)f} - P_{gf}V_{gf} - (L_{(l \to g)i} - P_{gi}V_{gi})]/(r_{gf} - r_{gi}) \qquad A.1.25$$

We can rewrite eqn A.1.25 as:

$$(U_{gf} - U_{gi})/(r_{gf} - r_{gi}) = -[L_{(l \to g)f} - L_{(l \to g)i} - (P_{gf}V_{gf} - P_{gi}V_{gi})]/(r_{gf} - r_{gi}) \qquad A.1.26$$

Rewriting eqn A.1.26 in terms of the delta function (Δ):

$$\Delta U_g / \Delta r_g = -(\Delta L_{(l \rightarrow g)} - \Delta(P_g V_g) / \Delta r_g \quad \text{A.1.27}$$

If all the changes are infinitesimally small ($\Delta \rightarrow 0$), then in differential form eqn A.1.27:

$$dU / dr = -d(L_{(l \rightarrow g)} - P_g V_g) / dr \quad \text{A.1.28}$$

Which can be expanded and rewritten as:

$$dU / dr = (V_g dP_g + P_g dV_g - dL_{(l \rightarrow g)}) / dr \quad \text{A.1.29}$$

Using eqn A.1.26, instead of eqn A.1.29. On a molecular basis eqn A.1.26, becomes:

$$X(u_{gf} - u_{gi}) / (r_{gf} - r_{gi}) = -X[L_{(l \rightarrow g)f} / X - L_{(l \rightarrow g)i} / X - \{(P_g v_g)_f - (P_g v_g)_i\}] / (r_{gf} - r_{gi}) \quad \text{A.1.30}$$

We can divide both sides by X, obtaining:

$$(u_{gf} - u_{gi}) / (r_{gf} - r_{gi}) = -[l_{(l \rightarrow g)f} - l_{(l \rightarrow g)i} - \{(P_g v_g)_f - (P_g v_g)_i\}] / (r_{gf} - r_{gi}) \quad \text{A.1.31}$$

Analyzing the empirical data

The Steam Table A.2 provides the empirically measured values for latent heat in relation to pressure. The latent heat of vaporization at 1 bar: $L_{(l \rightarrow g)} = X l_{(l \rightarrow g)} = 40676.39$ J/mole

Remember: A traditional way of writing latent heat in terms of a mole is: ΔH_{vap}. This author's preference is to write the latent heat in terms of a per molecule basis, i.e.: $\Delta H_{vap} = X l_{(l \rightarrow g)}$, when X is a mole of molecules.

Thus: $l_{(l \rightarrow g)}$ = 40676.39 (joules/mole) /6.022 x 10^{23} molecules/mole

$$= 6.7546 \times 10^{-20} \text{ joules/molecule}$$

The latent heat of vaporization at 1.1 bar is: $L_{(l \rightarrow g)}$ = 40547.42 joules/mole. Thus:

$l_{(l \rightarrow g)}$ = 40547.42 (joules/mole) /6.022 x 10^{23} molecules/mole

$$= 6.7332 \times 10^{-20} \text{ joules/molecule}$$

Therefore, the change in latent heat per molecule in going from 1 bar to 1.1 bar is:

$$L_{(l \rightarrow g)f} - L_{(l \rightarrow g)i} = 6.733 \times 10^{-20} - 6.755 \times 10^{-20} = -2.142 \times 10^{-22} \text{ j/molecule}$$

Next, we need to determine the intermolecular radii (r) and molecular volume (v). Instead of specific density (kg/m³), we will use the specific volume (m³/kg), which allows us to calculate: $\Delta P_g v_g = (P_g v_g)_f - (P_{lvl})_i$, as well as the change to intermolecular radius (r). (Note: Table A1.1 is located at the very end of this Appendix A).

I.e. consider the change in latent heat of vaporization from 1.0 to 1.1 bar. From Table A.1 we have: Steam's specific volume at 1.0 bar: 1.694 m³/kg; versus at 1.1 bar: 1.549 m³/kg

The mass of 1 water vapor molecule is: 18.015 (g/mole)/6.022x10^{23} (molecules/mole)

$$= 2.9915 \times 10^{-23} \text{ g/molecule} = 2.9915 \times 10^{-26} \text{ kg/molecule}$$

The molecular volume at 1.0 bar is: v_g = 2.9915x10^{-26} (kg/molecule) x 1.694 (m³/kg)

$$= 5.068 \times 10^{-26} \text{ m}^3/\text{molecule}$$

Thus at 1.0 bar the energy of the volume occupied by a solitary water vapor molecule is:

$(P_g v_g)_i = 1.0 \times 10^5$ (N/m^2) x 5.068×10^{-26} (m^3/molecule) = 5.068×10^{-21} joules/molecule

Molecular volume at 1.1 bar becomes: $v_g = 2.9915 \times 10^{-26}$ (kg/molecule) x 1.549 (m^3/kg)

$$= 4.634 \times 10^{-26} \text{ m}^3/\text{molecule}$$

Table A1.1's value is 4.640×10^{-26} m^3/molecule

Thus at 1.1 bar, the energy of the volume occupied by a solitary water vapor molecule is:

$(P_g v_g)_f = 1.1 \times 10^5$ (N/m^2) x 4.640×10^{-26} (m^3/molecule) = 5.097×10^{-21} joules/molecule

In going from 1.0 bar to 1.1 bar, there is a change in the PV occupied by each molecule of: $\Delta P_g v_g = (P_g v_g)_f - (P_{lvl})_i = 5.097 \times 10^{-21} - 5.068 \times 10^{-21} = 2.962 \times 10^{-23}$ joules/molecule: (Note: The value is from Table A1.1). Reconsider eqn A.1.31: $(u_{gf} - u_{gi})/(r_{gf} - r_{gi}) = -[l_{(l \to g)f} - l_{(l \to g)i} - \{(P_g v_g)_f - (P_g v_g)_i\}]/(r_{gf} - r_{gi})$

Our next step is to calculate: $(u_{gf} - u_{gi}) = -[l_{(l \to g)f} - l_{(l \to g)i} - \{(P_g v_g)_f - (P_g v_g)_i\}]$ Rewriting using the delta (Δ) function as:

$$\Delta u_g = -[\Delta l_{(l \to g)} - \Delta(P_g v_g)] \qquad \text{A.1.32}$$

Eqn A.1.32 can be rewritten as:

$$\Delta u_g = \Delta(P_g v_g) - \Delta l_{(l \to g)} \qquad \text{A.1.33}$$

Substituting in the previously obtained change when the pressure goes from 1.0 to 1.1 bar we obtain: $u_{gf} - u_{gi} = 2.962 \times 10^{-23} - (-2.142 \times 10^{-22}) = 2.438 \times 10^{-22}$ j/molecule

The radius occupied by a molecule is obtained by starting with the molecular volume (v_g):

$$v_g = 4\pi r_g^{\,3}/3 \qquad \text{A.1.34}$$

Reorganizing the terms:

$$r_g^{\,3} = 3v_g/4\pi \qquad \text{A.1.35}$$

And finally taking the cube root:

$$r_g = \sqrt[3]{3v_g/4\pi} \qquad \text{A.1.36}$$

Therefore, the mean intermolecular radius at 1.0 bar is obtained by:

$r_{gi} = [5.068 \times 10^{-26}/3/(4\pi)]^{1/3} = 2.296 \times 10^{-9}$ m

Similarly, at 1.1 bar: $r_{gf} = [4.634 \times 10^{-26}/3/(4\pi)]^{1/3} = 2.228 \times 10^{-9}$ m

Therefore, the change in intermolecular radius in going from 1.0 to 1.1 bar is:

$r_{gf} - r_{gi} = -6.746 \times 10^{-11}$ m

Before we can calculate C'', we need to calculate: $(1/r_{gf}^{\,2} - 1/r_{gi}^{\,2})$:

At 1.0 bar we have: $1/r_{gi}^{\,2} = 1/(2.296 \times 10^{-9})^2 = 1.898 \times 10^{17}$ m^{-2}

At 1.1 bar we have: $1/r_{gf}^2 = 1/(2.228 \times 10^{-9})^2 = 2.014 \times 10^{17}$ m^{-2}

Therefore: $(1/r_{gf}^2 - 1/r_{gi}^2) = 2.014 \times 10^{17} - 1.898 \times 10^{17} = 1.166 \times 10^{16}$ m^{-2}

Reconsider eqn A.1.20: $du/dr \approx C''(1/r_f^2 - 1/r_i^2)$. In terms of C'' eqn A.1.20:

$$C'' = (u_{gf} - u_{gi})/(r_{gf} - r_{gi})]/(1/r_{gf}^2 - 1/r_{gi}^2) \qquad \text{A.1.37}$$

Where: $(u_{gf} - u_{gi}) = -[l_{(l \rightarrow g)f} - l_{(l \rightarrow g)i} - (P_g v_g)_f - (P_g v_g)_i]$

Substituting in gives:

$$C'' = [2.438 \times 10^{-22} \text{ (j/molecule)}]/[-6.746 \times 10^{-11} \text{ (m)}]/ [1.166 \times 10^{16} \text{ (m}^{-2})]$$

$$C'' = -3.1 \times 10^{-28} \text{ joules*meter/molecule}$$

Similar calculations are performed for various pressures in Table A1.1. Averaging the value for C'' from all 65 pressure changes, as is given in Table A1.1, gives:

$$C''_{avg} = -3.86 \times 10^{-28} \text{ j*m/molecule}$$

With a standard deviation of: 1.34×10^{-28}.

The range of values is relatively large i.e.: $-8 \times 10^{-28} > C'' > -7 \times 10^{-29}$. Certainly, for a cloud of dipoles we would not expect our so-called constant (C'') to be a true constant, as the expectation is that there would be some sort or ordering of dipole moments, as the pressure increases. Note: In Table A1.1 the values of C'' were calculated while maintaining 4 decimal places for every calculation, even though only two decimal places are shown. Note: More decimal places, lower standard deviation should be.

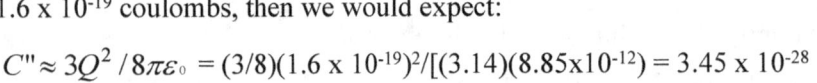

Graph A.1.3: Constant (C") vs Pressure
Pressure (Bar)
Constant C"

We considered a cloud of charged particles and found the following result eqn A.1.16:
$dU_e/dr = (3Q^2/8\pi\varepsilon_0)(1/R_f^2 - 1/R_i^2)$ defining how the potential energy of a cloud of similarly charged particles increases with pressure. Now consider: $3Q^2/8\pi\varepsilon_0$. Assuming that the charge is of the magnitude of an electron: 1.6×10^{-19} coulombs, then we would expect:

$$C'' \approx 3Q^2/8\pi\varepsilon_0 = (3/8)(1.6 \times 10^{-19})^2/[(3.14)(8.85 \times 10^{-12}) = 3.45 \times 10^{-28}$$

Obviously, the absolute value for our calculated value for $C''_{avg} = 3.86 \times 10^{-28}$ is of the same order of magnitude as the theoretical value: 3.45×10^{-28}, for a system of similar (repulsive) point charges. This is what one would expect if our theory were valid.

Conclusions

We must bear in mind that all we wanted to accomplish in this mammoth procedure was to show that our analysis for latent heat in Chapters 10 & 11 warrants consideration, e.g. both our descriptions for latent heat, and the kinematic number based theory.

Graph A.1.3 shows so-called "Constant (C'') vs Pressure". Its downward slope is what we might expect for dipoles. The saw-tooth pattern is due to the scale along the pressure axis. Certainly, nothing was proven beyond a reasonable doubt here. Simply showing the changes in latent heat with pressure   as being on the scale that we would expect for changes in the radius of a cloud of oppositely charged point charges, is not undeniable proof. Certainly it seemingly validates eqn 10.4! But it is food for thought, rendering the conceptualizations given herein, as very plausible! Hopefully, someone adept at dealing with a cloud of dipoles will take this on.

To reiterate: There are various ways of viewing this remarkable result:

1) Changes to the internal energy of a non-ideal gas can certainly be explained in terms of changes in electromagnetic potential within gases.

2) Eqn 10.4: $L(l \rightarrow g) = U_g - U_l + P_g V_g$, functions as an improvement over traditional writing of latent heat of vaporization.

3) The reason non-ideal gases do not adhere to the kinematic numbers as previously described in Chapter 11, may very well be due to complex variations in the electromagnetic potential within gases. How exactly, well that requires much more complex contemplation and theorizing followed by lots of experimentation.

Note: This author appreciates that some may disagree with publishing the following data tables (Table A1.1). However this author feels that it's the importance of the above result warrants the publication herein, so others can improve on the result.

References:

1. "The Electromagnetic Problem Solver", Edited by M. Fogiel, Research and Education Association, New York, 1983

Tables A.1: The following is the data set that goes with the above analysis. Due to its size the following Table is divided into two halves. Table A1.1A, and Table A1.1B

Table A1.1A: First half

P	Pressure	Specific V	$L(l \rightarrow g)$	$l(l \rightarrow g)$	$\Delta l(l \rightarrow g)$	v_g	$P_g v_g$	$\Delta(P_g v_g)$	$u_{gf} - u_{gi}$
Bar	N/m²	m³/kg	j/mole	j/molecule	j/molecule	m³/molec	Joule	Joule	Joule
0.02	2.00E+03	67.006	44320.15	7.36E-20		2E-24	4.01E-21		
0.03	3.00E+03	45.667	44039.57	7.31E-20	-4.66E-22	1.37E-24	4.1E-21	8.94E-23	5.55E-22
0.04	4.00E+03	34.802	43832.15	7.28E-20	-3.44E-22	1.04E-24	4.16E-21	6.6E-23	4.1E-22
0.05	5.00E+03	28.194	43664.71	7.25E-20	-2.78E-22	8.43E-25	4.22E-21	5.27E-23	3.31E-22
0.06	6.00E+03	23.741	43523.66	7.23E-20	-2.34E-22	7.1E-25	4.26E-21	4.42E-23	2.78E-22
0.07	7.00E+03	20.531	43407.23	7.21E-20	-2.02E-22	6.14E-25	4.3E-21	3.8E-23	2.4E-22
0.08	8.00E+03	18.105	43294.37	7.19E-20	-1.79E-22	5.42E-25	4.33E-21	3.36E-23	2.13E-22
0.09	9.00E+03	16.204	43197.08	7.17E-20	-1.62E-22	4.85E-25	4.36E-21	2.98E-23	1.91E-22
0.1	1.00E+04	14.675	43108.07	7.16E-20	-1.48E-22	4.39E-25	4.39E-21	2.73E-23	1.75E-22
0.2	2.00E+04	7.65	42486.58	7.06E-20	-1.03E-21	2.29E-25	4.58E-21	1.87E-22	1.22E-21
0.3	3.00E+04	5.229	42084.57	6.99E-20	-6.68E-22	1.56E-25	4.69E-21	1.16E-22	7.83E-22
0.4	4.00E+04	3.993	41780.61	6.94E-20	-5.05E-22	1.19E-25	4.78E-21	8.53E-23	5.9E-22
0.5	5.00E+04	3.24	41531.70	6.9E-20	-4.13E-22	9.69E-26	4.85E-21	6.82E-23	4.82E-22
0.6	6.00E+04	2.732	41319.76	6.86E-20	-3.52E-22	8.17E-26	4.9E-21	5.74E-23	4.09E-22
0.7	7.00E+04	2.365	41133.46	6.83E-20	-3.09E-22	7.07E-26	4.95E-21	4.88E-23	3.58E-22
0.8	8.00E+04	2.087	40966.78	6.8E-20	-2.77E-22	6.24E-26	4.99E-21	4.22E-23	3.19E-22

0.9	9.00E+04	1.869	40815.17	6.78E-20	-2.52E-22	5.59E-26	5.03E-21	3.74E-23	2.89E-22
1	1.00E+05	1.694	40676.39	6.75E-20	-2.3E-22	5.07E-26	5.07E-21	3.56E-23	2.66E-22
1.1	1.10E+05	1.549	40547.42	6.73E-20	-2.14E-22	4.63E-26	5.1E-21	2.96E-23	2.44E-22
1.2	1.20E+05	1.428	40426.74	6.71E-20	-2E-22	4.27E-26	5.13E-21	2.9E-23	2.29E-22
1.3	1.30E+05	1.325	40313.60	6.69E-20	-1.88E-22	3.96E-26	5.15E-21	2.66E-23	2.14E-22
1.4	1.40E+05	1.236	40206.50	6.68E-20	-1.78E-22	3.7E-26	5.18E-21	2.36E-23	2.01E-22
1.5	1.50E+05	1.159	40105.43	6.66E-20	-1.68E-22	3.47E-26	5.2E-21	2.42E-23	1.92E-22
1.6	1.60E+05	1.091	40008.89	6.64E-20	-1.6E-22	3.26E-26	5.22E-21	2.12E-23	1.82E-22
1.7	1.70E+05	1.031	39916.11	6.63E-20	-1.54E-22	3.08E-26	5.24E-21	2.12E-23	1.75E-22
1.8	1.80E+05	0.9775	39827.87	6.61E-20	-1.47E-22	2.92E-26	5.26E-21	2.03E-23	1.67E-22
1.9	1.90E+05	0.9294	39742.64	6.6E-20	-1.42E-22	2.78E-26	5.28E-21	1.9E-23	1.61E-22
2	2.00E+05	0.8857	39661.18	6.59E-20	-1.35E-22	2.65E-26	5.3E-21	1.66E-23	1.52E-22
2.2	2.20E+05	0.8097	39505.80	6.56E-20	-2.58E-22	2.42E-26	5.33E-21	2.97E-23	2.88E-22
2.4	2.40E+05	0.7463	39360.99	6.54E-20	-2.4E-22	2.23E-26	5.36E-21	2.93E-23	2.7E-22
2.6	2.60E+05	0.6925	39223.72	6.51E-20	-2.28E-22	2.07E-26	5.39E-21	2.81E-23	2.56E-22
2.8	2.80E+05	0.646	39093.99	6.49E-20	-2.15E-22	1.93E-26	5.41E-21	2.48E-23	2.4E-22
3	3.00E+05	0.6057	38970.29	6.47E-20	-2.05E-22	1.81E-26	5.44E-21	2.48E-23	2.3E-22
3.5	3.50E+05	0.5241	38684.43	6.42E-20	-4.75E-22	1.57E-26	5.49E-21	5.16E-23	5.26E-22
4	4.00E+05	0.4623	38424.97	6.38E-20	-4.31E-22	1.38E-26	5.53E-21	4.44E-23	4.75E-22
4.5	4.50E+05	0.4137	38186.63	6.34E-20	-3.96E-22	1.24E-26	5.57E-21	3.72E-23	4.33E-22
5	5.00E+05	0.3747	37964.88	6.3E-20	-3.68E-22	1.12E-26	5.6E-21	3.54E-23	4.04E-22
5.5	5.50E+05	0.3425	37757.47	6.27E-20	-3.44E-22	1.02E-26	5.64E-21	3.07E-23	3.75E-22
6	6.00E+05	0.3155	37561.36	6.24E-20	-3.26E-22	9.44E-27	5.66E-21	2.77E-23	3.53E-22
6.5	6.50E+05	0.2925	37375.82	6.21E-20	-3.08E-22	8.75E-27	5.69E-21	2.47E-23	3.33E-22
7	7.00E+05	0.2727	37199.33	6.18E-20	-2.93E-22	8.16E-27	5.71E-21	2.29E-23	3.16E-22
7.5	7.50E+05	0.2554	37030.37	6.15E-20	-2.81E-22	7.64E-27	5.73E-21	1.97E-23	3E-22
8	8.00E+05	0.2403	36862.18	6.12E-20	-2.79E-22	7.19E-27	5.75E-21	2.06E-23	3E-22
8.5	8.50E+05	0.2268	36711.33	6.1E-20	-2.5E-22	6.78E-27	5.77E-21	1.62E-23	2.67E-22
9	9.00E+05	0.2148	36561.23	6.07E-20	-2.49E-22	6.43E-27	5.78E-21	1.62E-23	2.65E-22
9.5	9.50E+05	0.204	36414.91	6.05E-20	-2.43E-22	6.1E-27	5.8E-21	1.44E-23	2.57E-22
10	1.00E+06	0.1943	36271.61	6.02E-20	-2.38E-22	5.81E-27	5.81E-21	1.5E-23	2.53E-22
11	1.10E+06	0.1774	36003.85	5.98E-20	-4.45E-22	5.31E-27	5.84E-21	2.51E-23	4.7E-22
12	1.20E+06	0.1632	35746.65	5.94E-20	-4.27E-22	4.88E-27	5.86E-21	2.09E-23	4.48E-22
13	1.30E+06	0.1511	35502.28	5.9E-20	-4.06E-22	4.52E-27	5.88E-21	1.77E-23	4.23E-22
14	1.40E+06	0.1407	35268.46	5.86E-20	-3.88E-22	4.21E-27	5.89E-21	1.65E-23	4.05E-22
15	1.50E+06	0.1316	35042.94	5.82E-20	-3.74E-22	3.94E-27	5.91E-21	1.26E-23	3.87E-22
16	1.60E+06	0.1237	34826.47	5.78E-20	-3.59E-22	3.7E-27	5.92E-21	1.56E-23	3.75E-22
17	1.70E+06	0.1166	34616.04	5.75E-20	-3.49E-22	3.49E-27	5.93E-21	8.97E-24	3.58E-22
18	1.80E+06	0.1103	34413.15	5.71E-20	-3.37E-22	3.3E-27	5.94E-21	9.57E-24	3.46E-22
19	1.90E+06	0.1046	34215.54	5.68E-20	-3.28E-22	3.13E-27	5.95E-21	5.98E-24	3.34E-22
20	2.00E+06	0.09953	34023.96	5.65E-20	-3.18E-22	2.98E-27	5.95E-21	9.57E-24	3.28E-22
21	2.10E+06	0.09489	33836.15	5.62E-20	-3.12E-22	2.84E-27	5.96E-21	6.25E-24	3.18E-22
22	2.20E+06	0.09065	33653.62	5.59E-20	-3.03E-22	2.71E-27	5.97E-21	4.82E-24	3.08E-22
23	2.30E+06	0.08677	33474.87	5.56E-20	-2.97E-22	2.6E-27	5.97E-21	4.22E-24	3.01E-22
24	2.40E+06	0.0832	33299.88	5.53E-20	-2.91E-22	2.49E-27	5.97E-21	3.26E-24	2.94E-22
25	2.50E+06	0.0799	33128.67	5.5E-20	-2.84E-22	2.39E-27	5.98E-21	2.09E-24	2.86E-22
26	2.60E+06	0.07685	32961.23	5.47E-20	-2.78E-22	2.3E-27	5.98E-21	1.79E-24	2.8E-22
27	2.70E+06	0.07402	32796.05	5.45E-20	-2.74E-22	2.21E-27	5.98E-21	1.32E-24	2.76E-22
28	2.80E+06	0.07139	32633.89	5.42E-20	-2.69E-22	2.14E-27	5.98E-21	1.14E-24	2.7E-22
29	2.90E+06	0.06893	32474.74	5.39E-20	-2.64E-22	2.06E-27	5.98E-21	1.5E-25	2.64E-22
30	3.00E+06	0.06662	32317.86	5.37E-20	-2.61E-22	1.99E-27	5.98E-21	-1.11E-24	2.59E-22

Table A1.1B : Second Half

P Bar	$r_g{}^3$ m³	r_g M	Δr_g M	$r_g{}^2$ m²	$1/r_g{}^2$	$\Delta(1/r_g{}^2)$	dU/dr	C"
0.02	4.79E-25	7.82E-09		6.12E-17	1.63E+16			
0.03	3.26E-25	6.88E-09	-9.38E-10	4.74E-17	2.11E+16	4.76E+15	-5.92E-13	-1.24E-28
0.04	2.49E-25	6.29E-09	-5.96E-10	3.95E-17	2.53E+16	4.19E+15	-6.89E-13	-1.64E-28
0.05	2.01E-25	5.86E-09	-4.26E-10	3.44E-17	2.91E+16	3.81E+15	-7.76E-13	-2.04E-28
0.06	1.7E-25	5.53E-09	-3.26E-10	3.06E-17	3.26E+16	3.53E+15	-8.53E-13	-2.41E-28
0.07	1.47E-25	5.27E-09	-2.62E-10	2.78E-17	3.6E+16	3.32E+15	-9.16E-13	-2.76E-28
0.08	1.29E-25	5.06E-09	-2.16E-10	2.56E-17	3.91E+16	3.14E+15	-9.83E-13	-3.12E-28
0.09	1.16E-25	4.87E-09	-1.84E-10	2.37E-17	4.21E+16	3E+15	-1.04E-12	-3.47E-28
0.1	1.05E-25	4.71E-09	-1.58E-10	2.22E-17	4.5E+16	2.88E+15	-1.11E-12	-3.84E-28
0.2	5.46E-26	3.79E-09	-9.2E-10	1.44E-17	6.95E+16	2.45E+16	-1.32E-12	-5.41E-29
0.3	3.73E-26	3.34E-09	-4.52E-10	1.12E-17	8.95E+16	2.01E+16	-1.73E-12	-8.64E-29
0.4	2.85E-26	3.06E-09	-2.87E-10	9.33E-18	1.07E+17	1.76E+16	-2.05E-12	-1.16E-28
0.5	2.31E-26	2.85E-09	-2.06E-10	8.12E-18	1.23E+17	1.6E+16	-2.34E-12	-1.46E-28
0.6	1.95E-26	2.69E-09	-1.57E-10	7.25E-18	1.38E+17	1.48E+16	-2.6E-12	-1.75E-28
0.7	1.69E-26	2.57E-09	-1.26E-10	6.58E-18	1.52E+17	1.39E+16	-2.83E-12	-2.03E-28
0.8	1.49E-26	2.46E-09	-1.05E-10	6.06E-18	1.65E+17	1.32E+16	-3.05E-12	-2.31E-28
0.9	1.33E-26	2.37E-09	-8.89E-11	5.63E-18	1.78E+17	1.26E+16	-3.25E-12	-2.58E-28
1	1.21E-26	2.3E-09	-7.63E-11	5.27E-18	1.9E+17	1.2E+16	-3.48E-12	-2.89E-28
1.1	1.11E-26	2.23E-09	-6.75E-11	4.96E-18	2.01E+17	1.17E+16	-3.61E-12	-3.1E-28
1.2	1.02E-26	2.17E-09	-5.96E-11	4.7E-18	2.13E+17	1.12E+16	-3.85E-12	-3.43E-28
1.3	9.46E-27	2.12E-09	-5.34E-11	4.47E-18	2.24E+17	1.09E+16	-4.01E-12	-3.69E-28
1.4	8.83E-27	2.07E-09	-4.85E-11	4.27E-18	2.34E+17	1.06E+16	-4.16E-12	-3.92E-28
1.5	8.28E-27	2.02E-09	-4.38E-11	4.09E-18	2.44E+17	1.03E+16	-4.38E-12	-4.27E-28
1.6	7.79E-27	1.98E-09	-4.04E-11	3.93E-18	2.54E+17	1.01E+16	-4.5E-12	-4.48E-28
1.7	7.36E-27	1.95E-09	-3.7E-11	3.78E-18	2.64E+17	9.78E+15	-4.73E-12	-4.84E-28
1.8	6.98E-27	1.91E-09	-3.43E-11	3.65E-18	2.74E+17	9.55E+15	-4.87E-12	-5.1E-28
1.9	6.64E-27	1.88E-09	-3.19E-11	3.53E-18	2.83E+17	9.37E+15	-5.04E-12	-5.38E-28
2	6.33E-27	1.85E-09	-2.99E-11	3.42E-18	2.92E+17	9.24E+15	-5.07E-12	-5.49E-28
2.2	5.78E-27	1.79E-09	-5.45E-11	3.22E-18	3.1E+17	1.8E+16	-5.28E-12	-2.93E-28
2.4	5.33E-27	1.75E-09	-4.81E-11	3.05E-18	3.28E+17	1.73E+16	-5.6E-12	-3.23E-28
2.6	4.95E-27	1.7E-09	-4.3E-11	2.9E-18	3.44E+17	1.68E+16	-5.95E-12	-3.55E-28
2.8	4.61E-27	1.66E-09	-3.9E-11	2.77E-18	3.61E+17	1.63E+16	-6.16E-12	-3.77E-28
3	4.33E-27	1.63E-09	-3.54E-11	2.65E-18	3.77E+17	1.58E+16	-6.51E-12	-4.11E-28
3.5	3.74E-27	1.55E-09	-7.67E-11	2.41E-18	4.15E+17	3.81E+16	-6.86E-12	-1.8E-28
4	3.3E-27	1.49E-09	-6.36E-11	2.22E-18	4.51E+17	3.62E+16	-7.47E-12	-2.07E-28
4.5	2.95E-27	1.43E-09	-5.41E-11	2.06E-18	4.86E+17	3.47E+16	-8E-12	-2.31E-28
5	2.68E-27	1.39E-09	-4.66E-11	1.93E-18	5.19E+17	3.31E+16	-8.67E-12	-2.61E-28
5.5	2.45E-27	1.35E-09	-4.1E-11	1.82E-18	5.51E+17	3.2E+16	-9.16E-12	-2.86E-28
6	2.25E-27	1.31E-09	-3.64E-11	1.72E-18	5.82E+17	3.1E+16	-9.71E-12	-3.13E-28
6.5	2.09E-27	1.28E-09	-3.27E-11	1.63E-18	6.12E+17	3.01E+16	-1.02E-11	-3.38E-28
7	1.95E-27	1.25E-09	-2.95E-11	1.56E-18	6.41E+17	2.93E+16	-1.07E-11	-3.66E-28
7.5	1.82E-27	1.22E-09	-2.7E-11	1.49E-18	6.7E+17	2.86E+16	-1.11E-11	-3.89E-28
8	1.72E-27	1.2E-09	-2.46E-11	1.43E-18	6.98E+17	2.78E+16	-1.22E-11	-4.4E-28
8.5	1.62E-27	1.17E-09	-2.30E-11	1.38E-18	7.2E+17	2.71E+16	-1.17E-11	-1.26E-28
9	1.53E-27	1.15E-09	-2.11E-11	1.33E-18	7.52E+17	2.60E+16	-1.20E-11	-4.7E-28
9.5	1.46E-27	1.13E-09	-1.97E-11	1.29E-18	7.78E+17	2.63E+16	-1.31E-11	-4.98E-28
10	1.39E-27	1.12E-09	-1.83E-11	1.24E-18	8.04E+17	2.57E+16	-1.39E-11	-5.39E-28
11	1.27E-27	1.08E-09	-3.33E-11	1.17E-18	8.54E+17	5.03E+16	-1.41E-11	-2.8E-28
12	1.17E-27	1.05E-09	-2.97E-11	1.11E-18	9.03E+17	4.88E+16	-1.51E-11	-3.09E-28
13	1.08E-27	1.03E-09	-2.67E-11	1.05E-18	9.51E+17	4.76E+16	-1.59E-11	-3.34E-28
14	1E-27	1E-09	-2.41E-11	1E-18	9.97E+17	4.63E+16	-1.68E-11	-3.63E-28
15	9.4E-28	9.8E-10	-2.21E-11	9.59E-19	1.04E+18	4.54E+16	-1.75E-11	-3.86E-28
16	8.83E-28	9.6E-10	-2E-11	9.21E-19	1.09E+18	4.39E+16	-1.87E-11	-4.27E-28

17	8.33E-28	9.41E-10	-1.87E-11	8.85E-19	1.13E+18	4.37E+16	-1.91E-11	-4.39E-28
18	7.88E-28	9.24E-10	-1.73E-11	8.53E-19	1.17E+18	4.26E+16	-2.01E-11	-4.71E-28
19	7.47E-28	9.07E-10	-1.62E-11	8.23E-19	1.21E+18	4.22E+16	-2.06E-11	-4.89E-28
20	7.11E-28	8.92E-10	-1.49E-11	7.96E-19	1.26E+18	4.09E+16	-2.2E-11	-5.38E-28
21	6.78E-28	8.78E-10	-1.41E-11	7.72E-19	1.3E+18	4.06E+16	-2.26E-11	-5.56E-28
22	6.47E-28	8.65E-10	-1.33E-11	7.48E-19	1.34E+18	4.01E+16	-2.32E-11	-5.78E-28
23	6.2E-28	8.53E-10	-1.25E-11	7.27E-19	1.38E+18	3.95E+16	-2.4E-11	-6.08E-28
24	5.94E-28	8.41E-10	-1.19E-11	7.07E-19	1.41E+18	3.91E+16	-2.48E-11	-6.34E-28
25	5.71E-28	8.29E-10	-1.13E-11	6.88E-19	1.45E+18	3.87E+16	-2.54E-11	-6.57E-28
26	5.49E-28	8.19E-10	-1.07E-11	6.7E-19	1.49E+18	3.82E+16	-2.62E-11	-6.85E-28
27	5.29E-28	8.09E-10	-1.02E-11	6.54E-19	1.53E+18	3.78E+16	-2.71E-11	-7.17E-28
28	5.1E-28	7.99E-10	-9.69E-12	6.38E-19	1.57E+18	3.73E+16	-2.79E-11	-7.47E-28
29	4.92E-28	7.9E-10	-9.28E-12	6.23E-19	1.6E+18	3.71E+16	-2.85E-11	-7.69E-28
30	4.76E-28	7.81E-10	-8.92E-12	6.09E-19	1.64E+18	3.69E+16	-2.91E-11	-7.89E-28

Table A.2: Steam Table: Part of table was taken with permission (email in 2011) from[42]: Thermexcel.com of France

(P) Bar	(T) Celsuis	Density Kg/m^3	Latent heat [$L(l \rightarrow g)$] Kcal/kg	Latent heat [$L(l \rightarrow g)$] Kj/kg	Latent heat [$L(l \rightarrow g)$] j/mole	$\Delta L(l \rightarrow g)$ j/mole	ΔP bar	Slope j/mole/bar
0.02	17.51	0.015	587.61	2460.18	44320.15			
0.03	24.1	0.022	583.89	2444.61	44039.57	-280.58	0.01	-28057.89
0.04	28.98	0.029	581.14	2433.09	43832.15	-207.42	0.01	-20741.72
0.05	32.9	0.035	578.92	2423.80	43664.71	-167.44	0.01	-16744.22
0.06	36.18	0.042	577.05	2415.97	43523.66	-141.04	0.01	-14104.37
0.07	39.02	0.049	575.44	2409.23	43402.23	-121.43	0.01	-12143.33
0.08	41.53	0.055	574.01	2403.24	43294.37	-107.86	0.01	-10785.69
0.09	43.79	0.062	572.72	2397.84	43197.08	-97.30	0.01	-9729.751
0.1	45.83	0.068	571.54	2392.90	43108.07	-89.00	0.01	-8900.082
0.2	60.09	0.131	563.3	2358.40	42486.58	-621.50	0.1	-6214.972
0.3	69.13	0.191	557.97	2336.08	42084.57	-402.01	0.1	-4020.122
0.4	75.89	0.25	553.94	2319.21	41780.61	-303.96	0.1	-3039.604
0.5	81.35	0.309	550.64	2305.40	41531.70	-248.90	0.1	-2489.006
0.6	85.95	0.366	547.83	2293.63	41319.76	-211.94	0.1	-2119.426
0.7	89.96	0.423	545.36	2283.29	41133.46	-186.30	0.1	-1862.983
0.8	93.51	0.479	543.15	2274.04	40966.78	-166.69	0.1	-1666.88
0.9	96.71	0.535	541.14	2265.62	40815.17	-151.60	0.1	-1516.031
1	99.63	0.59	539.3	2257.92	40676.39	-138.78	0.1	-1387.809
1.1	102.32	0.645	537.59	2250.76	40547.42	-128.98	0.1	-1289.758
1.2	104.13	0.7	535.99	2244.06	40426.74	-120.68	0.1	-1206.791
1.3	107.13	0.755	534.49	2237.78	40313.60	-113.14	0.1	-1131.366
1.4	109.32	0.809	533.07	2231.83	40206.50	-107.10	0.1	-1071.027
1.5	111.37	0.863	531.73	2226.22	40105.43	-101.07	0.1	-1010.687
1.6	113.32	0.916	530.45	2220.87	40008.89	-96.54	0.1	-965.4326
1.7	115.17	0.97	529.22	2215.72	39916.11	-92.77	0.1	-927.7204
1.8	116.93	1.023	528.05	2210.82	39827.87	-88.25	0.1	-882.4658
1.9	118.62	1.076	526.92	2206.09	39742.64	-85.23	0.1	-852.296
2	120.23	1.129	525.84	2201.56	39661.18	-81.46	0.1	-814.5838
2.2	123.27	1.235	523.78	2192.94	39505.80	-155.37	0.2	-776.8716

206

2.4	126.09	1.34	521.86	2184.90	39360.99	-144.81	0.2	-724.0745
2.6	128.73	1.444	520.04	2177.28	39223.72	137.27	0.2	-686.3623
2.8	131.2	1.548	518.32	2170.08	39093.99	-129.73	0.2	-648.65
3	133.54	1.651	516.68	2163.21	38970.29	-123.70	0.2	-618.4803
3.5	138.87	1.908	512.89	2147.35	38684.43	-285.86	0.5	-571.7171
4	143.63	2.163	509.45	2132.94	38424.97	-259.46	0.5	-518.92
4.5	147.92	2.417	506.29	2119.71	38186.63	-238.34	0.5	-476.6824
5	151.85	2.669	503.35	2107.40	37964.88	-221.75	0.5	-443.4956
5.5	155.47	2.92	500.6	2095.89	37757.47	-207.42	0.5	-414.8343
6	158.84	3.17	498	2085.00	37561.36	-196.10	0.5	-392.207
6.5	161.99	3.419	495.54	2074.71	37375.82	-185.54	0.5	-371.0882
7	164.96	3.667	493.2	2064.91	37199.33	-176.49	0.5	-352.9863
7.5	167.76	3.915	490.96	2055.53	37030.37	-168.95	0.5	-337.9014
8	170.42	4.162	488.73	2046.19	36862.18	-168.20	0.5	-336.3929
8.5	172.94	4.409	486.73	2037.82	36711.33	-150.85	0.5	-301.6977
9	175.36	4.655	484.74	2029.49	36561.23	-150.09	0.5	-300.1892
9.5	177.67	4.901	482.8	2021.37	36414.91	-146.32	0.5	-292.6468
10	179.88	5.147	480.9	2013.41	36271.61	-143.31	0.5	-286.6128
11	184.06	5.638	477.35	1998.55	36003.85	-267.76	1	-267.7567
12	187.96	6.127	473.94	1984.27	35746.65	-257.20	1	-257.1973
13	191.6	6.617	470.7	1970.71	35502.28	-244.38	1	-244.3751
14	195.04	7.106	467.6	1957.73	35268.46	-233.82	1	-233.8157
15	198.28	7.596	464.61	1945.21	35042.94	-225.52	1	-225.519
16	201.37	8.085	461.74	1933.19	34826.47	-216.47	1	-216.4681
17	204.3	8.575	458.95	1921.51	34616.04	-210.43	1	-210.4341
18	207.11	9.065	456.26	1910.25	34413.15	-202.89	1	-202.8917
19	209.79	9.556	453.64	1899.28	34215.54	-197.61	1	-197.612
20	212.37	10.047	451.1	1888.65	34023.96	-191.58	1	-191.578
21	214.85	10.539	448.61	1878.22	33836.15	-187.81	1	-187.8068
22	217.24	11.032	446.19	1868.09	33653.62	-182.53	1	-182.5271
23	219.55	11.525	443.82	1858.17	33474.87	-178.76	1	-178.7559
24	221.78	12.02	441.5	1848.45	33299.88	-174.98	1	-174.9847
25	223.94	12.515	439.23	1838.95	33128.67	-171.21	1	-171.2134
26	226.03	13.012	437.01	1829.65	32961.23	-167.44	1	-167.4422
27	228.06	13.509	434.82	1820.49	32796.05	-165.18	1	-165.1795
28	239.04	14.008	432.67	1811.48	32633.89	-162.16	1	-162.1625
29	231.96	14.508	430.56	1802.65	32474.74	-159.15	1	-159.1455
30	233.84	15.009	428.48	1793.94	32317.86	-156.88	1	-156.8828

Appendix B.1: **Blackbody Radiation**

This is an elaboration of Chapter 8. Different portions of the electromagnetic spectrum can be produced numerous ways i.e. blackbody radiation is emitted by any substance for $T>0$ K. The spectrum of emitted radiation depends upon the emitting body's temperature and composition as was predicted by Maxwell's electromagnetic theory due to oscillations of the body's internal electromagnetic charges.

There is a class of hot bodies whose spectra are universal in character, i.e. Planck's spectrum for blackbody radiation. Fig B.1.1 shows the blackbody radiation from four different oven temperatures that being 3000, 4000, 5000 and 6000 K. P_λ is the emitted power per unit area per unit wavelength (λ). Fig B.1.1 depicts ideal blackbody spectrums, whereas a real blackbody spectrum consists of a series of valleys, and peaks, dependent upon frequency. Our sun approximates a blackbody radiator at 5,700 K.

A true blackbody absorbs all thermal radiation incident upon it, hence its name black, implying that no light is reflected from its surface. An object, with no internal heat source, is in thermal equilibrium with its surroundings, when the object emits as much thermal energy as it absorbs. When blackbody radiation strikes a body, then that energy is either absorbed, reflected or transmitted, in such a manner that the total energy is conserved. The *absorptivity* is the total energy absorbed by the body per unit area per unit time. It follows that an object which is a good absorber must also be a good emitter of thermal radiation hence a blackbody is a perfect absorber e.g. carbon black whose absorptivity approaches unity

A method for studying blackbody radiation is to analyze the radiation emitted out through a small hole in an oven. Actually, any small opening inside of a larger container will radiate blackbody radiation through said opening, as long as the container's walls are isothermal. Hence the spectrum of the blackbody radiation represents the wall's temperature. Interestingly, the blackbody radiation spectrum is independent of the containers wall's material, i.e. it only depends upon the wall's temperature.

The *spectral radiancy* [$R_T(\upsilon)$] equals the energy emitted per unit time per unit surface area, for each frequency (υ). I.e. energy flux at each frequency. The *radiancy* (R_T) is the spectral radiancy integrated over the range of all possible frequencies (υ), i.e. the total flux of emitted energy:

$$R_T = \int_0^\infty R_T(\upsilon)d\upsilon \quad \text{B.1.1}$$

The curves, as depicted in Fig B.1.1 represents a blackbody's spectral radiancy [$R_T(\upsilon)$], at given temperatures. The total area enclosed between the spectral radiancy curve and its frequency axis defines the radiancy (R_T) i.e. the total power emitted per unit area:

$$Power \propto T^4 \quad \text{B.1.2}$$

Often, the radiancy is referred to as the substance's "emissive power". The power/area in eqn B.1.2 is the energy flux that being the total energy emitted from a surface of a blackbody per square meter per second. In 1879, the Stefan law was written:

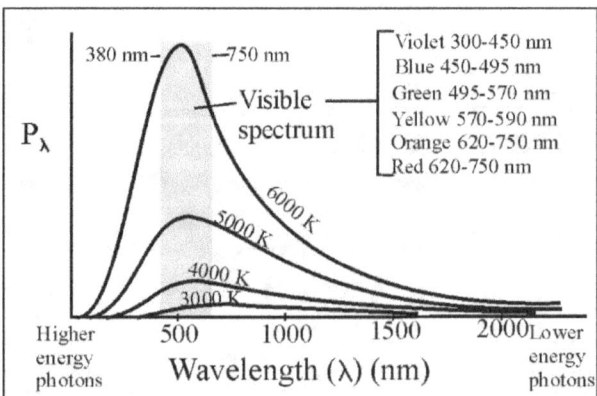

Fig B.1.1 Shows the spectrum signifying power per wavelength for Blackbody radiation. (Sketched from: "Contemporary College Physics")[9] Notice that for 6000°K: Rayleigh-Jeans approximation is valid for frequencies between approx.: 75nm and 300nm

$$R_T = \sigma T^4 \qquad\qquad \text{B.1.3}$$

where σ is the Stefan-Boltzmann constant[2]: $\sigma = 2\pi^5 k^4 / 15c^2 h^3$

$$\sigma = 5.670 \times 10^{-8} \text{ J/m}^2\text{s}^\circ\text{K}^4 = 5.670 \times 10^{-8} \text{ W/m}^{2\circ}\text{K}^4$$

The maximum wavelength (λ_{max}) in the spectral radiancy curves for a blackbody is related to the temperature by the Wein's displacement law, as illustrated in Fig B.1.2:

$$\lambda_{max} \propto 1/T \qquad\qquad \text{B.1.4}$$

In 1896 Wilhelm Wein (1864-1928) arrived at his experimentally determined proportionality constant, which is now accepted to be: 2.898×10^{-3} mK.: ($0.2014hc/k$.) Therefore eqn B.1.4 is often written: $\lambda_{max} = 2.898 x 10^{-3} / T$. In terms of frequency ($\upsilon = c/\lambda$), eqn B.1.4 can be written:

$$\upsilon_{max} \propto T \qquad\qquad \text{B.1.5}$$

Eqn B.1.5 can be written in terms of the proportionality constant: $0.2014h\upsilon_{max} = kT$. This was calculated by setting: $d\rho(\lambda)/d\lambda = 0$ to Plank's radiation spectrum[1pg27]. Note: It had been realized for centuries that actual colors represented temperature, e.g. kilns & blacksmith's temperatures were based upon color, i.e. measured with an optical pyrometer.

Now returning to the main discussion: Fig B.1.2 shows the blackbody radiation for three temperatures, namely 1000 K, 1500 K and 2000 K. To calculate the number of photons (N) at each frequency (υ) per second per unit surface area, we would divide the power at each frequency by: $h\upsilon$. Therefore, obtaining the flux of the number of photons emitted at each frequency, which is a decreasing function of frequency, as is illustrated in Fig B.1.3. Although, the plots seem to be linearly decreasing functions, a more accurate plot would include values measured where the energy per photon approaches zero ($\upsilon \to 0$) wherein the plots becomes exponentially decreasing functions.

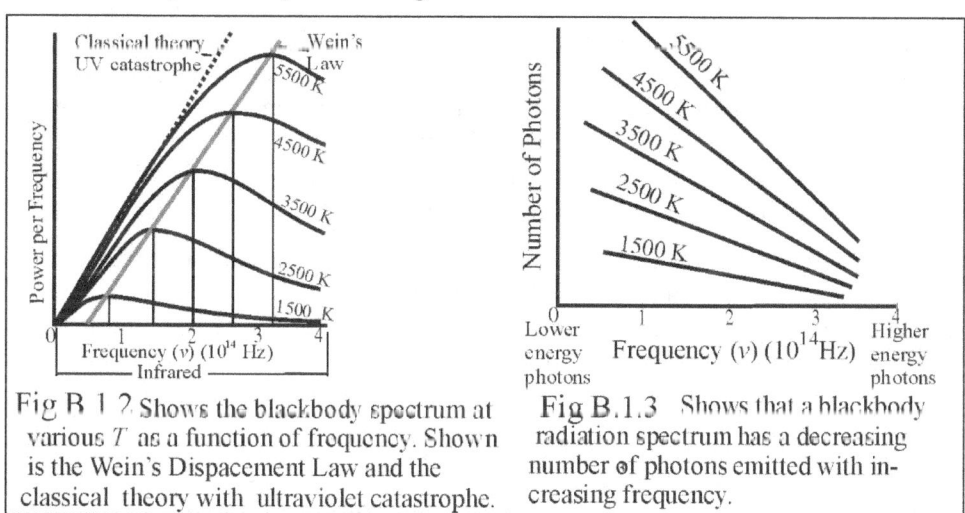

Fig B.1.2 Shows the blackbody spectrum at various T as a function of frequency. Shown is the Wein's Dispacement Law and the classical theory with ultraviolet catastrophe.

Fig B.1.3 Shows that a blackbody radiation spectrum has a decreasing number of photons emitted with increasing frequency.

Ultimately, blackbody radiation represents the radiation energy emitted by a body that absorbs and radiates all frequencies, whereby the number of photons emitted at each frequency is a decreasing function of frequency. This covers the basics in understanding the radiation emitted by a solid or liquid body (condensed matter).

Let's consider the science's history. Early in the 20th century, Rayleigh and Jeans wanted to explain the blackbody radiation energy density inside a cavity with isothermal walls, e.g. the blackbody radiation density within the said cavity. The analysis is detailed in various texts[1]. Within the cavity they only considered standing waves, that being waves whose wavelength is a multiple of the cavity's dimension. After an extensive analysis, they arrived at the Rayleigh-Jeans formula, which defines the energy density in classical terms, that being:

$$\rho_\lambda(\upsilon)d\upsilon = (8\pi\upsilon^2 kT/c^3)d\upsilon \qquad \text{B.1.6}$$

Please note that each standing wave was considered to have an energy of kT. This was based upon the equipartition theory, wherein each standing wave has a kinetic energy of $kT/2$, and a total energy equaling twice its kinetic energy.

The Rayleigh-Jeans formula is plotted in Fig B.1.2, as the classical formula dotted line, on the left. Obviously, the classical formulation only approximated the measured spectral radiancy for lower frequencies. For higher frequencies, say above the ultraviolet, the measured spectral radiancy is much lower than classical theory dictates. This became known as the "ultraviolet catastrophe".

Why does the catastrophe occur at the ultraviolet (UV)? To answer, start by looking at the blackbody spectrum for 6000 K, in the Fig. B.1.1. Is it mere coincidence that the thermal radiation density from our Sun (T=5,700 K) is not proportional to temperature for frequencies greater than UV? No it cannot be! Accordingly, both what we see as light, and/or feel as heat, e.g. infrared, tends to be proportional relations to our Sun's energy densities. As was stated in Chapter 8; this certainly explains why the thermal energy density on Earth is proportional to temperature. It is equally interesting that we evolved with our vision focused on the upper frequencies, whose energy density from our sun may still be approximated as linear relations. Note: Both visible light: 430 to 790 THz (370nm $<\lambda<$700 nm) & infrared: freq. <430THz (λ>700 nm), densities from our Sun can be approximated as being proportional to T, here on Earth.

So the energy density from our Sun, which is absorbed and then radiated by matter here on Earth, is proportional to T. Furthermore, Maxwell's speed distribution which defines the distribution of gaseous molecules speeds, at fixed pressure and temperature are rather similar to the spectral radiancy curves. It is no coincidence that the speed of gas molecules mimics the curves of the Sun's radiation that condensed matter (walls), absorbs. Max Planck (1858-1947) understood that the Maxwell speed distribution is a consequence of Boltzmann's distribution, which is based upon the quantization of energy. But like so many, he too failed to realize it is all a consequence of our Sun's output.

Thinking in terms of quantization of energy, Planck treated Boltzmann's distribution for standing waves as a discrete instead of a continuous quantity. He found that the spectral radiancy (continuous spectrum of electromagnetic radiation) for blackbody radiation could be understood by assuming that emitted radiation was in bundles whose size is defined by:

$$\Delta E = h\upsilon \qquad \text{B.1.7}$$

where υ = frequency, and h = Planck's constant = 6.626176x10^{-34} Js

The Rayleigh-Jeans formula correlates for small frequencies, hence Planck assumed that:

$$\Delta E \rightarrow kT \text{ as } \upsilon \rightarrow 0 \qquad \text{B.1.8}$$

Planck further realized that for blackbody radiation the ultraviolet catastrophe could be avoided by considering:

$$\Delta E \rightarrow 0 \text{ as } \upsilon \rightarrow \infty \qquad \text{B.1.9}$$

Planck envisioned that the mean energy change ($\Delta \overline{E}$) would be given by:

$$\Delta \overline{E}(\upsilon) = h\upsilon/(e^{h\upsilon'kT} - 1) \qquad \text{B.1.10}$$

Eqn B.1.10 is based upon the photon occupation number: $\eta = (e^{h\upsilon'kT} - 1)^{-1}$, with each photon having a discrete energy of: $h\upsilon$. For eqn B.1.10: $e^{h\upsilon'kT} \rightarrow 1 + h\upsilon/kT$ when, $h\upsilon/kT \rightarrow 0$, which satisfies condition set by B.1.8. Similarly: $h\upsilon/kT \rightarrow \infty$ then $e^{h\upsilon'kT} \rightarrow \infty$ satisfies the condition set by eqn B.1.9.

By realizing that kT in eqn B.1.6 was obtained from the equipartition theory and replacing it with the mean value ($\Delta \overline{E}$) in eqn B.1.10, Planck arrived at his blackbody radiation formula:

$$\rho_\lambda(\upsilon)d\upsilon = (8\pi\upsilon^2/c^3)[h\upsilon/(e^{h\upsilon/kT} - 1)]d\upsilon \qquad \text{B.1.11}$$

Planck's formula for blackbody radiation correlates with empirical data. Planck wrongly attributed his formulation to oscillations of matter rather than consisting of photons. Remember, Planck was still thinking in terms of energy quantization being restricted to points along the walls of the blackbody cavity. Interestingly, to Planck this meant that once radiated, the electromagnetic waves would spread out like waves from a rock tossed into a pond i.e. the classical wave theory, as first envisioned by Maxwell.

It was not until Einstein's successful interpretation, in 1905, of the photoelectric effect that the idea of radiation being quantized bundles of energy (photons) was first realized. The photons existence was experimentally verified in 1916 by Millikan's photoelectric effect, wherein the ejection of electrons from a metal surface, due to energy being transferred to the electrons from incident light. Einstein's photoelectric equation is:

$$K = h\nu - \Phi_e \qquad \text{B.1.12}$$

where Φ_e = constant = emitter work function, K = emitted electron's kinetic energy and ν = frequency

All warm matter radiates energy, whose spectral radiancy curves define the energy at each frequency, which equals the number of photons multiplied by energy of each photon at that frequency.

Our real concern is energy density i.e. the amount of energy contained within a given volume. The spectral radiancy curves are really a measurement in two dimensions, representing the energy at each frequency, passing through a unit surface area per unit time. It seems logical that the thermal radiation energy density (ρ_γ) must be proportional to the radiancy (R_T).

$$\rho_\gamma \propto R_T \qquad \text{B.1.13}$$

Combining eqn B.1.13 with eqn B.1.3:

$$\rho_\gamma \propto \sigma T^4 = R_T \qquad \text{B.1.14}$$

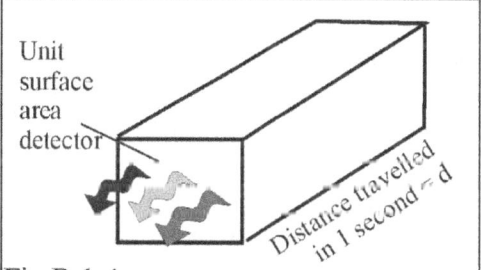

Fig B.1.4 Shows a unit surface area detector measuring the amount of energy contained within some volume

Imagine a unit surface area detector, as is illustrated in Fig B.1.4. If we were to now measure the number and type of each photon that passes through the detector in one second, it would be the same as measuring the number and type of each photon within a box traveling along some path, whose length was determined by the distance (d), wherein d equals one second (s) multiplied by the speed of light (c): $d=cs$. The thermal radiation energy density ($\rho_{\lambda path}$) traveling towards our detector is:

$$\rho_{\lambda path} = R_T / xyd \qquad \text{B.1.15}$$

where xy is the surface area of our detector, R_T is the radiancy and

d = distance traveled by light in one second along the z axis = cs.

The thermal energy density (ρ_λ) must be proportional to the radiancy (R_T): Eqn B.1.14. Instead of a rectangular box, consider a unit cube where: $x = y = z = 1$ meter. Since the speed of light is: c=3.0x10^8 m/s, the duration of the measurement of radiancy would be 1/(3x10^8) seconds. Therefore, the values for power emitted at each frequency would be 1/(3x10^8) its magnitude when measured over a 1 second duration. The radiancy would be 1/(3x10^8) of its value as measured over 1 second and the said value of radiancy should be proportional to the thermal energy density (ρ_λ). Expanding eqn B.1.14 further:

$$\rho_\lambda \propto \sigma T^4 / c = R_T / c \qquad \text{B.1.16}$$

Eqn B.1.15 only measured the photon's energy along a given path, but all possible paths needs consideration. For our cube, the radiancy was the total energy emitted through one of the cube's one square meter faces, in one second. Realizing that a cube has 6 such faces, then one might be inclined to argue that the energy density should be 6 times the radiancy divided by speed of light, c. Such a simple argument is incorrect, as the accepted value for the energy density is 4 times the radiancy divided by speed of light, c. I.e., the energy density[1] becomes:

$$\rho_\lambda = 4\sigma T^4 / c = 4R_T / c \qquad \text{B.1.17}$$

Note: This author has not spent any real duration on this issue, i.e. a good reference for this was not found. The accepted value of 4 (instead of 6) is based upon consideration of all components of radiation over all angles (See appendix B-7 for similar mathematical conceptualization). Eqn B.1.17 can be rewritten as eqn 8.4 $\rho_\lambda = \rho_B = aT^4$ where $a = 4\sigma/c$, which can also be written as: $a = 8\pi^5 k^4 / 15c^3 h^3$.

To further complicate things, when considering the radiancy, it is the emission of photons from the surface of matter. Hence, in Fig B.1.4: $d \to 0$. The reason being that the photon was previously trapped in the vibrations of matter, bouncing back and forth between various molecules prior to being emitted away from the matter. Since, the volume ov0er which the energy density is measured loses its depth (z), all we can say is that the matter behaves as if it had such and such, a thermal energy density (ρ_λ).

Reconsider thermal radiation. As previously stated our Sun's thermal radiation, as felt here on Earth, is proportional to T, i.e. Rayleigh Jeans approximation: eqn 1.4.27: $\rho_R = a'T$. Next consider a blast furnace here on Earth. Looking at Fig B.1.2, we can see that for blackbodies at 2000 K, that what we consider as thermal energy density is no longer simply a linear function of T. Instead the thermal radiation density i.e. the infrared spectrum density would be best defined by eqn 1.4.23. Thus, for close proximities to a blast furnace then certainly the thermal energy density would be best defined by: Eqn 1.4.23.

Closing Remarks/Summary

The energy associated with molecules in condensed matter can be considered in terms of phonons/photons. The number of photons, at a given temperature, is a linearly decreasing function the photons frequency (υ), as shown in Fig B.1.3. The energy is quantized ($h\upsilon$), hence, the energy flux as a function of frequency (spectral radiancy) is a Maxwell type distribution, as is illustrated in Fig B.1.2.

Wein's displacement law gives the peak frequency of the spectral radiancy. The radiancy is proportional to the energy density within the condensed matter. A more thorough classical approach to dealing with blackbody radiation can be found in text[12], which includes a more comprehensive derivation of eqn B.1.6 through eqn B.1.11. Formulated by Rayleigh and Jeans, this classical approach is based upon standing waves within a volume and their average energy. And most importantly, it can all readily be explained in terms of what we witness from of our Sun's output.

References:

1) Eisberg, R. Resnick R. "Quantum Physics", John Wiley & Sons   Toronto 1974

Appendix B.2: **Binomial, Gaussian & Boltzmann's Distributions**

This section is a simplified discussion of some relevant distributions utilized in thermodynamics. More intense discussion can be found in various texts on the subject[1,2].

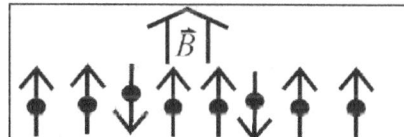

Fig B.2.1 Shows the magnetic moments for eight adjacent molecules (N=8). Six magnetic moments point up (n) and two point down (n'). The external magnetic field ($\vec{B}$) is upward.

Binomial Distribution

Consider an ideal system of N spins (each ½) having a magnetic moment (μ_o). Each magnetic moment can either be:

1) Pointed up (parallel to $\vec{B}$)
2) Pointed down (anti-parallel to $\vec{B}$)

If the system of spins is in equilibrium, then independent of time there is always a certain number of spins up, versus down. Now consider any one spin and say that the probability of the spin:

1) Pointing up is given by p.
2) Pointing down is given by q.

Two possibilities exist (up vs down) thus the total of the two probabilities is unity:

$$p + q = 1 \qquad \text{B.2.1}$$

If there were no external magnetic field ($\vec{B} = 0$) then we would expect that:

$$p = q = 1/2 \qquad \text{B.2.2}$$

In the presence of an external magnetic field ($\vec{B}$), more magnetic moments would be pointed up then down ($p' > q'$), as illustrated in Fig B.2.1. Let n be the number of magnetic moments (μ_o) pointed up and n' be the number of magnetic moments (μ_o) pointed down. Then, in terms of the number of spins:

$$n + n' = N \qquad \text{B.2.3}$$

What is the probability [$P'(n)$] that n molecules have an upward pointing μ_0? Since the probabilities are statistically independent, then the probability of any one configuration is:

$$P'(n) = C_N(n)p^n q^{n'} = C_N(n)p^n q^{N-n} \qquad \text{B.2.4}$$

where $C_N(n)$ is the number of possibilities having n magnetic moments pointing upward.

Fig B.2.2 shows all the possible configurations for four molecules whose magnetic moments are either aligned parallel or anti-parallel to each other. The total number of possible configurations is 16 i.e. when there are two possible choices (one up, one down), then the number of possible configurations is: 2^N

How many distinct configurations are there for two magnetic moments pointing up? Fig B.2.2 shows 6 possible configurations. How many distinct configurations are there for only one magnetic moments pointing up? Fig. B.2.2 shows 4 possible configurations. The general form for the total number of possible configurations $C_N(n)$ is based upon factorials (!):

$$C_N(n) = N!/n!(N-n)! \qquad \text{B.2.5}$$

Combining eqn B.2.5 with eqn B.2.4, we obtain:

$$P'(n) = [N!/n!(N-n)!]p^n q^{N-n} \quad \text{B.2.6}$$

For the special case wherein there is no external magnetic field ($\vec{B}=0$), then $p'=q'=1/2$ and eqn B.2.6 becomes:

$$P'(n) = [N!/n!(N-n)!](1/2)^N \qquad \text{B.2.7}$$

Which often is written in the following format:

$$P'(n) = [N!/n!n'!](1/2)^N \qquad \text{B.2.8}$$

⇑⇑⇑⇑	4 up	1 configuration
⇓⇑⇑⇑ ⇑⇓⇑⇑ ⇑⇑⇓⇑ ⇑⇑⇑⇓	3 up 1 down	4 configurations
⇓⇓⇑⇑ ⇑⇑⇓⇓ ⇑⇓⇓⇑ ⇓⇑⇑⇓ ⇑⇓⇑⇓ ⇓⇑⇓⇑	2 up 2 down	6 configurations
⇑⇓⇓⇓ ⇓⇑⇓⇓ ⇓⇓⇑⇓ ⇓⇓⇓⇑	1 up 3 down	4 configurations
⇓⇓⇓⇓	4 down	1 configuration

Fig. B.2.2 Shows the number of possible configurations for each possible state of total magnetic moments, for 4 independent molecules.

Eqn B.2.8 is the binomial distribution, which is symmetric about the mean value as graphically analyzing concepts like the tossing of a coin ($q=p=1/2$), or the rolling of a die illustrated in Fig. B.2.3. In the absence of a magnetic field, the plotted probability is perfectly symmetrical about the mean value. The binomial distribution is particularly useful when analyzing concepts like the tossing of a coin ($q=p=1/2$), or the rolling of a die ($q=1/6, p=5/6$).

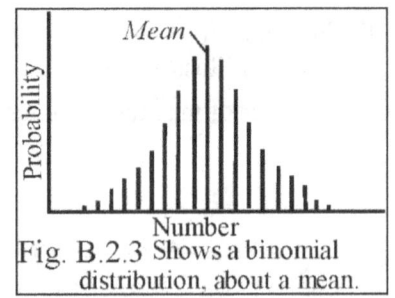

Fig. B.2.3 Shows a binomial distribution, about a mean.

Gaussian Distribution

Consider a large ensemble of molecules, where N becomes too large to handle using the binomial distribution. The problem is alleviated by the Gaussian distribution, which as an approximation simplifies the task.

As N increases, the probability tends to a maximum value, which becomes increasingly pronounced. Hence the probability becomes increasingly small as the value (n) differs appreciably from its mean value. Therefore, we are only concerned with the value close to that mean value. Based upon the concept of a large number around a mean value, the probability, as given by eqn B.2.6, can be approximated by the following Gaussian distribution:

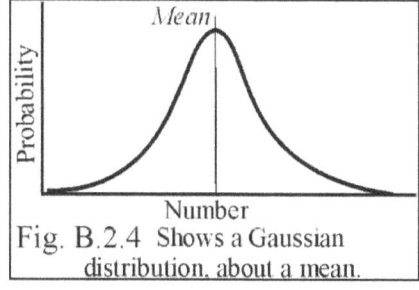

Fig. B.2.4 Shows a Gaussian distribution, about a mean.

$$P'(n) = (1/\sqrt{2\pi Npq})e^{-(n-Np)^2/2Npq} \qquad \text{B.2.9}$$

Boltzmann's Distribution

Envision a system of a large number of molecules isothermally exchanging energy with each other. Consider a small group of four molecules exchanging a quantized energy, ΔE amongst each other, such that the ensemble maintains a constant total energy of: $3\Delta E$. At any instant the total energy ($3\Delta E$) can be distributed amongst the four molecules numerous ways, with the number of possible choices being: 4x3x2x1= 4! Consider three cases:

Case 1: Three molecules have no energy, while the fourth molecule has energy: $3\Delta E$. For this to happen, four indistinguishable different paths exist, hence the probability of finding this energy distribution is: 4/4! = 4/20

Case 2: Two molecules have zero energy, while a third molecule has energy: $2\Delta E$ and the forth molecule has energy: ΔE. Twelve indistinguishable different ways exit for this energy to be distributed amongst the four molecules. Hence, the probability of finding the energy distribution is: 12/4! = 12/20

Case 3: One molecule has zero energy, while the other three molecules have energy: ΔE. There are four indistinguishable ways for this energy to be distributed amongst the four molecules. Hence, the probability of finding this energy distribution is: 4/4! = 4/20

To calculate the probable numbers of molecules in a given energy state e.g. zero energy. For case 1, there were three molecules with zero energy and a probability of: 4/20. For case 2, there were two molecules with zero energy and a probability of: 12/20. For case 3, there was one molecule with zero energy and a probability of: 4/20. Accordingly, the probable number of molecules [$n(E)$] being in energy state zero, is:

$n(0) = 3\text{x}(4/20)+2\text{x}(12/20)+1\text{x}(4/20) = 40/20$
$n(\Delta E) = 0\text{x}(4/20)+1\text{x}(12/20)+3\text{x}(4/20) = 24/2$
$n(2\Delta E) = 0\text{x}(4/20)+1\text{x}(12/20)+0\text{x}(4/20) = 12/20$
$n(3\Delta E) = 1\text{x}(4/20)+0\text{x}(12/20)+0\text{x}(4/20) = 4/20$

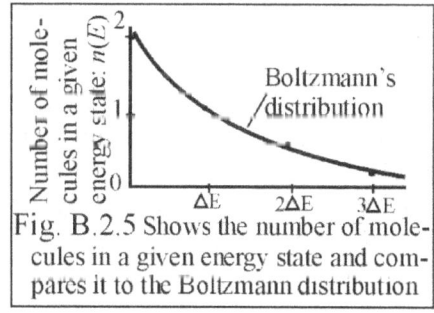

Fig. B.2.5 Shows the number of molecules in a given energy state and compares it to the Boltzmann distribution

Plotting the number of molecules in a given energy state versus that energy state we obtain Fig B.2.5. Thus, the number of molecules in each state is approximately:

$$n(E) = Ae^{E/kT} \qquad \text{B.2.10}$$

Eqn B.2.10 is the Boltzmann distribution with A being some constant. Basically, eqn B.2.10 states that the probable number of molecules in a system at temperature (T) that will be in the state of energy E is proportional to: $e^{-E/kT}$. In other words: The probability that a given molecule is in energy state E is proportional to: $e^{-E/kT}$. Accordingly, the probability of finding a molecule in a given energy range ($E + dE$) becomes:

$$P'(E) = \beta e^{-E/kT} \quad \text{B.2.11}$$

The constant β is determined by realizing that the probability integrated over all possibilities must be unity:

$$\int_0^\infty P'(E)dE = \int_0^\infty \beta e^{-E/kT}dE = \beta \int_0^\infty e^{-E/kT}dE = 1 \quad \text{B.2.12}$$

Solving eqn B.2.12, we obtain $\beta = 1/kT$. Therefore eqn B.2.11 can be written in the special form for the Boltzmann distribution:

$$P'(E) = e^{-E/kT}/kT \quad \text{B.2.13}$$

Eqn B.2.13 can be plotted i.e. Fig B.2.6. Multiplying the probability of each molecule having a particular energy, by that energy gives a plot of the total energy, as is illustrated in Fig B.2.7. Basically, the total energy between; $E + dE$ is the probability of the molecule having an energy between E and $E + dE$, multiplied by the energy E.

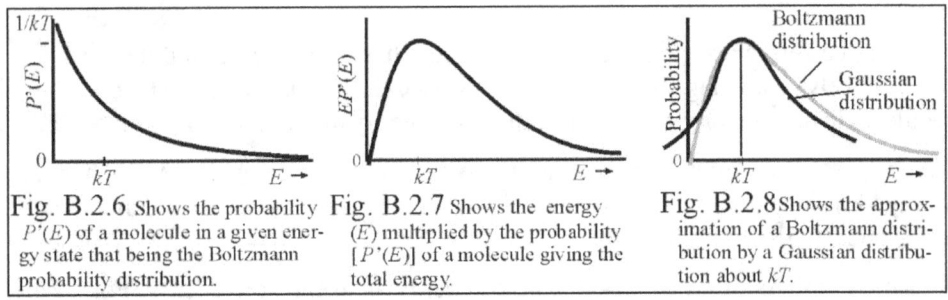

Fig. B.2.6. Shows the probability $P'(E)$ of a molecule in a given energy state that being the Boltzmann probability distribution.

Fig. B.2.7 Shows the energy (E) multiplied by the probability $[P'(E)]$ of a molecule giving the total energy.

Fig. B.2.8 Shows the approximation of a Boltzmann distribution by a Gaussian distribution about kT.

The mean energy ($\overline{E}$) is obtained by:

$$\overline{E} = \int_0^\infty EP'(E)dE / \int_0^\infty P'(E)dE \quad \text{B.2.14}$$

Solving, we obtain:

$$\overline{E} = kT \quad \text{B.2.15}$$

Although Fig B.2.7 does represent the energy of a system of molecules in thermal contact we can see that the curve can be approximated as some Gaussian curve about a mean value of kT, as is illustrated in Fig B.2.8. For the sake of simplicity in many applications a Gaussian distribution, rather than the more correct Boltzmann's distribution, can be used to approximate the energy of neighboring molecules.

References

1.Reif. F. "Fundamentals of Statistical and Thermal Physics", McGraw-Hill, New York, 1965
2. Carey V. "Statistical Thermodynamics and Microscale Thermophysics", Cambridge U 1999

Appendix B.3: **Equipartition Theorem**

This appendix is a discussion of equipartition theorem, as was briefly discussed in Chapter 2. Equipartition forms the basis of the traditional understanding of kinetic theory that being something that was only partially agreed with in our new perspective. This appendix is based purely upon traditional theory, showing the accepted general applicability of the equipartition theorem. The equipartition theorem is a useful result in statistical mechanics, based upon the concept that the energy of a system of f generalized coordinates is based upon some generalized coordinates ($q_1....q_f$) and the corresponding generalized momentum ($p_1...p_f$). Using generalized coordinates the energy is written:

$$E = E(q_1,...,q_f, p_1,..., p_f) \qquad \text{B.3.1}$$

Basically equipartition states that each independent quadratic term has a mean energy of $kT/2$. Examining this for a system of particles whose total energy splits into two additive parts:

$$E = \varepsilon(p_i) + E'(q_1...p_f) \qquad \text{B.3.2}$$

The most common situation for eqn B.3.2, is the energy term, ε , involving momentum i.e. kinetic energy, while the other energy term, E', involves a location dependent energy i.e. potential energy. The energy (ε) associated with momentum for the ith particle with mass (m) is:

$$\varepsilon_i = |\vec{p}_i|^2 /2m \qquad \text{B.3.3}$$

which leads to in terms of kinetic energy (E_k) of the ith particle:

$$\varepsilon = E_k = (|\vec{p}_x|^2 + |\vec{p}_y|^2 + |\vec{p}_z|^2)/2m \qquad \text{B.3.4}$$

where $\vec{p}_x$ = momentum along the x-axis, $\vec{p}_y$ is along the y-axis, $\vec{p}_z$ is along the z-axis, Since eqn B.3.4 contains three quadratic terms, equipartition states that the mean translational kinetic energy of an ideal gaseous molecule becomes[1,2]:

$$\overline{E}_k = 3kT/2 \qquad \text{B.3.5}$$

Eqn B.3.5 is equivalent to eqn 2.11: $\overline{E}_{(k,r)} = 3kT/2$, except eqn B.3.5 wrongly considers all the energy as being purely translational while eqn 2.11 considers it to be a combination of translational & rotational. We can rewrite eqn B.3.3 in a more general quadratic equation form that being:

$$\varepsilon = b|\vec{p}_i|^2 \qquad \text{B.3.6}$$

where b is a constant The system is considered to be in equilibrium at absolute temperature T, therefore it is distributed in accordance with the canonical distribution. In which case the mean energy ($\overline{\varepsilon}_i$) of the ith particle is:

$$\overline{\varepsilon}_i = [\int_{-\infty}^{\infty} e^{-\beta E(q_1...p_f)} \varepsilon dq_1...dp_f] / [\int_{-\infty}^{\infty} e^{-\beta E(q_1...p_f)} dq_1...dp_f] \qquad \text{B.3.7}$$

Now substituting in for E as given by eqn B.3.1, we obtain:

$$\overline{\varepsilon}_i = [\int_{-\infty}^{\infty} e^{-\beta(\varepsilon+E')} \varepsilon dq_1...dp_f] / [\int_{-\infty}^{\infty} e^{-\beta(\varepsilon+E')} dq_1...dp_f] \qquad \text{B.3.8}$$

Kent W. Mayhew

The multiplicative property of the exponential function allows us to rewrite eqn B.3.8:

$$\bar{\varepsilon}_i = [\int_{-\infty}^{\infty} e^{-\beta\varepsilon}\varepsilon_i dp_i \int_{-\infty}^{\infty} e^{-\beta E'} dq_1...dp_f]/[\int_{-\infty}^{\infty} e^{-\beta\varepsilon} dp_i \int_{-\infty}^{\infty} e^{-\beta E'} dq_1...dp_f] \qquad \text{B.3.9}$$

The integrals cancel leaving:

$$\bar{\varepsilon}_i = [\int_{-\infty}^{\infty} e^{-\beta\varepsilon}\varepsilon_i dp_i]/[\int_{-\infty}^{\infty} e^{-\beta\varepsilon} dp_i] \qquad \text{B.3.10}$$

Eqn B.3.10 simplifies by reducing the integral in the numerator to that in the denominator:

$$\bar{\varepsilon}_i = [-(\partial/\partial\beta)\int_{-\infty}^{\infty} e^{-\beta\varepsilon} dp_i]/[\int_{-\infty}^{\infty} e^{-\beta\varepsilon} dp_i] \qquad \text{B.3.11}$$

Which can be rewritten as:

$$\bar{\varepsilon}_i = -\partial\ln(\int_{-\infty}^{\infty} e^{-\beta\varepsilon} dp_i)/\partial\beta \qquad \text{B.3.12}$$

Now we must apply the fact that the energy term, ε_i is a quadratic as given by eqn B.3.3. Now, analyze the integral in eqn B.3.12, we obtain:

$$\int_{-\infty}^{\infty} e^{-\beta\varepsilon} dp_i = \int_{-\infty}^{\infty} e^{-\beta b p_i^2} dp_i \qquad \text{B.3.13}$$

Now let: $y = \beta^{1/2} p_i$. Substituting into eqn B.3.13 we obtain:

$$\int_{-\infty}^{\infty} e^{-\beta b p_i^2} dp_i = \beta^{-1/2}\int_{-\infty}^{\infty} e^{-by^2} dy \qquad \text{B.3.14}$$

Taking the natural logarithm:

$$\ln\int_{-\infty}^{\infty} e^{-\beta\varepsilon} dp_i = -\ln\beta/2 + \ln\int_{-\infty}^{\infty} e^{-by^2} dy \qquad \text{B.3.15}$$

Since the integral on the right does not involve β, then eqn B.3.12 simplifies into:

$$\bar{\varepsilon}_i = -\partial l(-\ln\beta/2)/\partial\beta = 1/2\beta = kT/2 \qquad \text{B.3.16}$$

Eqn B.3.16 states that for a system of gas molecules in thermal equilibrium, the mean kinetic energy per degree of freedom is: $kT/2$ i.e. mathematical conjecture! This was applied to any classical system consisting of a large number of identical entities. Just because degrees of freedom exists mathematically, does not mean molecules adhere to it i.e. a mathematical construct without logic!

References

1.Reif. F. "Fundamentals of Statistical and Thermal Physics", McGraw-Hill, New York, 1965
2. Carey V. "Statistical Thermodynamics and Microscale Thermophysics", Cambridge U 1999

Appendix B.4: **Debye's Model**

In Chapter 2, the classical approach for the heat capacity of a crystalline solid was briefly discussed. Each molecule was considered to be a simple harmonic oscillator about its lattice site in 3 dimensions (x, y and z). Each dimension allows for one degree of freedom, therefore the three dimensions allows for 3 degrees of freedom. And then by the equipartition theory, each degree of freedom has an average energy defined by kT, therefore for N molecules of condensed matter the total energy is approximated by eqn 2.5: $E_T = 3NkT$.

Einstein realized that the factor kT, which was derived from classical equipartition theory, had to be replaced by a simple harmonic oscillator, in much the same way as Planck had done for blackbody radiation. Einstein replaced kT with: $hv/(e^{hv/kT} - 1)$, which is a combination of Planck's energy quantization and Boltzmann's distribution. And then determined for a crystalline substance that eqn 2.5 becomes:

$$E = 3Nhv/(e^{hv/kT} - 1) \quad \text{B.4.1}$$

Now consider eqn B.4.1 in the following limits:

i) For high temperatures: $e^{hv/kT} \to 1 + hv/kT$, as $hv/kT \to 0$, therefore: $E \to 3NkT$

ii) For low temperatures: $e^{hv/kT} \to \infty$, as $hv/kT \to \infty$, therefore: $E \to 0$

Einstein's explanation did not properly explain the empirically known result that the molar heat capacity varied as T^3 for extremely low temperatures. In other words the above may also simply be a mathematical result lacking constructive logic. Based upon discussion throughout this text, we can now speculate that systems of extremely low temperatures are not systems whose temperatures are related to the thermal energy density from our Sun, as most other systems are here on Earth. Accordingly, they would not correlate very well with the energy density of our mother of all heat baths e.g. atmosphere & oceans.

Based upon eqn 2.5, the Law of Dulong and Petit is attained, which defines the heat capacity for a mole of molecules, at a constant volume, by eqn 2.7: $C_V = dQ/dT = 3R = 6$cal/mole/K $= 25.10$ joules/mole/K

What is interesting about the Law of Dulong and Petit (eqn 2.7) is that the energy needed to raise a mole of molecules by one degree (K) is independent of the element. Hence, within reason eqn 2.12 is applicable to most solid substances, except when we are dealing with extremely low temperatures. i.e. eqn 2.12 being: $E_{Tk(t,r)} = 3NkT/2$

In fact, the heat capacity of all solid substances tends to zero, as the temperature tends toward absolute zero. And this may be best explained by eqn B.4.1.

Appendix B.5: **Vibrational Energy (Polyatomic) & Motion**

Consider that the center of mass of the two molecules is motionless with respect to the wall and they are not affected by gravity. They start at rest and are then compressed towards each other, and allowed to oscillate. The position coordinates of the two molecules, and masses are respectively: m_1, m_2, x_1 and x_2, i.e. Fig. B.5.1. The length of the spring is: $x_2 - x_1$. Consider that in the relaxed position the distance between the two molecules is "d", then the extension/compaction (x) of the spring is[1]:

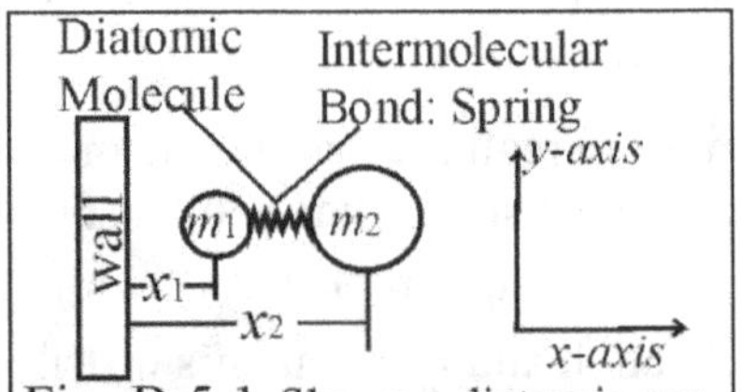

$$x = x_2 - x_1 - d \qquad \text{B.5.1}$$

The force upon, m_1 is given by Hooke's law as:

$$F_1 = k'x \qquad \text{B.5.2}$$

where k' is the spring constant. Conversely, the force on m_2 is:

$$F_2 = -k'x \qquad \text{B.5.3}$$

Fig. B.5.1 Shows a diatomic gas molecule vibrating along the x-axis. Intermolecular bonds are analogous to springs.

The equation for motion of m_1 is:

$$m_1 d^2 x_1 / dt^2 = k'x \qquad \text{B.5.4}$$

The equation for motion of m_2 is:

$$m_2 d^2 x_2 / dt^2 = -k'x \qquad \text{B.5.5}$$

Multiply eqn B.5.4 by m_2 and eqn B.5.5 by m_1, we obtain:

$$m_2 m_1 d^2 x_1 / dt^2 = m_2 k'x \qquad \text{B.5.6}$$

$$m_1 m_2 d^2 x_2 / dt^2 = -m_1 k'x \qquad \text{B.5.7}$$

Subtract eqn B.5.6 from eqn B.5.7 to obtain:

$$m_1 m_2 (d^2 x_2 / dt^2 - d^2 x_1 / dt^2) = -k'x(m_1 + m_2) \qquad \text{B.5.8}$$

Which becomes:

$$m_1 m_2 (d^2 x / dt^2)/(m_1 + m_2) = -k'x \qquad \text{B.5.9}$$

The movement of the spring is similar to a spring fixed to a wall whose reduced mass (M_R) is:

$$M_R = m_1 m_2 /(m_1 + m_2) \qquad \text{B.5.10}$$

The characteristic angular frequency (ω_0) would be the same for both molecules that being:

$$\omega_0 = \sqrt{k'/M_R} \qquad \text{B.5.11}$$

The energy of oscillation can be expressed in terms of the maximum potential energy (U'_m) when the spring is fully extended. Let "x_m" be the maximum extended position, then the energy is:

$$U'_m = k' x_m^2 / 2 \qquad \text{B.5.12}$$

The energy of oscillation can also be expressed in terms of the total kinetic energy (E'_{kT}) of both molecules when the spring is in relaxed position e.g. molecules are separated by distance d. Expressed in terms of momentum ($\vec{p}$) with respect to the center of mass, we would write:

$$E'_{kT} = E'_{k1} + E'_{k2} = |\vec{p}_1|^2 / 2m_1 + |\vec{p}_2|^2 / 2m_2 \quad \text{B.5.13}$$

The magnitude of the momentum of both molecules must be the same, therefore: $|\vec{p}| = |\vec{p}_1| = |\vec{p}_2|$, when the molecules are separated by d. The reasoning is that there is no net motion of the diatomic molecule, as we assumed its center of mass remained stationary in space. Since there is no net motion of the center of mass, then the momentum must be the same. Therefore, eqn B.5.13 becomes:

$$E'_{kT} = (|\vec{p}|^2 / 2)(1/m_1 + 1/m_2) \quad \text{B.5.14}$$

Since: $(1/m_1 + 1/m_2) = (m_2 + m_1)/m_1 m_2 = M_R$, then eqn B.5.14 can be rewritten as:

$$E'_{kT} = |\vec{p}|^2 / 2M_R \quad \text{B.5.15}$$

Rather than equate eqn B.5.15 to kinetic energy often in mechanics we use the Hamiltonian (H), in which case based upon eqn B.5.15 and eqn B.5.12 we would write:

$$H = |\vec{p}|^2 / 2M_R = k' x_m^2 / 2 \quad \text{B.5.16}$$

Based upon eqn B.5.14, we can see that ratio between the kinetic energies of molecule 1 and molecule 2, when they are separated by distance d would be. Note : d is the relaxed position, A.K.A: zero potential energy position:

$$E'_{k1} / E'_{k2} = m_2 / m_1 \quad \text{B.5.17}$$

Ratio of Kinetic Energies versus Momentums

When deriving eqn B.5.14 we realized that the magnitudes of the momentum are the same. Thus:
$$\vec{p}_1 + \vec{p}_2 = m_1 \vec{v}_1 + m_2 \vec{v}_2 = 0 \quad \text{B.5.18}$$
Therefore
$$\vec{v}_1 / \vec{v}_2 = -m_2 / m_1 \quad \text{B.5.19}$$

It seems strange that the ratio of velocities as given by eqn B.5.19 and ratio of kinetic energies as given by eqn B.5.17 are the same. In order to demonstrate that they are: Square both sides of B.5.19

$$\vec{v}_1^2 / \vec{v}_2^2 = m_2^2 / m_1^2 \quad \text{B.5.20}$$

Multiply both sides by: $(m_1/2)/(m_2/2)$ in order to make the L.H.S. of eqn B.5.20 into kinetic energy

$$(m_1 \vec{v}_1^2 / 2)/(m_2 \vec{v}_2^2 / 2) = (m_1 m_2^2 / 2)/(m_2 m_1^2 / 2) \quad \text{B.5.21}$$

The L.H.S. of eqn B.5.21 is kinetic energy and canceling the terms on the R.H.S. returns eqn B.5.17.

Diatomic Momentum/Impulse

Assume that the center of mass of the diatomic gas molecule had no initial velocity. Consider that an impulse ($\vec{J}$ =transfer of momentum) is passed onto the diatomic gas molecule by a wall molecule and is directed along the x-axis, as illustrated in Fig B.5.2. The two molecules must share the impulse. Therefore, the center of mass of a diatomic

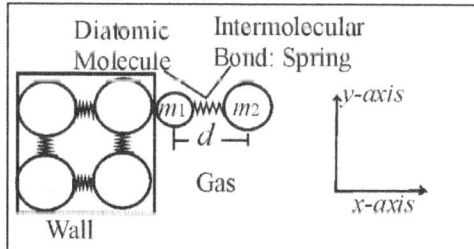

Fig. B.5.2 Shows a diatomic gas molecule in contact with a vibrating wall, which passes an impulse (J) onto the gas molecule.

molecule would be moving with a velocity ($\vec{v}_c$). Therefore:

$$\bar{J} = (m_1 + m_2)\vec{v}_c \qquad \text{B.5.22}$$

Therefore, the velocity of the center of mass is:

$$\vec{v}_c = \bar{J}/(m_1 + m_2) \quad \text{B.5.23}$$

The diatomic molecule's kinetic energy (E_k) is:

$$E_k = (m_1 + m_2)\vec{v}_c{}^2/2 = \bar{J}^2/(m_1 + m_2) \qquad \text{B.5.24}$$

The diatomic molecule colliding with a wall would have vibrational energy (E_v) (Chapter 2). Accordingly, when contemplating an ensemble of diatomic gas molecules, the mean diatomic molecule would not extract vibrational energy from the collision with the wall. In other words, some diatomic gas molecules would give up some of their vibrational while others will receive vibrational energy due to collisions with the walls. Therefore, the velocity as defined by eqn B.5.23 must be the mean velocity for our ensemble of gas molecules. Which implies (wrongly?) that all the momentum imparted to the diatomic atom goes into its translational energy as defined by eqn B.5.24: Actually should be rotational plus translational!! Consider the unrealistic case there is no initial vibrational energy between the two m_1 and m_2 of the diatomic molecule. Assume that they start off being separated by a distance d i.e. the relaxed position. We realize that the impulse must be felt by m_1 first. Therefore, at the moment the impulse is transferred:

$$\bar{J} = m_1\vec{v}_1 \qquad \text{B.5.25}$$

Since m_1 is moving while m_2 is not, then the spring/bond is compressing. Accordingly, some energy from the impulse is transformed into the spring's potential energy, hence the magnitude of $\vec{v}_c$ would be less than that defined by eqn B.5.23. Thus, some of the energy that would become part of the translational energy of the diatomic molecule gets transferred into vibrational energy within the diatomic molecule.Let the Hamiltonian (H) represent the spring's potential energy which equals the vibrational energy imparted onto the diatomic molecule by the initial impulse. Conservation of energy allows us to write:

$$H = m_1\vec{v}_1{}^2/2 - (m_1 + m_2)\vec{v}_c{}^2/2 \text{ B.5.26}$$

where $m_1\vec{v}_1{}^2/2$, is the initial kinetic energy imparted onto m_1, while: $(m_1 + m_2)\vec{v}_c{}^2/2$, is the kinetic energy of the center of mass of the diatomic molecule after the center of mass attains its maximum velocity. Reorganizing eqn B.5.26, gives:

$$(m_1 + m_2)\vec{v}_c{}^2/2 = m_1\vec{v}_1{}^2/2 - H \text{ B.5.27}$$

Therefore:

$$\vec{v}_c{}^2 = 2[m_1\vec{v}_1{}^2/2 - H]/(m_1 + m_2) \qquad \text{B.5.28}$$

Therefore:

$$\vec{v}_c = \sqrt{2[m_1\vec{v}_1{}^2/2 - H]/(m_1 + m_2)} \qquad \text{B.5.29}$$

Eqn B.5.29 illustrates a diatomic molecule with no initial vibrational energy would have a lower final velocity after wall collision. Note: The molecule's rotational energy was not considered!

Reference: 1)"The Mechanics Problem Solver", Director of editing M. Fogiel, Research and Education Association New York N.Y. 1983

Appendix B.6: **Mechanical Work**

In Chapters 3 and 4, work and its application to thermodynamics was discussed. Herein is a mathematical elaboration of the principles of work. From a mechanical perspective, work signifies the energy transferred by a force onto an object in motion. Work is a scalar quantity that can be positive or negative depending upon how it is directed with respect to an object's motion. Work is measured in the same units as energy that being the joule (*J*) in SI units.

A practical way to express work is in terms of *c*; the curve/path traveled by the object, $\vec{F}$; the force vector, and $\vec{s}$; is the position vector:

$$W = \int_C \vec{F} \bullet d\vec{s} \qquad \text{B.6.1}$$

For those not familiar with the above line integral, work can be expressed in terms of force and distance as vector quantities, hence work (*W*) becomes the following scalar product:

$$W = \vec{F} \bullet \vec{d} \qquad \text{B.6.2}$$

Consider a simple example of the work required to lift some mass. In order to lift a block of a given mass (*m*) up into the air, the force exerted has to be sufficient to elevate the mass's weight against the force of gravity ($\vec{g}$), as is illustrated in the free-body diagram Fig. B.6.1.

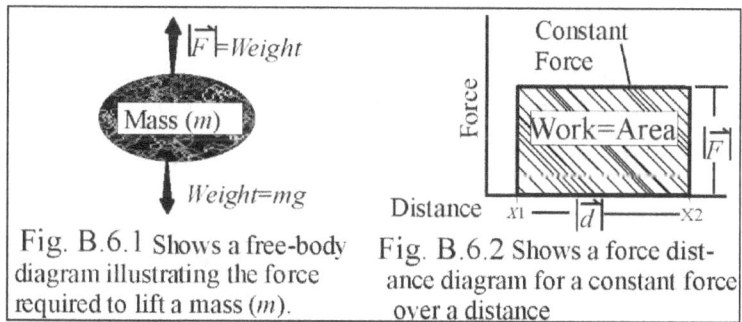

Fig. B.6.1 Shows a free-body diagram illustrating the force required to lift a mass (*m*).

Fig. B.6.2 Shows a force distance diagram for a constant force over a distance

Weight is equals the mass multiplied by the magnitude of the force of gravity ($\vec{g}$).

$$Weight = mg \qquad \text{B.6.3}$$

Therefore, the magnitude of force required must be mass times gravity.

$$|\vec{F}| = m|g| \qquad \text{B.6.4}$$

Understandably, in order to lift this mass, the force must be directed opposite to the weight, as illustrated in Fig B.6.1. Hence in terms of force vectors eqn B.6.4 becomes:

$$\vec{F} = -m\vec{g} \qquad \text{B.6.5}$$

Combining the above, the work (*W*) required for lifting the block becomes:

$$W = m|\vec{g} \bullet \vec{d}| \qquad \text{B.6.6}$$

Kent W. Mayhew

If the object's motion changes over time, then the more general formulation, which employs differentials, is needed. I.e., work is contemplated in terms of infinitesimal work (dw), thus the total work (W) is obtained by the integration of the dw. Integrating:

$$W = \int dw = \int_{x1}^{x2} F dx = F\int_{x1}^{x2} dx = F(x_2 - x_1) \quad \text{B.6.7}$$

A more general consideration of work occurs when the force vector is at some angle with respect to the displacement, as shown in Fig B.6.3. Now the work is:

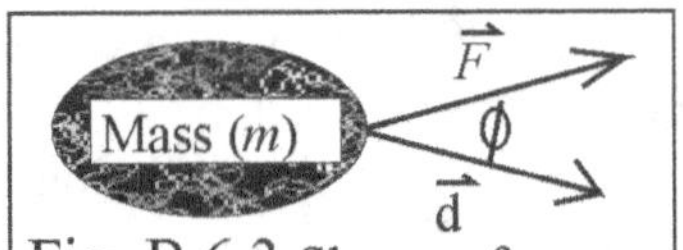

Fig. B.6.3 Shows a force an angle to a mass's displacement

$$W = |F||d|\cos\phi \quad \text{B.6.8}$$

where ϕ represents the angle between the two vectors. As is illustrated in Fig. B.6.3, when $\phi = 0$, then $\cos\phi = 1$, and we arrive at eqn B.6.3. Conversely, when $\phi = 90^0$, then $\cos\phi = 0$ and the work performed by that force becomes zero. Finally when $\phi = 180^0$, then $\cos\phi = -1$, which was actually the case for our example depicted in Fig B.6.1.

Consider what happens when the force varies with displacement, such as is the case for a spring. Springs require increasing amounts of force to either expand or compress them, the further they are from their neutral position. Generally, the required force is linearly proportional to the spring's displacement (d) from its neutral position, i.e.:

$$\vec{F} = b\vec{d} \quad \text{B.6.9}$$

Calculating the force ($\vec{F_s}$) required as the spring is stretched, we would need to realize that the spring's force is in the opposite direction to the spring's displacement. This is nothing more than a variation of eqn B.6.9, when $\phi = 180^0$. Therefore, we can write:

$$|\vec{F_s}| = -b|\vec{d}| \quad \text{B.6.10}$$

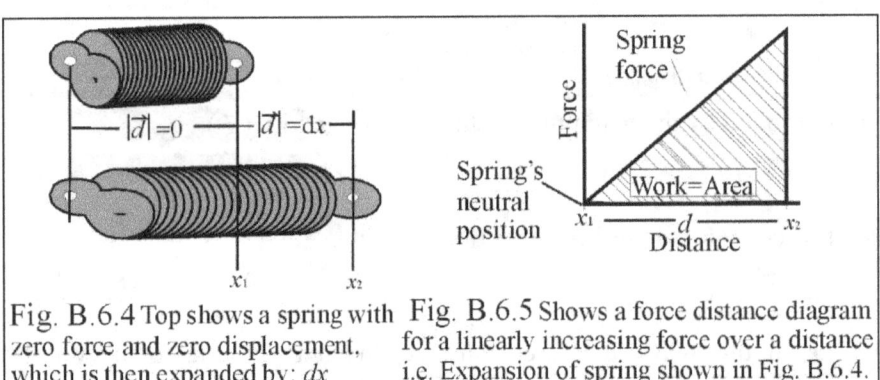

Fig. B.6.4 Top shows a spring with zero force and zero displacement, which is then expanded by: dx

Fig. B.6.5 Shows a force distance diagram for a linearly increasing force over a distance i.e. Expansion of spring shown in Fig. B.6.4.

Fig B.6.4 illustrates the expansion of a spring a distance (d), where: $d = \vec{x}_2 - \vec{x}_1$. It shows the spring's original position as being located at the spring's neutral position. A more general analysis would have x_1 located at some expanded distance from the spring's original position and then the spring is stretched further to the more expanded position located at x_2. The work required is simply the area under the graph between x_1 and x_2, i.e.:

$$W = (\vec{x}_2 \bullet \vec{F}_2 - \vec{x}_1 \bullet \vec{F}_1)/2 \quad \text{B.6.12}$$

Realizing that the applied forces and displacements are in the same direction and substituting in for eqn B.6.10 (i.e. $|\vec{F}_2| = bx_2$), gives the work required for expansion:

$$W = b(x_2{}^2 - x_1{}^2)/2 \qquad \text{B.6.13}$$

We could have similarly used integration to calculate eqn B.6.13 as follows:

$$W = \int dw = \int_{x_1}^{x_2} F dx = b \int_{x_1}^{x_2} x dx = b(x_2{}^2 - x_1{}^2)/2 \quad \text{B.6.14}$$

Obviously the integration approach is best when the object's motion changes over time, wherein work concerns infinitesimally small change, which is written (dw):

$$dw = \vec{F} \bullet d\vec{s} \qquad \text{B.6.15}$$

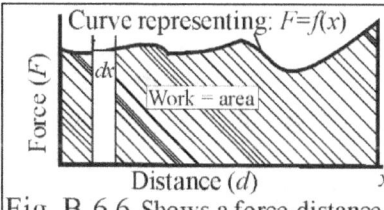

Fig. B.6.6 Shows a force-distance wherein the area defines the work in some process where the force is a function of location [$F=f(x)$].

The integration of eqn B.6.15 allows us to determine the total work (W) as defined by eqn B.6.1: $W = \int dw = \int_C \vec{F} \bullet d\vec{s}$

The above line integral, considers that some process experiences/requires a force that changes with location, around curve/path c, i.e. the force is a function of location, i.e.: $F=f(x)$.

An arbitrary example of such a situation is depicted in Fig B.6.6. The component of the position vector (in this case: $d\vec{s}$) that is affected by the force vector ($\vec{F}$) is considered to be dx. The multiplication of the magnitude of the force vector with the component of the position vector (dx) gives the work required/experienced over the distance dx.

The total work is the summation of all the forces multiplied by distances over f(x), i.e.:

$$W = \sum F dx \qquad \text{B.6.16}$$

Eqn B.6.16 is simply the approach that in terms of the integration of differentials would be written.

$$W = \int dw \qquad \text{B.6.17}$$

When this work simply relies on the endpoints, then it is simplest to use equations like eqn B.6.3. Otherwise we must use equations like eqn B.6.17 or B.6.14.

Appendix B.7: **Explaining 66.67% for Work vs Energy**

The following analysis concerning the number of gaseous molecules striking a surface is based upon an analysis found in Reif[1] (pg 271). The total molecular "flux" is the number of molecules that strike a unit surface area per unit time. A crude analysis arriving at the flux along any one axis in a given direction is:

$$\Phi_0 = n\bar{\bar{v}}_x/6 \qquad 1.4.5$$

More exacting analysis of Flux

Draw an elemental area dA of some system's wall in the x,y plane with the z-axis being perpendicular to said wall, i.e. Fig B.7.1.

Consider the molecules in the immediate vicinity of the wall's elemental area (dA) whose velocities lay between: $v+dv$. ; its direction specified by its polar angle θ with respect to z-axis and its azimuthal angle φ is such that these angles lie between $\theta+d\theta$ and $\varphi+d\varphi$ respectively.

Over a time interval dt the above molecules experience a displacement of vdt. Therefore all the molecules that lay in a infinitesimal cylinder whose cross sectional area dA and have a length of vdt making an angle θ with the z-axis will strike the wall, The infinitesimal cylinder's volume is:

$$dAvdt\cos\theta \qquad B.7.1$$

While the number of molecules is:

$$f(v)d^3v \qquad B.7.2$$

Therefore the number of molecules that strike the walls surface area dA over a time period of dt becomes the multiplication of eqn B.1.7.1 and B.1.7.2, i.e.:

$$[f(v)d^3v](dAdt\cos\theta) \qquad B.7.3$$

Let $\Phi(v)d^3dv$ be the number of molecules with velocity between v and $v+dv$ which strike elemental area dA over time frame dt. Dividing eqn B.7.3 by elemental surface area (dA) and elemental time interval (dt) gives:

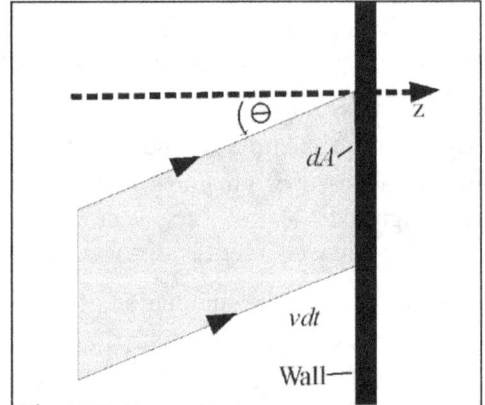

Fig B.7.1 Shows molecules with velocity between v and $v+dv$ striking an elemental area (dA) of some wall

$$\Phi(v)d^3dv = d^3vf(v)v\cos\theta \quad B.7.4$$

Let Φ_0 represent the total number of molecules that strike dA over time dt. This would be obtained by summing over:

    a) all possible speeds: $0 < v < \infty$
    b) all possible azimuth angles: $0 < \varphi < 2\pi$
    c) all possible angles: $0 < \theta < \pi/2$
Note: Molecules where $\pi/2 < \theta < \pi$ are directed away from the wall's elemental area dA.

Therefore:

$$\Phi_0 = \int\limits_{v(z)>0} d^3v f(v) v \cos\theta \quad \text{B.7.5}$$

However:

$$d^3v = v^2 dv(\sin\theta d\theta d\varphi) = v^2 dv d\Omega \qquad \text{B.7.6}$$

Note: eqn B.7.6 is based upon the elemental solid angle being: $d\Omega = \sin\theta d\theta d\varphi$.

Therefore eqn B.7.5 becomes:

$$\Phi_0 = \int\limits_{v(z)>0} d^2 dv \sin\theta d\theta d\varphi f(v) v \cos\theta \qquad \text{B.7.7}$$

Which becomes:

$$\Phi_0 = \int\limits_0^\infty f(v) v^3 dv \int\limits_0^{\pi/2} \sin\theta \cos\theta d\theta \int\limits_0^{2\pi} d\varphi \qquad \text{B.7.8}$$

Since the integration over the azimuth angle (φ) gives 2π, while the integration over all possible angles (θ) gives ½, thus:

$$\Phi_0 = \pi \int\limits_0^\infty f(v) v^3 dv \qquad \text{B.7.9}$$

The mean speed (which is intrinsically positive) is given by:

$$\bar{v} = (1/n)\iiint f(v) v d^3v \qquad \text{B.7.10}$$

Thus

$$\bar{v} = (1/n)\int d^3v f(v) v \qquad \text{B.7.11}$$

Hence:

$$\bar{v} = (1/n)\int\limits_0^\infty \int\limits_0^\pi \int\limits_0^{2\pi} (v^2 dv \sin\theta d\theta d\varphi) f(v) v \qquad \text{B.7.12}$$

Or:

$$\bar{v} = (4\pi/n)\int\limits_0^\infty f(v) v^3 dv \qquad \text{B.7.13}$$

227

Since the integration over the angles (θ) and azimuth angles (φ) is just the total solid angle about a point a point. Hence eqn B.7.9 can be rewritten as:

$$\Phi_0 = (1/4)n\bar{v} \qquad \text{B.7.14}$$

We can see that this exact result differs from the simple initial assumption that the flux along a given axis is 1/6 of the total flux. So our initial rudimentary thought was based upon eqn 1.4.5: $\Phi_0 = (1/6)n\bar{v}$. Hence our rudimentary conceptualization (used in kinetic theory) was 2/3 e.g. 66.667% of the actual result.

What really concerns us is this. The energy of a gas enclosed by a box is due to the contribution of energy from along 3 orthogonal axis thus six directions, as was discussed in Chapter 2 on kinetic theory, and this was really based upon energy contributions from all directions (an offshoot of $\Phi_0 = (1/6)n\bar{v}$ if you prefer). In kinetic theory the total translational plus rotational energy of a monatomic gas was due to energy being pumped/exchanged into the gas along all three orthogonal axis, with that energy being determined to be: $3NkT/2$. (Note our analysis differs with the traditionally accepted theory wherein the energy of a monatomic gas is only translational)

But the actual energy of this gas that can be used as work i.e. being passed onto a given wall, so that it can move in a given direction e.g. along one axis in one direction, is based upon eqn B.7.14: $\Phi_0 = (1/4)n\bar{v}$. Accordingly, only 2/3 e.g. $(1/4)n\bar{v}/(1/6)n\bar{v} = 2/3$, of a monatomic gas's total energy can be used for work, that being: NkT. Remember gas molecules passes both their translational and rotational energy onto a wall, as part of its kinetic energy exchange with the wall molecules.

References:

1) Reif. F "Fundamentals of Statistical and Thermal Physics", McGraw-Hill, New York, 1965

Appendix B.8: **Elastic Collisions: Analysis of**

Considering an elastic collision where two different masses moving at different velocities both before and after the collision is not easily solved because we have two equations with two unknowns. Consider mass (M_1) moving with initial velocity ($\vec{v}_{1i}$) experience an elastic collision with mass (M_2) moving at velocity ($\vec{v}_{2i}$). The problem is to find the final velocity ($\vec{v}_{1i}$) of mass (M_1) and of mass (M_2) i.e. $\vec{v}_{2f}$, after the collision. To keep the analysis fairly simple, we shall consider that the trajectory of the two masses is along a solitary plane that passes through their center of masses. For any elastic collision the kinetic energy is conserved therefore:

$$(M_1\vec{v}_{1i}^2 + M_2\vec{v}_{2i}^2)/2 = (M_1\vec{v}_{1f}^2 + M_2\vec{v}_{2f}^2)/2 \qquad \text{B.8.1}$$

Multiplying both of eqn B.8.1 sides by 2 gives:

$$M_1\vec{v}_{1i}^2 + M_2\vec{v}_{2i}^2 = M_1\vec{v}_{1f}^2 + M_2\vec{v}_{2f}^2 \qquad \text{B.8.2}$$

Rearranging the variables gives:

$$M_1(\vec{v}_{1i}^2 - \vec{v}_{1f}^2) = M_2(\vec{v}_{2f}^2 - \vec{v}_{2i}^2) \qquad \text{B.8.3}$$

Eqn (B.8.3) can be rewritten in terms of difference of squares:

$$M_1(\vec{v}_{1i} + \vec{v}_{1f})(\vec{v}_{1i} - \vec{v}_{1f}) = M_2(\vec{v}_{2f} + \vec{v}_{2i})(\vec{v}_{2f} - \vec{v}_{2i}) \qquad \text{B.8.4}$$

For any collision conservation of momentum means:

$$M_1\vec{v}_{1i} + M_2\vec{v}_{2i} = M_1\vec{v}_{1f} + M_2\vec{v}_{2f} \qquad \text{B.8.5}$$

Rearranging Eqn B.8.5 gives:

$$M_1(\vec{v}_{1i} - \vec{v}_{1f}) = M_2(\vec{v}_{2f} - \vec{v}_{2i}) \qquad \text{B.8.6}$$

Dividing equation B.8.4 by eqn B.8.6 yields:

$$\vec{v}_{1i} + \vec{v}_{1f} = \vec{v}_{2i} + \vec{v}_{2f} \qquad \text{B.8.7}$$

Rearranging this equation gives an interesting result:

$$\vec{v}_{1i} - \vec{v}_{2i} = \vec{v}_{2f} - \vec{v}_{1f} \qquad \text{B.8.8}$$

As William Layton (physics teacher) explains it; "This equation says the relative velocity before the collision equals minus the relative velocity after the collision. This is always true in all perfectly elastic collisions. A "super" ball should bounce up with the same speed it had just before striking the floor. This equation also makes it easy to solve the original problem". Note: Eqn B.8.8 is in other texts[1]. Dividing eqn B.8.4 by B.8.6 seems strange as we divided conservation of kinetic energy by conservation of momentum. Even so it is doable because of the equalities. Rectifying: Rewrite Eqn B.8.8:

$$\vec{v}_{1i} = \vec{v}_{2f} + \vec{v}_{2i} - \vec{v}_{1f} \qquad \text{B.8.9}$$

Inserting eqn B.8.9 into eqn B.8.5 gives:

$$M_1(\vec{v}_{2f} + \vec{v}_{2i} - \vec{v}_{1f}) + M_2\vec{v}_{2i} = M_1\vec{v}_{1f} + M_2\vec{v}_{2f} \qquad \text{B.8.10}$$

Which can be rewritten:

$$M_1(\vec{v}_{2f} + \vec{v}_{2i} - \vec{v}_{1f} - \vec{v}_{1f}) = M_2(\vec{v}_{2f} - \vec{v}_{2i}) \qquad \text{B.8.11}$$

Collecting the terms gives:

$$M_1(\vec{v}_{2f} + \vec{v}_{2i} - 2\vec{v}_{1f}) = M_2(\vec{v}_{2f} - \vec{v}_{2i}) \qquad \text{B.8.12}$$

Which can be rewritten as:

$$M_1/M_2 = (\vec{v}_{2f} - \vec{v}_{2i})/(\vec{v}_{2f} + \vec{v}_{2i} - 2\vec{v}_{1f}) \qquad \text{B.8.13}$$

Eqn (B.8.13) can be rewritten as:

$$M_2/M_1 = (\vec{v}_{2f} + \vec{v}_{2i} - 2\vec{v}_{1f})/(\vec{v}_{2f} - \vec{v}_{2i}) \qquad \text{B.8.14}$$

Which can be rewritten:

$$M_2/M_1 = 1 - |2\vec{v}_{1f}/(\vec{v}_{2f} - \vec{v}_{2i})| \qquad \text{B.8.15}$$

Certainly eqn 1.10.38 is not some simple discernible equation that easily determines what an elastic collision looks like. There is one solution to (14) that is readily envisioned that being set: $M_1=M_2$, then:

$$(\vec{v}_{2f} + \vec{v}_{2i} - 2\vec{v}_{1f}) = (\vec{v}_{2f} - \vec{v}_{2i}) \qquad \text{B.8.16}$$

Collecting the variable gives:

$$-2\vec{v}_{1f} = -2\vec{v}_{2i} \qquad \text{B.8.17}$$

Thus the simple solution is:

$$\vec{v}_{1f} = \vec{v}_{2i} \qquad \text{B.8.18}$$

A solution to our problem is that an elastic collision involves two identical masses where all the momentum is transferred from M_1 to M_2, with M_2 having an initial velocity of zero. The classic example being in a game of billiards when the cue ball hits another ball through their center of mass, with the only rotation being along the plane through the center of their masses. In this case all the cue ball momentum will be passed onto the other ball. Other solutions remain plausible but these are not simple to visualize.

Inelastic Collision

Certainly inelastic collisions are the norm. Consider a bullet hitting a block of wood where the bullet lodges itself in the wood. Such a collision is not an elastic one hence kinetic energy is not conserved, although energy is conserved with the lost energy being associated with deformation and/or heat. For such an inelastic collision where the two masses become one, then based upon eqn B.8.5:

$$M_1\vec{v}_{1i} + M_2\vec{v}_{2i} = (M_1 + M_2)\vec{v}_f \qquad \text{B.8.19}$$

Commentary

In this book, our real concern concerning elastic collisions is one of, is it logical that inter molecular collisions are all elastic as prescribed by traditional kinetic theory, or were we right to question this supposition, as was done in Chapter 7. Nothing was prove but we know most collisions are inelastic.

References:

1)"University Physics" Sears, Francis, Hugh,Young, Zemansky, Mark, Addison Wesley Publishing company . Reading Massechussettes, 1987

Appendix B.9: **Gas Kinematics**

Traditional kinetic theory starts by considering the energetics of a gas as given here. The translational momentum ($\vec{p}$) of a gas molecule in terms of the gas molecule's mass (m), and velocity ($\vec{v}$) is[1,2]:

$$\vec{p} = m\vec{v} \qquad \text{B.9.1}$$

The translational energy of a gas molecule is defined by its the kinetic energy (E_k)[1,2]:

$$E_k = m\vec{v}^2 / 2 \qquad \text{B.9.2}$$

The following analysis is based upon the analysis in a text by Reif[2]. Assume that the:

1) Number of molecules is large.
2) Separation between molecules is large.
3) Molecules move randomly.
4) Distribution of molecular velocities does not change.
5) Gas molecules undergo elastic collision between each other and with the walls.
6) Molecules obey Newton's laws.

Fig. B.9.1 shows an ideal gas consisting of N molecules, within a box of volume V, which is in thermal equilibrium with its surrounding, at T. We want to calculate how many molecules collide with the wall's surface area (A) during time interval dt. The gas molecules move with a mean velocity ($\bar{\bar{v}}$) in a random motion. Assume that one-third of the molecules will be moving along each of the x, y and z axis/directions.

Let the concentration of gaseous molecules be $n = N/V$, where N signifies the total number of molecules in our system. One-third of gaseous molecules will be moving in the x-direction ($n_x = n/3$), half in the positive x-direction, while the other half are moving in the negative x-direction. Thus, the average number of gaseous molecules moving in the positive x-direction must be $n_{x+} = n/6$.

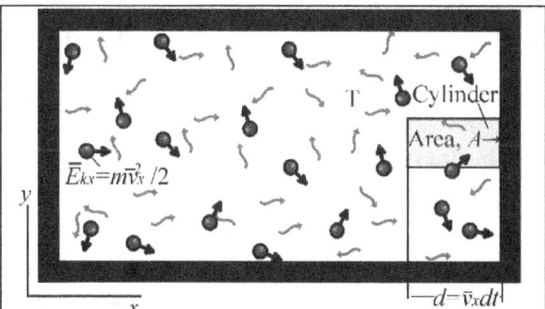

Fig.B.9.1 Shows an ideal gas in a parallelepiped. We are interested in how the pressure is calculated & how it relates to the kinetic energy of the gas.

Consider that any molecule whose mean velocity ($\bar{\bar{v}}$) is predominately in the positive x-direction ($\bar{\bar{v}}_x$) travels a distance ($d = \bar{\bar{v}}_x dt$), in the time interval (dt). For a molecule within a distance (d) from the wall's surface area (A) moving in the positive x-direction, then that molecule will strike the wall's surface, in that time interval. Fig B.9.2 shows a cylinder with surface area is A, has a volume (V_{cyl}). Accordingly:

$$V_{cyl} = Ad = A\bar{\bar{v}}_x dt \qquad \text{B.9.3}$$

Calculating the number of gaseous molecules ($N_{strikeA}$) that strike the surface area, A, in the time interval dt, can be obtained by multiplying the concentration of molecules moving in the positive x-direction ($n_x = n/6$) by the volume of the cylinder, as given by eqn B.9.3:

$$N_{strikeA} = nA\bar{\bar{v}}_x dt / 6 \qquad \text{B.9.4}$$

The *flux density* (Φ_0) is the average number of molecules striking a unit wall surface area per unit time, which is calculated by dividing eqn B.9.4 by the surface area (A) and time interval (dt) giving:

$$\Phi_0 = n\bar{\bar{v}}_x / 6 \qquad \text{B.9.5}$$

Note: Eqn B.9.5 is based upon an overly simplified analysis. A more precise analysis can be obtained by considering all the molecules impacting area A from all possible angles as was done in Appendix B.7. I.e., by considering that all molecules do not simply move directly along one of the three orthogonal axis i.e. along x, y, or z. In this case the flux density become 3/2 that given by eqn B.9.5 [1page273] i.e. $\Phi_0 = n\bar{\bar{v}}_x / 4$. Does this mean that this traditional derivation requires a rethink? Seemingly yes!

Continuing; the mean kinetic energy ($\bar{E}_{kx}$) of a gaseous molecule moving along the positive x-axis is:

$$\bar{E}_{kx} = m\bar{\bar{v}}_x^2 / 2 \qquad \text{B.9.6}$$

Similarly, the mean momentum ($\bar{\bar{p}}$) along the positive x-axis must be:

$$\bar{\bar{p}}_x = m\bar{\bar{v}}_x \qquad \text{B.9.7}$$

Note: Be aware of the similarity between symbols e.g. momentum ($\bar{\bar{p}}_x$) and pressure (P). The momentum imparted onto the wall during the collision must be twice the momentum, as given by eqn B.9.7 because, on average, the molecules travel in the negative x-direction with the same magnitude of velocity, as it had in the positive *x*-direction. Thus:

$$\Delta \bar{\bar{p}}_x = 2m\bar{\bar{v}}_x \qquad \text{B.9.8}$$

The pressure (*P*) exerted upon the walls must equate to the average momentum ($\bar{\bar{p}}$) imparted upon the wall by each molecule, multiplied by the flux density, that being:

$$P = \Delta \bar{\bar{p}}_x \Phi_0 \qquad \text{B.9.9}$$

Substituting both eqn B.9.5 and B.9.8 into eqn B.9.9, the pressure of a gas becomes:

$$P = (2m\bar{\bar{v}}_x)(n\bar{\bar{v}}_x/6) = nm\bar{\bar{v}}_x^2/3 \qquad \text{B.9.10}$$

Eqn B.9.10 can be rewritten in the form originally derived by Daniel Bernoulli (in 1738). In terms of the mass density of the gas, i.e. $\rho = nm = Nm/V$, that being:

$$P = \rho \bar{\bar{v}}_x^2/3 \qquad \text{B.9.11}$$

Eqn B.9.10 represents the pressure exerted by an ideal gas along the positive *x*-axis. Its magnitude equally represents the pressure exerted along any axis. Note: Although, eqn B.9.6 was defined as representing the mean kinetic energy, it actually is the root mean square, meaning it can be obtained from velocity distribution, i.e. traditional Maxwell's velocity distribution, which is not exactly correct! Remember the vibrating wall molecules impart both rotational and translational energy onto the gas molecules that they surround.

Kinetic Energy of an Ideal Gas

When developing eqn B.9.4, it was stated that the concentration of gaseous molecules moving predominately along the positive *x*-direction is $n_x = n/6$. Multiplying $n/6$ by the system's volume (*V*) gives the number of gaseous molecules moving along the positive *x*-axis that being $N_x = Vn_x$. Next multiply N_x by the kinetic energy of each molecule, as defined by eqn B.9.6, gives the total kinetic energy of the gas molecules that are moving in the predominately positive *x*-direction (E_{kxT}), i.e.:

$$E_{kxT} = (nV/6)(m\bar{\bar{v}}_x^2/2) = nVm\bar{\bar{v}}_x^2/12 \qquad \text{B.9.12}$$

Note: Again be aware of the nomenclature. In order that we have no misunderstanding for the above subscripts: "*k*" signifies kinetic energy, while the subscript "*x*" means along the positive *x*-axis, and the subscript "*T*" signifies total energy. Realizing that $n = N/V$, therefore $nV = N$. We can rewrite eqn B.9.12, in terms of the total number of gas molecules (*N*) within our system:

$$E_{kxT} = Nm\bar{\bar{v}}_x^2/12 \qquad \text{B.9.13}$$

In order to calculate the *N* gas molecule system's total kinetic energy, we simply multiply eqn B.9.13 by six, thus covering all the molecules moving in all six possible directions. Thus, the total kinetic energy (E_{kT}) of the gas becomes:

$$E_{kT} = Nm\bar{\bar{v}}^2/2 \qquad \text{B.9.14}$$

where $|\bar{\bar{v}}| = |\bar{\bar{v}}_x| = |\bar{\bar{v}}_y| = |\bar{\bar{v}}_z|$, which simply states that the magnitudes of the mean velocities in all directions are considered to be equal.

Pressure versus Kinetic Energy

Realizing that $n = N/V$, eqn B.9.10: $P = nm\overline{\overline{v}}_x^2/3$ can be rewritten as:

$$P = Nm\overline{\overline{v}}_x^2/3V \qquad \text{B.9.15}$$

Comparing eqn B.9.15 to eqn B.9.14: $E_{kT} = Nm\overline{\overline{v}}^2/2$, it is obvious that:

$$2E_{kT} = 3PV \qquad \text{B.9.16}$$

Eqn B.9.16 can be rewritten to give the total kinetic energy (E_{kT}) of a gas in terms of the pressure and volume of that gas, as follows:

$$E_{kT} = 3PV/2 \qquad \text{B.9.17}$$

Eqn B.9.17 gives the relationship between pressure and kinetic energy. In the 19th century, the theory was based upon the *equation of Clausius*[13] when the gas experienced no attraction/repulsion between the molecules, i.e. an ideal gas. Eqn B.9.16 can be rewritten to give the pressure of a gas in terms of its total kinetic energy, and the volume, of that gas:

$$P = 2E_{kT}/3V \qquad \text{B.9.18}$$

Based upon eqn B.9.18, the pressure exerted by a gas is two thirds of its total kinetic energy density:

$$PV = 2E_{kT}/3 \qquad \text{B.9.19}$$

The above relationships between the pressure, volume and total kinetic energy for an ideal gas, were derived using classical considerations of motion and are traditionally accepted as being valid.

Commentary:

The traditional consideration of starting with a gas's momentum was reviewed here. Part of the new perspective presented in this book was that the kinematics of a gas is imposed upon the gas via the vibrational energies of wall molecules and larger polyatomic gas molecules. Now consider what happens if you use the true flux ($\Phi_0 = n\overline{v}_x/4$) rather than the approximate flux used in eqn B.9.5? Another traditionally accepted convenient oversight? To what end?

References:

1. Reif. F "Fundamentals of Statistical and Thermal Physics", McGraw-Hill, New York, 1965
2. Reif. F. "Statistical Physics", McGraw-Hill, New York, 1967
3. J.C. Maxwell, J. Chem. Soc (London), **28**, 493-508, (1875) [facsimile published in Mary Jo Nye, The Question of the Atom (Los Angeles: Tomash 1984)]

Appendix B.10: **The Polytropic Equation**

The polytropic equation is a useful tool employed by technologists, engineers and cosmologists. Although useful, its traditional derivation is troublesome as we are about to witness. Basically the ideal gas law applies to quasi-static isothermal processes wherein thermal energy (free energy) can either freely enter (expansion) or freely escape (compression) a system. If you think about it, this has a lot to do with our atmosphere being the mother of all heat baths for experiments on Earth.

Polytropic equations are employed in adiabatic processes is given by eqn 6.4: $PV^n = C$, where n is the polytropic exponent and C is the proportionality constant.

The following is the traditional problematic approach to the development of the polytropic equation. Starting with a traditional interpretation of the first law:

$$dQ + dW = d\varepsilon \qquad \text{B.10.1}$$

where Q is the heat, W is the work and ε is the internal energy.

Since the process is adiabatic then $dQ = 0$. Traditionally, reversible work is $dW = PdV$, which can be problematic. We now realize that for useful expanding systems, that $W = PdV$ often signifies lost work, which certainly cannot be part of some reversible process.

Continuing with tradition, therefore:

$$dQ = d\varepsilon - dW = d\varepsilon + PdV = 0 \quad \text{B.10.2}$$

For n moles the traditional ideal gas law gives: $PV = nRT$ thus: $P = nRT/V$. Substituting in gives:

$$d\varepsilon + nRTdV/V = 0 \qquad \text{B.10.3}$$

Again, we must caution. As was discussed in Chapter 6, adiabatic expansion is not necessarily an isothermal process, as is often traditionally contemplated. Accordingly, the validity of using eqn B.10.3, wherein adiabatic volume change is contemplated in terms of being isothermal, must be questioned. Furthermore, as the volume increases the pressure decreases, which questions the validity of eqn B.10.2 wherein the work is considered in terms of constant pressure. Ignoring the apparent problems, let us continue with the traditional derivation.

Traditional thermodynamics considers the internal energy (ε) as being: $d\varepsilon = nc_vdT$. Note that in this nomenclature, n is number of moles and c_v is the isometric molar heat capacity.

$$nc_vdT + nRTdV/V = 0 \quad \text{B.10.4}$$

Another inherent problem, based upon the application of the above equation to an adiabatically expanding system. The isometric molar heat capacity (c_v) strictly applies to constant volume systems. Accordingly, the traditional application of c_v to a system whose volume is not necessarily constant, as implied by B.10.4, is illogical. Continuing with the traditional approach, we now divide through by temperature to obtain:

$$nc_vdT/T + nRdV/V = 0 \quad \text{B.10.5}$$

Dividing through by the number of moles (n) and realizing that the ratio of specific volume change divided by specific volume must equate to the volume change divided by volume: $dv/v = dV/V$. We can now write:

$$c_v dT / T + R dv / v = 0 \quad \text{B.10.6}$$

Integrating, traditional thermodynamics obtains:

$$c_v InT + R Inv = C' \qquad \text{B.10.7}$$

where C' is a constant.

In which case we obtain:

$$c_v In(pv / R) + R Inv = C' \quad \text{B.10.8}$$

Dividing through by: c_v, traditional thermodynamics then writes:

$$In(pv / R) + (R / c_v) Inv = C' / c_v \quad \text{B.10.9}$$

Now consider eqn 1.35, when: $c_v = R / (\gamma - 1)$, where $\gamma = c_p / c_v$: Substituting eqn 1.35 into eqn, traditional thermodynamics obtains:

$$In(pv / R) + (\gamma - 1) Inv = C' / c_v \quad \text{B.10.10}$$

Which can be rewritten:

$$In(pv / R) + Inv^{\gamma - 1} = C' / c_v \qquad \text{B.10.11}$$

Which becomes:

$$In(pvv^{\gamma - 1} / R) = C' / c_v \quad \text{B.10.12}$$

Therefore:

$$In(pv^{\gamma} / R) = C' / c_v \qquad \text{B.10.13}$$

Taking the exponent of both sides, we obtain:

$$pv^{\gamma} / R = e^{C'/c_v} \qquad \text{B.10.14}$$

Assuming that the isometric heat capacity is constant, then the R.H.S. of eqn B.10.14 is a constant, in which case: $C = e^{C'/c_v}$, where: C is a proportionality constant.

$$pv^{\gamma} / R = C \qquad \text{B.10.15}$$

Since R and C are taken as being constants, Eqn B.10.15 can be rewritten as the following version of eqn 6.4:

$$PV^{\gamma} = C \qquad \text{B.10.16}$$

Please note that the constant C in eqn B.10.16 is not the same constant C in eqn B.10.15.

How valid is eqn B.10.16? As previously stated the use of c_v in an expanding or contracting system, is a weak link. The concept that we have an adiabatic system whose volume changes whilst its temperature does not, is perplexing. Furthermore, look at all the substitutions that were required to obtain eqn 1.10.7. Something is amiss. The reason that the polytropic equation applies to empirical data is more a consequence of how: $PV^n = C$ plots rather than it being enshrined in some strong logically constructed theoretical deduction.

Continuing on with the traditional interpretation of work in a polytropic process Consider the ideal gas law: $PV = NkT$

The more general form for work would be:

$$W = \int dw = \int_{x1}^{x2} Fdx = \int_{x1}^{x2} PAdx = \int_{V_1}^{V_2} Pdv \qquad \text{B.10.16}$$

For an ideal gas we can rewrite eqn B.10.16 in terms of volume and volume change, by realizing that for our ideal gas: $P = V/C$. Therefore:

$$W = \int_{V_1}^{V_2} Pdv = C\int_{V_1}^{V_2} dv/v \qquad \text{B.10.17}$$

Performing the integration, we then arrive at:

$$W = C\ln(V_2/V_1) \qquad \text{B.10.18}$$

The more general case for real a gas can be approximated the $P\text{-}V$ relationship using the polytropic equation with the polytropic exponent being: $1<n<2$. For such a real gas:

$$W = \int_{V_1}^{V_2} Pdv = C\int_{V_1}^{V_2} v^{-n}dv \qquad \text{B.10.19}$$

Performing the integration we obtain:

$$W = [C/(1-n)](V_2^{1-n} - V_1^{1-n}) \qquad \text{B.10.20}$$

Substituting in for the polytropic equation ($PV^n = C$), we obtain:

$$W = [1/(1-n)](P_2V_2 - P_1V_1) \qquad \text{B.10.21}$$

Closing Remarks

One could argue that the traditional derivation shows the beauty of math. Others may argue that it is mathematical conjecture, e.g. most steps seem to be pure math lacking any real logical construct. This author leans to the side of conjecture. It really is not that much different to how the Clausius-Clapeyron equation is traditionally derived i.e. a series of eloquent mathematical equations based upon hard to fathom logic, that may not exists at all. Note: I was going to put the Clausius-Clapeyron derivation in this book but decided, why bother. And of course the ultimate justification bestowed upon such equations is that we were able to make them fit empirical data. "Make" being the provocative word in sledge hammer math. Such arguments equally apply to both Gibbs and Helmholtz free energies (Chapter 15).

ABOUT THE AUTHOR

I was never a great student nor did I ever embrace the education system when I attended Brock University in the 1980's. So as happened when I was the first to figure out how much energy is required to nucleate a bubble (2004 paper : *Energetics of Nucleation*) there will always be those who snobbishly (rightfully in their eyes) declare who the **** am I to do this. Human arrogance rarely hides.

The reality is if I were a better student, then I too might have been equally fully blinded by my education. As a partially indoctrinated human, it still took me several years to accept that entropy and the second law were the biggest blunders in the history of the sciences. It then another equal time span to collect and hone my thoughts into this book .

The fact remains that only a true outsider could write a book like this at this time. And that is exactly who this author is. A nobody from way outside of the system. One struggling to survive who at this rate will work for pennies till the day I make the grave. Luckily, I am able to do most of my own carpentry, mechanics as well as deal with most of whatever else life throws at me. I remain: A true blue common sense man continually willing to push my own boundaries thus generally failing yet willing to get back up onto my knees and then find my feet.

Someone scientists could learn from, if they only can toss their educational blinders. Then I too may learn from them once they learn to abridge their thoughts.

www.ingramcontent.com/pod-product-compliance
Lightning Source LLC
Chambersburg PA
CBHW081556220526

45468CB00010B/2673